Master Math:

ESSENTIAL PHYSICS

**Master everything from motion, force, heat and work
to energy, fluids, waves, optics and electricity**

By

Debra Anne Ross Lawrence

with contributions by

David Allen Lawrence

Course Technology PTR
A part of Cengage Learning

COURSE TECHNOLOGY
CENGAGE Learning·

Australia · Brazil · Japan · Korea · Mexico · Singapore · Spain · United Kingdom · United States

ii

COURSE TECHNOLOGY
CENGAGE Learning·

Master Math:
Essential Physics
Debra Anne Ross
Lawrence

Publisher and General
Manager,
Course Technology
PTR:
Stacy L. Hiquet

Associate Director of
Marketing:
Sarah Panella

Manager of Editorial
Services:
Heather Talbot

Senior Marketing
Manager:
Mark Hughes

Senior Acquisitions
Editor:
Emi Smith

Technical Reviewer:
David A. Lawrence

Cover Designer:
Jeff Cooper

Proofreader:
Sue Boshers

For product information and technology assistance, contact us at
Cengage Learning Customer & Sales Support, 1-800-354-9706
For permission to use material from this text or product,
submit all requests online at cengage.com/permissions
Further permissions questions can be emailed to
permissionrequest@cengage.com

Library of Congress Control Number: 2012930934

ISBN-10: 1-4354-5888-5

ISBN-13: 978-1-4354-5888-8

Course Technology, a part of Cengage Learning
20 Channel Center Street
Boston, MA 02210
USA

Cengage Learning is a leading provider of customized learning solutions with office locations around the globe, including Singapore, the United Kingdom, Australia, Mexico, Brazil, and Japan. Locate your local office at:
international.cengage.com/region

Cengage Learning products are represented in Canada by Nelson Education, Ltd.
For your lifelong learning solutions, visit courseptr.com
Visit our corporate website at cengage.com

Printed by RR Donnelley, Crawfordsville, IN. 1st Ptg. 03/2012
Printed in the United States of America
1 2 3 4 5 6 7 14 13 12

TABLE OF CONTENTS

Dedication

To the Designer

vii

Acknowledgments

First, I deeply appreciate my brilliant husband, David A. Lawrence, for creating the end-of-chapter problem sets, for serving as technical editor, and for engaging in countless valuable discussions about the book's content.

I am eternally grateful to Dr. Channing R. Robertson, Professor of Chemical Engineering at Stanford University, for reviewing this book, for his insightful comments, and for his generous endorsement. I am very thankful for all his guidance and for his friendship.

I am also very grateful to Maggie Ross for the adorable duck drawings in Section 1.7 and for carefully editing the manuscript.

I thank George Skladal for reading the early manuscript and for his helpful comments.

Very special thanks to Emi Smith, Senior Acquisitions Editor, for facilitating a myriad of details concerning the book's publication and for her infinite patience with me.

Many thanks to Sue Boshers for proofreading this book and improving its presentation. It was a pleasure working with her!

I am also thankful to Stacy Hiquet, Publisher and General Manager, for publishing the *Master Math* books and providing me the opportunity to write this addition to the series. I also thank Heather Talbot, Manager of Editorial Services, for all her help. Additional special thanks to Sarah Panella, Associate Director of Marketing; Mark Hughes, Senior Marketing Manager; and Jeff Cooper, cover designer.

I want to recognize Ron Fry and the staff of Career Press for their work in publishing and launching my original *Master Math* books as a successful series.

Finally, I especially thank my wonderful agent, Sidney B. Kramer, and the staff of Mews Books. Thank you, Sidney!

About the Author

Debra Anne Ross Lawrence is the author of six other books in the *Master Math* series: *Basic Math and Pre-Algebra, Algebra, Geometry, Pre-Calculus, Trigonometry,* and *Calculus.* Debra earned a double Bachelor of Arts degree in Biology and Chemistry with honors from the University of California at Santa Cruz and a Master of Science degree in Chemical Engineering from Stanford University.

Debra's research experience encompasses investigating the photo-synthetic light reactions using a dye laser, studying the eye lens of diabetic patients, creating a computer simulation program of physiological responses to sensory and chemical disturbances, genetically engineering bacteria cells for over-expression of a protein, and designing and fabricating biological reactors for in-vivo study of microbial metabolism using nuclear magnetic resonance spectroscopy.

Debra's work history includes: developing and bringing to market the first commercial biosensor system with a small team of scientists and engineers; managing an engineering group responsible for the scale-up of combinatorial synthesis for pharmaceutical development; and managing intellectual property for a scientific research and development company. Debra's research and work has been published in scientific journals and/or patented.

Debra is also the author of *The 3:00 PM Secret: Live Slim and Strong Live Your Dreams* and *The 3:00 PM Secret 10-Day Dream Diet.* She is the coauthor with her husband, David A. Lawrence, of *Arrows Through Time: A Time Travel Tale of Adventure, Courage, and Faith.* Debra is President of GlacierDog Publishing (visit glacierdogpublishing.com or GlacierDog.com). When Debra is not engaged in all-season mountaineering with David near her Alaska home, she is endeavoring to understand the seemingly incomprehensible workings of the universe.

David Allen Lawrence holds Bachelor and Master of Science degrees from the Massachusetts Institute of Technology and a Juris Doctor *cum laude* from the University of Minnesota. David has served as Director of Law for a Fortune 500 energy company, as Senior Vice President and General Counsel for a telecommunications engineering and consulting company, and as an energy and utility Law Judge.

x

IMPORTANT DEFINITIONS

(Skim this section before reading the book and refer back as needed.)

1. **Pay attention to units**. *When working on a problem or equation, all the values must be in, or converted to, one system of units*. All units must agree. Those units may be SI (or MKS), CGS, or British, but you cannot mix units in a given equation or problem. We will mostly use the MKS system, but will also use British for practice. In this book we will generally use abbreviations such as: seconds s, hours h, minutes min, meters m, centimeters cm, miles mi, feet ft, inches in, miles/hour mph, pounds lb, grams g, kilograms kg, slugs sl, Newtons N, Joules J, volts V.

Units can assist you in developing or solving an equation since the remaining units after simplifying must be the units of your desired result. When you work a problem, ask yourself if the answer has the correct units. For example, if you are calculating a velocity, do you get velocity units (m/s, or mph) after you solve and simplify or cancel?

Physicists often use the *metric* **SI (Système International) units**, also referred to as the **International System of Units** or the **MKS system**. This system uses m for length, kg for mass, and s for time (hence, "MKS"). *Note that the terms SI units and MKS units are used interchangeably*. Another *metric system of units* is the **CGS system**, based on the cm, the g, and the s. Other values measured using these two metric systems include *work* and *energy*, which use the **Joule** (J) in MKS and the **erg** in CGS. Measuring **force** uses the **Newton** (N) in MKS and the **dyne** in CGS. While the **British System of units** is officially used in the USA, most countries use the MKS system. In the British System, length is in ft, mass is in sl, and time is in s.

Converting between systems of units for force, mass, and weight can be tricky because of gravity. Units of force represent mass times acceleration, or $F = ma = $ (mass)(length)/(time)^2. When we work in the SI (or MKS) system, force is the derived unit called a **Newton** N. A Newton is the force required to accelerate a 1-kg mass 1 m/s over a second's time. 1 N = 1 kg·m/s^2. In the CGS system, force is a derived unit called a **dyne**. A dyne is the force required to accelerate a 1-g mass 1 cm/s over a second's time. 1 dyne = 1 g·cm/s^2, and 1 N = 10^5 dynes. In the British system, a **pound-force** (lbf) is the force required to accelerate a 1-sl mass 1 ft/s over a second's time. 1 lbf = slug·ft/s^2.

There can be confusion whether *pounds* are a mass or a weight (force). Pound-force and Newtons are units that measure force or weight. Kilograms and slugs (or pound-mass [lbm]) are units of mass. The relationship between slugs and lbm is based on *acceleration of gravity* $g = 32$ ft/s^2: 1 slug = 32.17 lbm. *When you are on the Earth's surface, where gravity is about 32 ft/s^2, one lbm weighs one lbf, and the mass of*

1 slug weighs 32 lbf. This means a lbf is the force required to accelerate 1 lbm 32.17 ft/s over 1 second's time. For 1 lbm = 1 lbf, we need the acceleration of gravity g = 32 ft/s^2, which only works in the gravitational field on the surface of the Earth. The slug came about because in the British system force was defined to be one lbf, time to be s, and distance to be ft, and mass was derived and called a slug. Therefore, on Earth lbm = lbf, and you can use just lb.

Conversion factors between SI and British systems for force and mass are: Force: 1 N = 0.2248 lb, 1 lb = 4.448 N. Mass: 1 kg = 0.06852 sl, 1 sl = 14.59 kg. The conversion between kg and lb is often given as 1 kg = 2.2 lb, which only applies when the acceleration due to gravity is 9.8 m/s^2 or 32 ft/s^2, as it is near Earth's surface. (See also Section 2.1.)

2. **Equations**: Once you verify that the units are all consistent within an equation or problem, think about whether your answer makes sense. *Does the order of magnitude of the answer seem right to you?* If you are calculating the speed of a plane and your answer is faster than the speed of light, there is error. Also, when you encounter any equation, consider what it is telling you about its fundamental values or variables, and how those variables depend on one another. For example, in the Ideal Gas equation PV = nRT, pressure P, volume V, and temperature T affect one another. We see that P and T are directly proportional (doubling P will double T, holding V constant), whereas P and V are inversely proportional (increasing P decreases V).

3. **Remember the basic trigonometric relationships**. (See *Master Math: Trigonometry* for details.) The **trigonometric functions** are sine (sin), cosine (cos), tangent (tan), cotangent (cot), secant (sec), and cosecant (csc). These relate to each other as: tan = sin/cos; cot = cos/sin = 1/tan; sec = 1/cos; and csc = 1/sin. The *trigonometric functions* can be defined using ratios of the sides of a right triangle where sin θ = opposite/hypotenuse, cos θ = adjacent/hypotenuse, and tan θ = opposite/adjacent. Remember, a right triangle has one right (90°) angle and two acute (<90°) angles that sum to 90°. The total sum of the angles in a planar triangle is 180°. (To remember the side ratios use SohCahToa: sin = opposite/hypotenuse, cos = adjacent/hypotenuse, tan = opposite/adjacent.) The *trigonometric functions* can also be described using the coordinates of points on a circle with a radius of one and, due to their periodic nature, can be depicted on a graph.

4. **Roots**: The **square root** of any number, x, can be written $\sqrt{x}$ or $[x]^{1/2}$ or $(x)^{1/2}$. Likewise for a cube root or nth root: $\sqrt[3]{x} = [x]^{1/3}$ and $\sqrt[n]{x} = [x]^{1/n}$.

5. **Vectors** possess both magnitude and direction. *Letters representing vectors are boldface when their direction is being recognized.*

INTRODUCTION

Master Math: Essential Physics presents, teaches, and explains the fundamental topics of **algebra-based physics**. It includes engaging, fun examples and applications throughout the book, as well as challenging practice problems with explanatory answers at the end of each chapter. *Master Math: Essential Physics* was written for you, the student, parent, teacher, tutor, or curious thinker.

This book covers the essentials of high school and *algebra-based* college curricula. It can serve as a supplement to your textbook, a handy reference, or a tutor for lifetime learners. Topics encompass motion, force, momentum, Newton's Laws, equilibrium, friction, forces in nature, energy, work, elasticity, harmonic motion, static and moving fluids, heat, temperature, gas, electric fields, electromagnetism, direct and alternating current, waves, sound, radiation, light and optics, and an introduction to relativity, quanta, the atom, dark matter, and dark energy.

Master Math: Essential Physics logically presents each topic, developing and explaining the algebraic equations that represent the physical concepts. If you want to know what physics is all about, this book will provide you with a broad picture of the amazing and surprising macro and micro realities of our world. I must warn you, however, you may fall in love with this fascinating subject and decide to dive in and pursue a deeper understanding.

This book is a new addition to this author's contributions to the *Master Math* series, which include *Master Math: Basic Math and Pre-Algebra, Master Math: Algebra, Master Math: Pre-Calculus, Master Math: Geometry, Master Math: Trigonometry,* and *Master Math: Calculus.*

Enjoy delving into this fascinating world of physics and remember:

"The important thing is not to stop questioning. Curiosity has its own reason for existing. One cannot help but be in awe when he contemplates the mysteries of eternity, of life, of the marvelous structure of reality. It is enough if one tries merely to comprehend a little of this mystery every day."
Attributed to Albert Einstein

Chapter 1

MOTION

"All science is either physics or stamp collecting."
Attributed to Ernest Rutherford

"I do not feel obliged to believe that the same God who has endowed us with senses, reason, and intellect has intended us to forgo their use."
Attributed to Galileo Galilei

1.1. Average Speed

- **Average speed** is simply: total distance traveled per total time.

Average speed = (total distance traveled) / (total elapsed time)

$$v_{Ave} = \Delta x/\Delta t = (x_1 - x_0)/(t_1 - t_0)$$

v_{Ave} is average speed.
Δx is total distance traveled from initial point x_0 to final point x_1.
Δt is total elapsed time from initial time t_0 to final time t_1.
Note: The Greek letter Δ (Delta) is often used to denote "the change in."

• **Example**: What is your **average speed** if you walk 12 miles in 6 hours?

Total distance traveled per total elapsed time

12 mi / 6 h

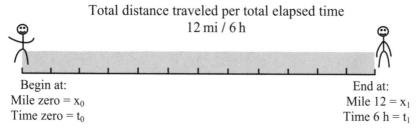

Begin at:
Mile zero = x_0
Time zero = t_0

End at:
Mile 12 = x_1
Time 6 h = t_1

Total distance traveled is the change in distance x from x_0 to x_1, or:

$$x_1 - x_0 = \Delta x$$
$$12 \text{ mi} - 0 \text{ mi} = 12 \text{ mi}$$

Total elapsed time is the change in time t between t_0 and t_1, or:

$$t_1 - t_0 = \Delta t$$
$$6 \text{ h} - 0 \text{ h} = 6 \text{ h}$$

Your average speed is:

$$v_{Ave} = \Delta x / \Delta t = (12 \text{ mi})/(6 \text{ h}) = 2 \text{ mi/h}$$

Next, suppose your friend walks the same 12 miles, but the first half she walks 2 mi/h and the second half she walks 1 mi/h. What is her **average speed**? (No, it's not one and one-half mi/h. Let's see why.)

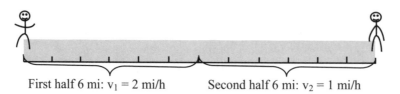

First half 6 mi: $v_1 = 2$ mi/h   Second half 6 mi: $v_2 = 1$ mi/h

Begin with average speed: $v_{Ave} = \Delta x / \Delta t$.
We know the total distance Δx is 12 mi.
6 mi are walked at 2 mi/h (mph) and 6 mi at 1 mi/h (mph).
To find average speed, solve $v_{Ave} = \Delta x / \Delta t$ for total time: $\Delta t = \Delta x / v_{Ave}$.
First segment: She walked 6 mi at 2 mph.

$$6 \text{ mi} / (2 \text{ mi/h}) = 3 \text{ h}$$

Second segment: She walked 6 mi at 1 mph.

$$6 \text{ mi} / (1 \text{ mi/h}) = 6 \text{ h}$$

Now substitute into equation for average speed:
average speed = (total distance traveled in both segments) / (total elapsed time)

$$v_{Ave} = (6 \text{ mi} + 6 \text{ mi})/(3 \text{ h} + 6 \text{ h}) = (12 \text{ mi})/(9 \text{ h}) = 1\tfrac{1}{3} \text{ mi/h}$$

Your friend's average speed was 1⅓ mph. Remember, she spent twice as long traveling at only 1 mph.

Average Speed Geometrically

• If an object is traveling at a **constant speed** over an interval of time, the speed is the same throughout that interval. The graph of distance traveled vs. time is a straight line, describing a linear function.

Distance vs. Time

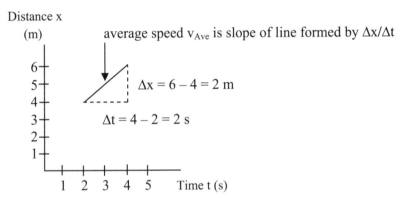

Distance x (m)

average speed v_{Ave} is slope of line formed by $\Delta x/\Delta t$

$\Delta x = 6 - 4 = 2$ m

$\Delta t = 4 - 2 = 2$ s

Time t (s)

Average speed v_{Ave} is the length of the vertical dashed line Δx divided by the length of the horizontal dashed line Δt. The average speed v_{Ave} is the slope of the line formed by $\Delta x/\Delta t$. The average speed in the interval shown in the above graph of an object moving at constant speed is:

$$v_{Ave} = \Delta x/\Delta t = (2\ m)/(2\ s) = 1\ m/s \quad \text{or about 2.2 mph}$$

• A greater (steeper) slope shows an object moving faster, since it covers more distance per time interval.

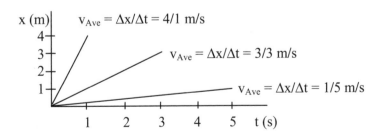

x (m)

$v_{Ave} = \Delta x/\Delta t = 4/1$ m/s

$v_{Ave} = \Delta x/\Delta t = 3/3$ m/s

$v_{Ave} = \Delta x/\Delta t = 1/5$ m/s

t (s)

1.2. Instantaneous Speed

• The **instantaneous speed** is the speed of an object at a specific time.

> $$v = \lim_{\Delta t \to 0}(\Delta x/\Delta t)$$
>
> The instantaneous speed v is:
>   the *limit* as Δt approaches zero of the average speed;
>   the *limit* as $\Delta t \to 0$ of $\Delta x/\Delta t$;
>   the slope of a line that is tangent to the distance vs. time
>   graph at a specified point.

• If speed is not constant, the graph of Distance vs. Time will NOT be a straight line. What if we want to find the speed at point A on a curve?

Distance vs. Time for Average Speed

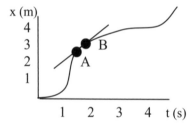

Slope of straight line
between A and B
is **average speed.**

The average speed is:
   v_{Ave} = (x value at B – x value at A) / (t value at B – t value at A) = $\Delta x/\Delta t$

The **average speed** between two points is the slope of the *straight line* between the two points.

Distance vs. Time for Instantaneous Speed

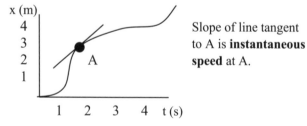

Slope of line tangent
to A is **instantaneous
speed** at A.

As point B moves toward A and time interval Δt approaches zero, the average speed between A and B becomes the instantaneous speed at A. The instantaneous speed v is:
$$v = \lim_{\Delta t \to 0}(\Delta x/\Delta t)$$

The **instantaneous speed** at a single point on a *curve* is the *slope of a tangent line at that point.*

To summarize: The slope of a line between two points on a graph of distance vs. time describes *average speed* during the time interval. Similarly, the slope of a tangent line at a single point on a curve describes the *instantaneous speed* at that point, so that *instantaneous speed* is the speed at a single point on a curve.

• Note: The equation for instantaneous speed uses the **limit** (abbreviated lim), which describes *closeness* to a value when the exact value cannot be identified. In calculus the *limit* is used in the development of the derivative. In fact, instantaneous speed is the first derivative of distance with respect to time (where dx/dt is the derivative of distance with time):

$$v = \lim_{\Delta t \to 0}(\Delta x/\Delta t) = dx/dt$$

For more on *limits* and the *derivative* see *Master Math: Pre-Calculus* and *Master Math: Calculus.*

• **Example**: What is the speed of an object in meters/second (m/s) at time t if the equation describing position is x = 7t + 3?

At time $t = 0$, $t = t_0$ and position is: $x_0 = 7t_0 + 3$.
At a time t_1, where $t_1 = t_0 + \Delta t$, position is: $x_1 = 7t_1 + 3 = 7(t_0 + \Delta t) + 3$.
To determine the **instantaneous speed** at any point use the equation:

$$v = \lim_{\Delta t \to 0}(\Delta x/\Delta t) = \lim_{\Delta t \to 0}[(x_1 - x_0) / (\Delta t)]$$

$$= \lim_{\Delta t \to 0}[((7(t_0 + \Delta t) + 3) - (7t_0 + 3)) / \Delta t]$$

$$= \lim_{\Delta t \to 0}[(7t_0 + 7\Delta t + 3 - 7t_0 - 3) / \Delta t] = \lim_{\Delta t \to 0}[(7\Delta t)/\Delta t] = 7 \text{ m/s}$$

This equation for v works out so that Δt cancels in the numerator and denominator leaving speed simply as $v = 7$ m/s.

1.3. Acceleration

• Just as **speed** $\Delta x/\Delta t$ is the rate of change of distance with time, **acceleration** $\Delta v/\Delta t$ is the rate of change of speed or velocity with time. Average acceleration is given by:

$$a_{Ave} = (v_1 - v_0) / (t_1 - t_0) = \Delta v/\Delta t$$

When you are traveling in a straight line, the speed or velocity changes during acceleration. If you push on the gas pedal in your car, the car accelerates and speed increases. When you press the brake, the car undergoes negative acceleration (or deceleration) and slows.

• **Units** of acceleration quantify: speed/time = (distance/time)/time = distance/time^2. Acceleration is often measured in m/s^2 or mi/h^2.

• Note that speed indicates magnitude while velocity is a vector which describes both magnitude and direction. Acceleration is also a vector and possesses both magnitude and direction. Vectors are discussed later in this chapter.

• Let's visualize constant speed with no acceleration and linearly increasing speed with constant acceleration.

Speed vs. Time

Speed is constant with time:
Motion with **no** acceleration

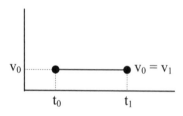

Speed increases linearly with time:
Motion with **constant** acceleration

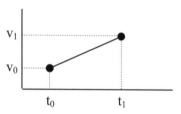

When **acceleration is constant,** a_{Ave} is a *constant* and can be written as "a". If time t_0 is zero, then the time interval is from 0 to t.

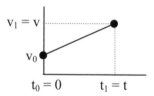

The equation for acceleration becomes:

$$a_{Ave} = a = (v_1 - v_0)/(t_1 - t_0) = (v - v_0)/(t - 0) = (v - v_0)/t = \Delta v/\Delta t$$

rearranging:

$$at = v - v_0$$

or:

$$\boxed{v = v_0 + at}$$

Speed v at time t is initial speed v_0 plus the additional speed caused by constant acceleration during time t.

1.4. Finding Distance Traveled at Constant Acceleration

• To find **distance traveled** *when acceleration is constant*, first remember that speed and distance are related:

$$v_{Ave} = \Delta x / \Delta t = (x_1 - x_0)/(t_1 - t_0)$$

If initial time is at $t_0 = 0$ and position begins at $x_0 = 0$, this equation can be rearranged as simply:

$$x = v_{Ave}t$$

This relationship is often referred to as *distance equals rate times time.*

• If *acceleration is constant* and the beginning and ending speeds v_0 and v are known, then the average speed between time $t_0 = 0$ and time t is the average of v_0 and v.

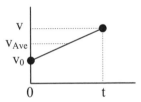

Average speed can be expressed as a simple average:

$$v_{Ave} = (1/2)(v_0 + v)$$

To find **distance** x, substitute v_{Ave} into $x = v_{Ave}t$:

$$x = (1/2)(v_0 + v)t = (1/2)v_0t + (1/2)vt$$

Substitute $v = v_0 + at$:

$$x = (1/2)v_0t + (1/2)(v_0 + at)t = (1/2)v_0t + (1/2)v_0t + (1/2)at^2$$

$$x = v_0t + (1/2)at^2$$

This equation for **distance traveled** involves the vt term which is the distance that would be traveled in the absence of acceleration, plus the acceleration-dependent term that is proportional to the square of elapsed time t^2.

If an object begins at rest so $v_0 = 0$, then the equation for distance traveled under constant acceleration becomes:

$$x = (1/2)at^2$$

• **To summarize**: Motion in one dimension involves distance x, time t, velocity or speed v, and acceleration a. Speed is the rate of change of distance, and acceleration is the rate of change of velocity. The average velocity and average acceleration are:

$$v_{Ave} = \Delta x/\Delta t \quad \text{and} \quad a_{Ave} = \Delta v/\Delta t$$

When **acceleration is constant**, the following equations describe motion.

$$a = (v - v_0)/t$$
$$v = v_0 + at$$
$$v_{Ave} = (1/2)(v_0 + v)$$
$$x = v_{Ave}t$$
$$x = v_0t + (1/2)at^2$$

We can develop an additional useful equation involving **velocity** by beginning with:

$$x = v_{Ave}t$$

Substitute $a = (v - v_0)/t$ or $t = (v - v_0)/a$:

$$x = v_{Ave}(v - v_0)/a$$

Substitute $v_{Ave} = (1/2)(v_0 + v)$:

$$x = (1/2)(v_0 + v)(v - v_0)/a$$
$$2ax = (v_0 + v)(v - v_0)$$
$$v^2 - v_0^2 = 2ax$$

$$v^2 = v_0^2 + 2ax$$

When acceleration is constant, motion uniformly accelerates, and the distance traveled will be proportional to t^2. When graphed, distance vs. time will form a parabola:

Distance vs. Time

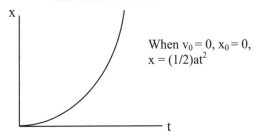

When $v_0 = 0$, $x_0 = 0$,
$x = (1/2)at^2$

• **Example Part 1**: Your friend was bragging about his recent adventures in his new Zodiac boat. He said he exited his local marina at 7.5 mph and then, with a constant acceleration, sped up to 30 mph. Just as his speed reached 30 mph, he arrived at his favorite little offshore island. It took him exactly 2 min, or 120 s, to reach the island from the marina. How many miles is the island from the marina?

This situation involves constant acceleration. We know speed and time and need to find distance and acceleration. The useful equations are:

$$a_{Ave} = \Delta v/\Delta t \quad \text{and} \quad x = v_0 t + (1/2)at^2$$

Change the miles/hour to feet/second:

$$7.5 \text{ mi/h} = (7.5 \text{ mi/1 h})(5280 \text{ ft/1 mi})(1 \text{ h/3600 s}) = 11 \text{ ft/s}$$

$$30 \text{ mi/h} = (30 \text{ mi/1 h})(5280 \text{ ft/1 mi})(1 \text{ h/3600 s}) = 44 \text{ ft/s}$$

The acceleration is:

$$a_{Ave} = \Delta v/\Delta t = (44 \text{ ft/s} - 11 \text{ ft/s}) / (120 \text{ s}) = 0.28 \text{ ft/s}^2$$

The distance from the marina to the island is:

$$x = v_0 t + (1/2)at^2$$

$$x = (11 \text{ ft/s})(120 \text{ s}) + (1/2)(0.28 \text{ ft/s}^2)(120 \text{ s})^2$$

$$= 1320 \text{ ft} + 2016 \text{ ft} = 3{,}336 \text{ ft}$$

The distance can be converted to miles:

$$3{,}336 \text{ ft} \times (1 \text{ mi})/(5280 \text{ ft}) \approx 0.63 \text{ mi}$$

The distance from the marina to the island is 0.63 mi (or about 1,014 m).

• **Example Part 2**: After spending a few hours on the island, your friend sped off the beach and at constant acceleration took his new Zodiac up to a speed of 50 mph at which point he passed a buoy which was 0.5 mi away. What was his acceleration, that is, how much did his speed increase each second?

We know initial and final speeds were zero and 50 mph, respectively, so average speed is:

$$v_{Ave} = (1/2)(v_0 + v) = (1/2)(0 + 50 \text{ mi/h}) = 25 \text{ mi/h}$$

We can use $x = v_{Ave}t$ to determine time:

$$t = x/v_{Ave} = (0.5 \text{ mi})/(25 \text{ mi/h}) = 0.02 \text{ h}$$

$$0.02 \text{ h} \times 3600 \text{ s/h} = 72 \text{ s}$$

Acceleration is $a_{Ave} = \Delta v/\Delta t$, where initial speed is zero, final speed is 50 mph, and the time interval is 72 s:

$$a = (50 \text{ mi/h}) / (72 \text{ s}) = 0.69 \text{ (mi/h)/s}$$

Therefore, during the 72 s of travel, the boat's speed increased by 0.69 mph each second.

1.5. Acceleration Due to Gravity—No Air Resistance

• An object in **free fall** accelerates downward due to the force of gravity. For now we will ignore the frictional effects of air resistance, which would slow the motion of a falling object. In the absence of air resistance, the acceleration experienced by any object falling near the surface of the Earth is approximately 32 ft/s^2 or 9.8 m/s^2. The **acceleration of gravity g** is approximately:

$$\boxed{g = 32 \text{ ft/s}^2 = 9.8 \text{ m/s}^2}$$

Note: Because Earth's density and terrain vary and it rotates, g is not exactly the same in all places. As an object flies further from the Earth, the gravitational force decreases inversely with the square of its distance from Earth's center. Equations for constant acceleration apply, with $a = g$.

• **Example**: Suppose you climb a mountain with a 1,000-ft (or 304.8-m) sheer drop-off on one side, and you stand at the edge and shove a smooth round rock over the edge. How fast will it be moving when it hits the bottom? How long does it take to hit bottom? (Note, if we ignore air resistance, the rock will be accelerated uniformly by gravity until it hits bottom.)

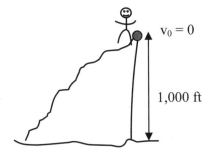

$v_0 = 0$

1,000 ft

$g = -32$ ft/s^2 = -9.8 m/s^2

The negative sign is used to show the downward direction of gravitational force.

We need equations involving distance and speed that account for the constant acceleration of gravity. (Note: acceleration a = gravity g.) When acceleration (of gravity) is constant, distance is:

$$x = v_0t + (1/2)at^2 = v_0t + (1/2)gt^2$$

Because initial speed $v_0 = 0$:

$$x = (1/2)gt^2$$

solve for t:

$$t^2 = 2x/g$$

or take the square root:

$$t = [2x/g]^{\frac{1}{2}}$$

Acceleration is:

$$a_{Ave} = \Delta v/\Delta t$$

We need to find the speed the rock was moving when it hits bottom. Rearrange to isolate v and note that initial speed $v_0 = 0$. Also substitute a = g since this describes constant acceleration due to gravity:

$$v = gt$$

Substitute t:

$$v = g[2x/g]^{\frac{1}{2}} = [g^2 2x/g]^{\frac{1}{2}} = [2gx]^{\frac{1}{2}}$$

or:

$$v^2 = 2gx$$

The downward direction is negative, so g = -32 ft/s^2 and x = $-1,000$ ft:

$$v^2 = (2)(-32 \text{ ft/s}^2)(-1,000 \text{ ft}) = 64,000 \text{ ft}^2/\text{s}^2$$

Take the square root:

$$v \approx \pm 253 \text{ ft/s}$$

Use the negative root so that -253 ft/s (about 77 m/s) is the speed of the rock in the downward (negative) direction when it hits the bottom.

To determine the time it takes to hit bottom, remember $v_0 = 0$ and a = g, and use v = gt. Rearrange to find t:

$$t = v/g = (-253 \text{ ft/s}) / (-32 \text{ ft/s}^2) \approx 7.9 \text{ s}$$

We can also find time using:

$$t = [2x/g]^{\frac{1}{2}} = [(2)(1{,}000 \text{ ft}) / (32 \text{ ft/s}^2)]^{\frac{1}{2}} \approx 7.9 \text{ s}$$

Therefore, when it hits bottom the rock will be moving 253 ft/s (or about 77 m/s), and it will take 7.9 s, or nearly 8 seconds.

• **Example**: Next you climb another higher mountain with a 2,000-ft drop-off and throw another smooth round rock vertically up in the air. It rises 20 ft before stopping and then falling, just clearing the edge of the cliff on the way down to the bottom. What was the rock's initial speed? What was its speed when it hit bottom, and how long did it take?

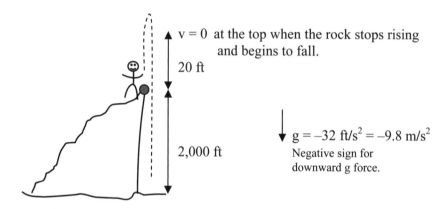

v = 0 at the top when the rock stops rising and begins to fall.

20 ft

2,000 ft

$g = -32 \text{ ft/s}^2 = -9.8 \text{ m/s}^2$
Negative sign for downward g force.

As the rock initially is rising, the acceleration of gravity is acting to reduce its speed until it stops its upward motion (v = 0) and begins to fall downward. The height the rock reaches when its speed decreases to zero is given as 20 ft. We can use the following equation for velocity we developed in section 1.4:

$$v^2 = v_0^2 + 2ax$$

At its high point, v = 0, x = 20 ft, and a = g = −32 ft/s². Solving for v_0^2 gives:

$$v_0^2 = -2(-32 \text{ ft/s}^2)(20 \text{ ft}) = 1{,}280 \text{ ft}^2/\text{s}^2$$

$v_0 \approx 36$ ft/s (about 11 m/s)   which is the rock's initial upward speed

Next, find the rock's speed when it hit bottom. There is often more than one way to determine an answer. Let's calculate the rock's final speed using two different methods. One method is to use the rock's entire trip and a second is to begin at the top of the rock's trajectory where its speed is zero. In either case, on the way down, the rock speeds up since its acceleration (due to gravity) is in the same direction as its speed.

First, calculate v using the entire trip of the rock. In this case, $v_0 = 36$ ft/s and $a = g = -32$ ft/s^2.

$$(v_{final})^2 = v_0^2 + 2ax$$

$$(v_{final})^2 = (36 \text{ ft/s})^2 + 2(-32 \text{ ft/s}^2)(-2{,}000 \text{ ft}) = 129{,}296 \text{ ft}^2/\text{s}^2$$

Take the square root (choose negative root since speed is downward):

$$v_{final} \approx -360 \text{ ft/s} \quad \text{the rock's speed when it hits bottom}$$

Alternatively, calculate v beginning at the highest point of the rock's path where $v_0 = 0$:

$$(v_{final})^2 = v_0^2 + 2ax$$

$$(v_{final})^2 = 0 + 2(-32 \text{ ft/s}^2)(-2{,}020 \text{ ft}) = 129{,}280 \text{ ft}^2/\text{s}^2$$

Take the square root (choose negative root since speed is downward):

$$v_{final} \approx -360 \text{ ft/s} \quad \text{the rock's speed when it hits bottom}$$

Both methods gave essentially the same answer for the rock's speed at the bottom (360 ft/s or 110 m/s), with the difference due to rounding errors. Finally, let's calculate how long it took for the rock to reach the bottom from its starting point when it was thrown upward from the cliff.

For the initial segment up, $v_0 = 36$ ft/s, $v_{final} = 0$, $a = g = -32$ ft/s^2.

$$a = (v_{final} - v_0)/t$$

$$t = (v_{final} - v_0)/a = (0 - 36 \text{ ft/s})/(-32 \text{ ft/s}^2) = 1.125 \text{ s}$$

For the down segment, $v_0 = 0$, $v_{final} = -360$ ft/s, $a = -32$ ft/s^2.

$$t = (v_{final} - v_0)/a = (-360 \text{ ft/s} - 0) / (-32 \text{ ft/s}^2) = 11.25 \text{ s}$$

Total time for the rock to reach bottom is up segment plus down segment:

$$1.125 \text{ s} + 11.25 \text{ s} = 12.375 \text{ s}$$

You can also calculate the total time for the rock's entire flight using one equation:

$$t = (v_{final} - v_0)/a = (-360 \text{ ft/s} - 36 \text{ ft/s})/(-32 \text{ ft/s}^2) = 12.375 \text{ s}$$

1.6. Acceleration Due to Gravity—With Air Resistance

• We have learned that an object in free fall accelerates due to gravity, often called the acceleration of gravity. Near Earth's surface:

$$g = 32 \text{ ft/s}^2 = 9.8 \text{ m/s}^2$$

Note: We sometimes insert a negative sign to denote the downward direction.

The acceleration of gravity affects all free-falling objects equally regardless of how long they have been falling, or whether they began from rest or were initially projected upward. ***In the absence of air resistance, all objects fall at the same rate regardless of size or mass.***

But doesn't this seem counterintuitive since we would imagine that a more massive object would fall or accelerate faster than a less massive one? In fact, a more massive object does **not** accelerate at a greater rate than a less massive one *in a vacuum in the absence of air resistance.*

In the presence of air resistance, however, a more massive object will fall faster, and the rates of falling of different objects are affected by their size and shape. Air creates friction on the falling object. **Friction** and other forces cause the actual motion of a falling object to deviate from its theoretical motion, creating what is known as **terminal velocity**.

Let's visualize this concept. If a person, a tiny animal, and a bug all fall from a third-story balcony, what will happen?

If you guess the person will experience serious injury and the tiny animal and the bug will survive, you are right. But why? If you answer that one reason is "**terminal velocity**" you are also right. But why?

Terminal velocity of a small bug is a few meters per second.
Insects and tiny animals have a large surface-area-to-volume ratio,
so the frictional resistance of the air limits the speed that they
fall to a certain speed, called "terminal velocity."
Terminal velocity of a human spread out is about 55–60 m/s
(120–135 mph), and tucked into a ball is about 90 m/s (200 mph).
The motion of a falling object through air is slowed by air resistance. An object's terminal velocity depends on its size, shape, and mass. The greater the area-to-volume or surface-to-mass of an object, the lower (slower) its terminal velocity. A smooth stone will fall faster than a cotton ball, because the cotton ball has more surface area relative to its weight with which to contact air.

Without air resistance an object falls faster and its speed increases the farther it falls. The equation for **speed ignoring air resistance**,

$$v = v_0 + at$$

suggests that the speed continues to increase. But when friction or air resistance is considered, a terminal speed for a falling object is reached, where the force of air resistance cancels the force of gravity, and acceleration ceases.

We can visualize the effect of terminal velocity on a graph:

Speed vs. Time

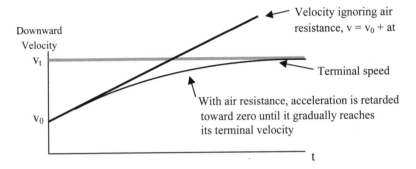

In real world situations, motion is retarded or impeded by frictional effects. Examples include air resistance on a falling object or friction slowing an object sliding down a sloped surface.

1.7. Vector and Scalar Excursion

• Both scalar and vector quantities are used in physics, so let's take a quick detour to understand the difference. In simple terms, **scalars** describe **magnitude** while **vectors** describe both **magnitude** (shown by vector's length) and **direction** (vector's pointing). This can be visualized by comparing the definitions of distance vs. displacement.

Distance vs. Displacement

• **Distance** is a **scalar** quantity indicating the length of the path between two points and is measured along that actual path (which could be curvy or circular) connecting the points. **Displacement** is the **vector** describing the magnitude of the separation between two points and the direction from the first to the second. The displacement is the separation between two points and is expressed as a vector whose magnitude is the straight line distance between them.

If a duck walks around the perimeter of a pond and returns to his starting point, the *distance* the duck traveled will be the total path length around the pond, but the duck's *displacement* is zero since he returned to his starting point. Displacement is independent of the path taken to get from one point to another.

• **Example**: If the duck then walks east 500 ft, turns around and walks west 200 ft, what is his distance and displacement?

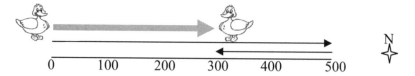

The duck's *distance* traveled was 700 ft (500 ft + 200 ft), but his *displacement* was 300 ft east (500 ft – 200 ft). (Note that when displacement has a value, it also has a direction.)

Summary of Properties of Scalars and Vectors

Scalars:

Quantities represent magnitudes.
Described by a real number that is positive, negative, or zero.
Can be compared with each other if they have the same units
  or physical dimensions such as walk 5 mi vs. walk 20 mi.
Examples include length, time, distance, speed, volume, temperature,
  work, density, and mass.
Do NOT indicate direction.

Vectors:

Describe both *magnitude* (vector's *length*) and *direction* (vector points).
Depicted as line segment or arrow with an initial point and
  an arrowhead at the terminal point.
The length of a vector represents its magnitude.
Examples of vectors include displacement, velocity, acceleration,
  force, weight, and momentum.

• A *displacement vector* represents the movement or displacement between two points in a *coordinate system*. The *length* of a displacement vector is the distance between the two points and the *direction* of a displacement vector is the direction it is pointing.

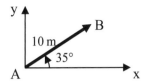

Displacement vector from point A to B is shown by the vector **AB** with magnitude (length) 10 m and direction 35° from the horizontal x direction.

• **Vectors** are written as one or two letters with an arrow over them, a boldface single letter (**A** or **a**), or two boldface letters with the first letter representing the initial point and second letter the terminal point (**AB**). Note: When just the magnitude of a vector is discussed, **boldface** type is not used.

• The **magnitude (or length)** of a vector is the absolute value of the displacement and therefore never a negative number. The direction of a vector can, however, be positive or negative. Remember: The absolute value of a number n is represented by $|n|$, where $|1| = 1$ and $|-1| = 1$.

• Note: Vectors that point in the *same direction* and have the *same length* (or *magnitude*) are ***equivalent vectors*** even if they are not in the same location. A vector can be relocated and still be considered the same vector as long as its length and direction remain the same. The ***negative of a vector*** is a vector with the same length but pointing in the opposite direction.

Addition and Subtraction of Vectors

• Two *vectors can be added or subtracted* if they have the same dimensions by adding or subtracting the corresponding components. For example, a two-dimensional vector can be added to another two-dimensional vector.

• In **vector addition** the initial point of the second vector is placed at the terminal point of the first vector. The sum is a third vector with its initial point at the initial point of the first vector and its final point at the final point of the second vector.

Illustration 1 shows **a** + **b** = **c** by placing the initial point of **b** at the final point of **a**. The sum is the vector joining the initial point of **a** to the final point of shifted **b**, or vector **c**.

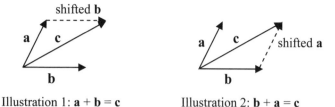

Illustration 1: **a** + **b** = **c**   Illustration 2: **b** + **a** = **c**

Illustration 2 shows **b** + **a** = **c** by placing the initial point of **a** at the final point of **b**. The sum is the vector joining the initial point of **b** to the final point of shifted **a**, or vector **c**.

Remember that the starting point of a vector can be moved as long as its length and direction stay the same. Note that vector addition is communicative, ie., **a** + **b** = **b** + **a**.

• ***Subtraction of two vectors*** is equivalent to adding the first vector to the negative of the second vector. The *negative of a vector* is a vector with the same length but pointing in the opposite direction.

To subtract two vectors, **a** − **b** = **c**, reverse the direction of the second vector, then add the first vector to the negative of the second vector. To do this, position the vectors so that the initial point of the negative **b** vector is at the final point of the first, **a**, vector. The sum is the vector **c** with its initial point at the initial point of the first vector and its final point at the final point of the second (negative) vector.

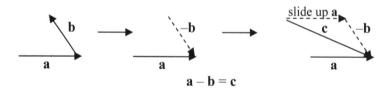

a − **b** = **c**

Just as two vectors can add to a single vector sum, a single vector can be resolved into two component vectors that sum to the original vector.

Vectors and Their Components in a Coordinate System

• A vector can be described by its horizontal and vertical components (which are also vectors) in a coordinate system.

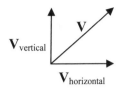

Vectors $V_{vertical}$ and $V_{horizontal}$ are each *component vectors* of vector **V**.

• **Right-Triangle Relations**: Remember from trigonometry that the *trigonometric functions* are defined according to the ratios of the three sides of a *right triangle*. A right triangle can be drawn alone or at the origin of a coordinate system. (See *Master Math: Trigonometry* for complete instruction of trigonometry.)

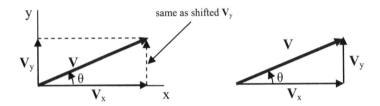

$$cosine \ \theta = \cos \theta = adjacent/hypotenuse = V_x/V$$
$$sine \ \theta = \sin \theta = opposite/hypotenuse = V_y/V$$
$$tangent \ \theta = \tan \theta = opposite/adjacent = V_y/V_x = \sin \theta / \cos \theta$$

Component vectors of vector **V** can be expressed as:

$$V_x = V \cos \theta$$
$$V_y = V \sin \theta$$

The **magnitude** of **V** is given by the Pythagorean Theorem:

$$|V| = [V_x^2 + V_y^2]^{1/2}$$

Remember, the **Pythagorean Theorem** states that the sum of the squares of the lengths of the sides of a right triangle equals the square of the length of its hypotenuse.

The **angle** θ is:

$$\tan \theta = V_y/V_x \quad or \quad \theta = \tan^{-1}(V_y/V_x) = \arctan(V_y/V_x)$$

• **A vector can be resolved into its components** that run along horizontal and vertical axes of a coordinate system.

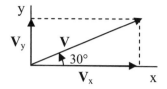

If vector **V** is at an angle θ of 30° from the x-axis:

Horizontal component is: $V_x = V \cos 30°$

Vertical component is: $V_y = V \sin 30°$

V is the vector sum of vectors V_x and V_y

• **Example**: If vector **V** is at an angle of 30° from the x-axis and its magnitude is 50, calculate vector components V_x and V_y. Show that the length of vector **V** is 50 using the component vectors V_x and V_y.

The vector components V_x and V_y are:

$$|V_x| = V \cos 30° = 50 \cos 30° \approx 43$$
$$|V_y| = V \sin 30° = 50 \sin 30° = 25$$

The **length or magnitude** of a vector **V** is the square root of the sum of the squares of its component vectors, V_x and V_y:

$$|V| = [V_x^2 + V_y^2]^{1/2}$$

where the magnitude, represented |n|, is the absolute value of number n. Using $V_x = 43$ and $V_y = 25$ calculated above, the length of vector **V** is:

$$|V| = [V_x^2 + V_y^2]^{1/2} = [43^2 + 25^2]^{1/2} \approx 50 \quad \text{(which was given initially)}$$

• Now that we have reviewed some basics of vectors, we can continue with our adventure into the world of motion.

1.8. Velocity vs. Speed

• When we study an object's speed or velocity, we look at how far it moves during a time interval. When considering how far the object moves, we are measuring either its distance or displacement. Remember: **distance** is a **scalar** indicating the length of the path between two points and is measured along that path (which could be curvy or circular), while **displacement** is the **vector** describing the magnitude of the separation between the two points and has direction from the first to the second. So how do **speed** and **velocity** differ?

Speed is the **rate of change of distance over time**.

Average speed = (total distance traveled) / (total elapsed time)

Velocity is the **rate of change of displacement over time**.

Average velocity = (displacement) / (time)

In English when we say *speed* or *velocity* we generally mean the same concept, but in physics, *speed represents the scalar quantity* and *velocity represents the vector quantity*.

Speed is a scalar quantity and describes how fast an object is moving. Speed is the rate at which an object covers distance.

Velocity is a vector quantity having both magnitude and direction and describes the rate at which an object changes its position. For example, if you quickly move around haphazardly but return to your starting position, you will have many instantaneous velocities, but a zero average velocity. Because you returned to the original starting position, your motion would not result in a net change in position. Since velocity is defined as the rate at which the position changes, this motion results in zero average velocity.

Velocity is a vector quantity which indicates both magnitude and direction. A *velocity vector* describes an object in motion and has a magnitude representing the speed of the object and a direction representing the direction of motion.

For motion in a straight line in the x direction, the **average velocity** is:

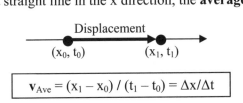

$$v_{Ave} = (x_1 - x_0) / (t_1 - t_0) = \Delta x / \Delta t$$

For any point along the path, the **instantaneous velocity** is found by taking the **limit** as the time interval becomes infinitesimal.

Instantaneous velocity = $v = \lim_{\Delta t \to 0}(\Delta x / \Delta t)$

Note: In calculus this *limiting* process defines the derivative:

$$v = \lim_{\Delta t \to 0}(\Delta x / \Delta t) = dx/dt$$

The **instantaneous speed** of an object is the magnitude of its instantaneous velocity (whether velocity is positive or negative).

$$v = |\mathbf{v}|$$

(The velocity is boldface because it is a vector.)

Since **displacement** is measured along the shortest path between two points, its magnitude is always less than or equal to the **distance** (which can be a curvy path). As the time interval becomes infinitesimal, the magnitude of the displacement approaches the value of the distance. Because speed is calculated using distance and velocity using displacement, speed and velocity will have essentially the same magnitude as the time interval approaches zero.

• **Units of both speed and velocity** are distance/time as measured in, for example, meters/second (m/s), miles/hour (mph), or feet/second (ft/s).

1.9. Acceleration Vector

• We learned that when an object is in free fall, it accelerates due to gravity in one direction—toward Earth. Position, velocity, and acceleration are related to each other and are vectors. *Velocity* is the rate of change of *displacement* with time, $\Delta x/\Delta t$. *Acceleration* is the rate of change of *velocity* with time, $\Delta v/\Delta t$. While vectors possess both magnitude and direction, there are situations when we are only concerned with magnitude. The magnitude of acceleration for an object in motion is the rate of change of velocity, and the direction of acceleration is the direction of the change in velocity.

• Units of acceleration are:

$$(\text{velocity})/(\text{time}) = (\text{distance/time})/(\text{time}) = (\text{distance})/(\text{time}^2)$$

This can be measured in: m/s^2, mi/h^2, or even $(mi/h)/(s)$.

• Let's look at some examples of velocity and acceleration vectors:

Suppose a car is moving 30 mph in a straight line in the +x direction at time $t_0 = 0$.

$t_0 = 0$ h, $v_0 = 30$ mi/h

At $t_1 = 1$ h, the car is moving
60 mph in the same +x direction.
The x-component of *velocity has
increased*, and the *acceleration vector
is in the +x direction* with:
$a = \Delta v/\Delta t = (30 \text{ mi/h})/(1 \text{ h}) = 30 \text{ mi/h}^2$

$t_1 = 1$ h, $v_1 = 60$ mi/h

30 mi/h^2

Now suppose that at time t_0, velocity
is 60 mph. The car is still in
the +x direction traveling in a
straight line. At time $t_1 = 1$, velocity
is $v_1 = 30$ mph. In this case
the *velocity has decreased* in the
x-direction and the *acceleration
vector is in the –x direction* with:
$a = \Delta v/\Delta t = (-30 \text{ mi/h})/(1 \text{ h})$
$= -30 \text{ mi/h}^2$

$t_0 = 0$ h, $v_0 = 60$ mi/h

$t_1 = 1$ h, $v_1 = 30$ mi/h

–30 mi/h^2

Next suppose at $t_0 = 0$ the car is
moving east at $v_0 = 30$ mph.
At $t_1 = 2$ min, the car is moving
north at $v_1 = 30$ mph.
Suppose you want to determine any
change in average acceleration.
You will need to find
$a_{Ave} = \Delta v/\Delta t = (v_1 - v_0)/\Delta t$, where
Δv is the change in velocity between v_1 and v_0.
Note: You can draw $v_1 - v_0 = \Delta v$ using vector
subtraction so that Δv has its initial point at the
initial point of v_1 and its final point at the final
point of negative v_0.

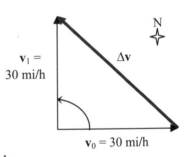

The magnitude of Δv is the square root of the
sum of the squares of its component vectors:
$$|\Delta v| = [v_1^2 + v_0^2]^{1/2} = [(30 \text{ mi/h})^2 + (30 \text{ mi/h})^2]^{1/2} \approx 42.4 \text{ mi/h}$$

The average acceleration is:
$$a_{Ave} = \Delta v/\Delta t = (42.4 \text{ mi/h})/(2 \text{ min}) = 21.2 \text{ mi/h per min}$$

Or in mi/h per second:
$$a_{Ave} = \Delta v/\Delta t = (42.4 \text{ mi/h})/(120 \text{ s}) \approx 0.35 \text{ mi/h per s}$$

Acceleration is about 21 mi/h per minute or 0.35 mi/h per second in the
northwesterly direction (because the change in velocity is toward the
northwest.

1.10. Two-Dimensional Motion: Projectiles

Horizontal Projectile

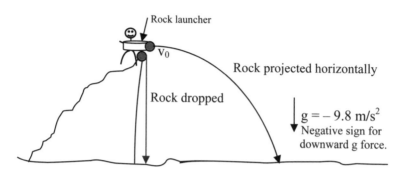

- If you drop an object it will accelerate downward due to gravity. If you project it sideways at some initial velocity in the horizontal x direction, its motion will be at an angle to the vertical and will have two components. Because velocity is a vector, it can be decomposed into its components.

The *vertical or y-component* of velocity undergoes acceleration due to gravity and increases linearly with time. The time that a projectile is in flight depends on the vertical component of its motion. We will generally choose the upward direction as positive y-direction and the downward direction as negative y-direction so that the acceleration of gravity will be in the downward, negative y-direction. Therefore, the vertical velocity component of an object beginning from rest is: $v_y = -gt$.

There is no force and therefore no acceleration in the x direction, so the *horizontal or x-component* of velocity proceeds at the constant initial velocity v_{0x}: $v_x = v_{0x}$.

The *horizontal displacement* of a projected object depends on the horizontal component of its motion and its time of flight. *The path of a horizontally projected object depends on the simultaneous effects of the horizontal and vertical components of its motion.* **The horizontal and vertical components act independently.**

Motion of Object Projected Horizontally Follows a Parabolic Path

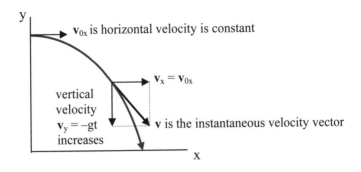

Equations for **object projected horizontally**:

	x dimension	y dimension
Displacement:	$x = v_{0x}t$	$y = v_{0y}t - (1/2)gt^2$
Velocity:	$v_x = v_{0x}$	$v_y = v_{0y} - gt$
Acceleration:	$a_x = 0$	$a_y = -g$

Note: Earlier in this chapter when we introduced acceleration, we *developed the equations for motion in one dimension.* At that time we did not consider the acceleration of gravity or using a negative sign to indicate direction. The above equations can be written without negative signs, but when we put in a value for gravity we would use a negative number. It is important to always keep track of positive and negative directions in physics models that involve motion or forces that move or act in a certain direction.

• **Example**: Suppose you hiked your rock launcher up to the top of a 300-m cliff and shot a rock horizontally at 10 m/s (22.4 mph). What is the horizontal distance traveled when the rock hits the ground? (Assume no air resistance.)

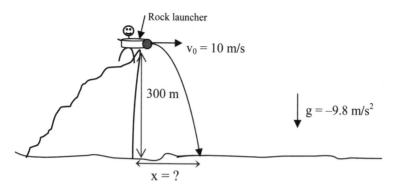

The displacement in the x direction is: $x = v_{0x}t$

To find t we can use: $y = v_{0y}t - (1/2)gt^2$

Because initial velocity $v_{0y} = 0$: $y = (1/2)gt^2$

Solve for t^2: $t^2 = 2y/g$

Substitute the vertical distance y, which is 300 m downward in the same negative direction as gravity, and gravity $g = -9.8 \text{ m/s}^2$:

$$t^2 = 2(-300 \text{ m})/(-9.8 \text{ m/s}^2) \approx 61 \text{ s}^2$$

Take the square root to find t: $t \approx 7.8 \text{ s}$

Now find the horizontal displacement:

$$x = v_{0x}t = (10 \text{ m/s})(7.8 \text{ s}) = 78 \text{ m} \quad \text{or about 256 ft}$$

Projectile Motion at An Angle

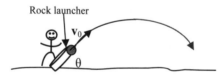

• The *velocity components of a projectile* shot at an **angle** to the horizontal can be derived from the right-triangle relationship for the components of a vector (discussed in Section 1.7). When the initial velocity is v_0 and the angle from horizontal is θ, the components of v_0 are:

$$v_{0x} = v_0 \cos\theta \quad \text{and} \quad v_{0y} = v_0 \sin\theta$$

The **displacement equations for a projected object** become:

$$x = v_{0x}t = (v_0 \cos\theta)t$$
$$y = v_{0y}t - (1/2)gt^2 = (v_0 \sin\theta)t - (1/2)gt^2$$

The equations for the **velocity components at some instant for a projectile** are found by combining the *initial velocity component* equations ($v_{0x} = v_0 \cos \theta$ and $v_{0y} = v_0 \sin \theta$) with the equations for an *object projected horizontally* ($v_x = v_{0x}$ and $v_y = v_{0y} - gt$) to give:

$$v_x = v_{0x} = v_0 \cos \theta$$

$$v_y = v_{0y} - gt = v_0 \sin \theta - gt$$

When discussing projectiles the ***maximum height*** and the ***horizontal distance in the x direction*** are often desired.

Parabolic Path of a Projectile

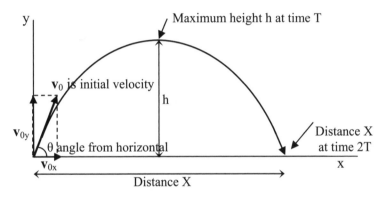

Maximum height h occurs when the projectile stops rising and its upward velocity slows to zero, $v_y = 0$, (just before it begins to fall). At the top, $y = h$, $v_y = 0$, and time $t = T$.

An equation for **time T at the top** can be found by substituting $v_y = 0$ and $t = T$ into the *y-component velocity equation,* $v_y = v_{0y} - gt$, and solving for T:

$$0 = v_{0y} - gT$$

$$T = v_{0y}/g$$

To express **time T at the top** in terms of initial velocity v_0 instead of v_{0y}, substitute the *y velocity component equation,* $v_{0y} = v_0 \sin \theta$:

$$T = (v_0 \sin \theta)/g$$

The **total time 2T** is twice the time to the top:

$$2T = 2(v_0 \sin \theta)/g$$

To find an equation for **height h at the top**, we substitute $y = h$ and $t = T$ into the *y-displacement equation,* $y = v_{0y}t - (1/2)gt^2$:

$$h = v_{0y}T - (1/2)gT^2$$

Substitute $T = v_{0y}/g$:

$$h = v_{0y}(v_{0y}/g) - (1/2)g(v_{0y}/g)^2$$

$$h = (v_{0y}^2/g) - (1/2)(v_{0y}^2/g) = (v_{0y}^2/g)(1 - 1/2) = (1/(2g))(v_{0y}^2)$$

To express **maximum height h** in terms of initial velocity v_0 instead of v_{0y}, substitute the *y velocity component equation,* $v_{0y} = v_0 \sin\theta$:

$$h = (1/(2g))(v_{0y}^2)$$

$$\boxed{h = (1/(2g))(v_0 \sin\theta)^2}$$

To find the **horizontal distance X traveled for a projectile** when time $t = 2T$ substitute $x = X$ and $t = 2T$ into the equation for *horizontal displacement,* $x = v_{0x}t$:

$$X = v_{0x}2T$$

To express X in terms of initial velocity v_0 instead of v_{0x}, substitute the *x velocity component equation,* $v_{0x} = v_0 \cos\theta$:

$$X = (v_0 \cos\theta)2T$$

Substitute total time $2T = 2(v_0 \sin\theta)/g$:

$$X = (v_0 \cos\theta)(2\,v_0 \sin\theta)/g$$

This may be rearranged as:

$$X = (2v_0^2/g) \cos\theta \sin\theta$$

or simplified using the trigonometric relation $2 \sin x \cos x = \sin 2x$:

$$\boxed{X = (v_0^2/g) \sin 2\theta}$$

Note that the maximum value for the sine function is at 1 which occurs when $2\theta = 90$ since $\sin 90 = 1$. When $2\theta = 90$, $\theta = 45$, which is the angle that produces the maximum distance. This was discovered by **Galileo** when he studied projectile motion and found that maximum distance occurred when the elevation angle was 45 degrees. Therefore, at $\theta = 45$, X_{max} occurs:

$$X_{max} = (v_0^2/g) \quad \text{at} \quad \theta = 45 \text{ degrees}$$

• **Example**: You upgrade your rock launcher into a mini cannon and shoot a cannon ball up from the ground at an angle of 60 degrees and an initial velocity of 40 m/s. What is the total time your cannon ball will be in the air? What is the time when it is at the high point? How high and far will it travel, and where will it be 1 s from launch?

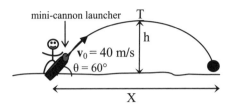

The total time is given by:

$$2T = 2(v_0 \sin \theta)/g = 2(40 \text{ m/s} \times \sin 60)/(9.8 \text{ m/s}^2) \approx 7.07 \text{ s}$$

Time at the highest point is: $T = (7.07 \text{ s})/2 \approx 3.5 \text{ s}$

The total horizontal distance is:

$$X = (v_0^2/g) \sin 2\theta = [(40 \text{ m/s})^2/(9.8 \text{ m/s}^2)] \sin 120 \approx 141 \text{ m}$$

Alternatively, you can calculate total horizontal distance using the T dependent equation:

$$X = (v_0 \cos \theta)2T = (40 \text{ m/s} \times \cos 60)(7.07 \text{ s}) \approx 141 \text{ m}$$

The maximum height is:

$$h = (1/(2g))(v_0 \sin \theta)^2 = (1/(2 \times 9.8 \text{ m/s}^2))(40 \text{ m/s} \times \sin 60)^2 \approx 61.22 \text{ m}$$

At $t = 1$ s, the location is given by the displacement equations:

$$x = (v_0 \cos \theta)t = (40 \text{ m/s} \times \cos 60)(1 \text{ s}) = 20 \text{ m}$$
$$y = (v_0 \sin \theta)t - (1/2)gt^2 = (40 \text{ m/s} \times \sin 60)(1 \text{ s}) - (1/2)(9.8 \text{ m/s}^2)(1 \text{ s})^2$$
$$\approx 29.74 \text{ m}$$

Therefore, at 1 s the cannon ball was at a horizontal distance of 20 m and a height of about 30 m.

1.11. Circular Motion

Angular Velocity

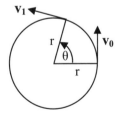

The *velocity vector* **v** of an object moving at a constant speed around a circle changes direction continually even though the speed may be constant.

• As an object moves along a circle, the angle θ changes. The rate at which the angle θ changes is the ***angular velocity*** ω. (The Greek letter omega usually denotes angular velocity.) The angular velocity is a measure of the change in angle θ as an object moves at a constant speed around a circle. The *angular velocity* ω is given by: $\omega = \Delta\theta/\Delta t$, where angle θ is in radians and is the measure of where object P is at time t.

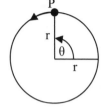

Angular velocity: $\omega = \Delta\theta/\Delta t$
Time required for one revolution or *period* is:
$\tau = 2\pi/\omega$
Speed of an object moving in a circular path
of radius r is: $v = r\omega$

An object moves at an *angular rate* of 1 revolution per second if it takes it one second to travel (at a constant speed) completely around a circular path. Because one revolution around a circle is 2π radians the object moves at an *angular velocity* of 2π radians per second. (Remember a full circle is 360° which equals 2π radians.)

The **time** required for one complete revolution or cycle of motion is the **period** τ of circular motion. (Note: We use the Greek letter τ, or tau, to denote period.) The period and the angular velocity are inversely related because the greater the angular velocity, the shorter the time required to make one revolution. The *period* τ is given by: $\tau = 2\pi/\omega$.

For an object moving at constant speed around a circle of radius r, the distance traveled in one period is the circumference of the circle, $2\pi r$. Given that the time required for one revolution is $\tau = 2\pi/\omega$, the *speed of the object moving in a circular path* of radius r is:

$$v = \text{distance/time} = (2\pi r)/(2\pi/\omega) = r\omega \quad \text{or just} \quad v = r\omega$$

If an object is moving at a constant speed around a circle, then the *distance traveled is an arc s*. Therefore, for an object moving around a circle at a constant velocity the distance along the arc is: $s = vt$.

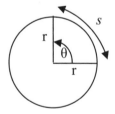

Solving $s = vt$ for v gives the *linear velocity* around the circle as: $v = s/t$. Combining the *angular velocity* $\omega = \Delta\theta/\Delta t$ with the equation for the *arc length s* of a circle, given by $s = r\theta$ we can relate the *linear and angular velocity* formulas as: $v = s/t = r\theta/t$.

Also, because $\omega = \theta/t$, then the following relations for *linear and angular velocity* can be written: $v = r\theta/t$, $v = r\omega$, $v = s/t$, $\omega = \theta/t$, $\omega = s/rt$, $\omega = v/r$.

Centripetal Acceleration

• In motion with no acceleration, the velocity vector is constant in magnitude and direction. If there is a change in velocity, then motion must have experienced acceleration. Acceleration may cause a change in either magnitude or direction. In fact, a change in the direction of the velocity vector requires acceleration even though the magnitude (speed) may remain constant. Because a change in the direction of the velocity vector requires acceleration, an object moving uniformly in a circular path is continually accelerated. The acceleration required to produce constant motion in a circle is *centripetal* (center seeking) *acceleration*.

• For those interested in math we show one way of developing the equations for *centripetal acceleration* in the following paragraphs and figures. Otherwise you can just examine the final equations.

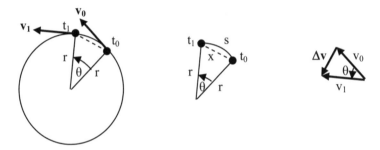

At time t_0 the tangent *velocity vector* of a moving object is v_0. At time t_1 the object has moved by $\Delta\theta$ and the velocity vector is v_1. (v_0 and v_1 are perpendicular to the radius lines r.)

As the tangent *velocity vector* moves around the circle, its direction changes toward the center of the circle so it remains tangent. If the motion is uniform the magnitude of velocity is $v_0 = v_1$, even though the direction of vectors v_0 and v_1 are different.

Between t_0 and t_1 the object moves along arc s by length Δs. The straight-line distance between t_0 and t_1 is Δx. For small time increments, $\Delta\theta$ gets small and $\Delta s \approx \Delta x$.

If we set the initial points of tangent vectors v_0 and v_1 together (right-hand figure), we see that $\Delta\theta$ about the circle's center equals $\Delta\theta$ between vectors v_0 and v_1. (The velocity vectors have to rotate the same $\Delta\theta$ to remain perpendicular to r.) A change from v_0 to v_1 requires acceleration. The change in the velocity vector from v_0 to v_1 is Δv. Component vectors of Δv are v_0 and v_1.

Note that as θ gets small, triangles r r s and $v_0 v_1 \Delta v$ above are similar isosceles triangles, so that:

$$\Delta x/r = \Delta v/v$$

where v can be either v_0 or v_1 (since magnitude $v_0 = v_1$).

Alternatively we can reason this because the length of arc s is $s = r\theta$ (or $\theta = s/r$) and the velocity vector changes by the same angle θ as the central angle through which the object moves. Therefore: $\Delta\theta = \Delta s/r = \Delta v/v$.
As $\Delta\theta$ and Δs become small, $\Delta s \approx \Delta x$, so that: $\Delta x/r = \Delta v/v$,
where v can be either v_0 or v_1 (since magnitude $v_0 = v_1$).

Solving $\Delta x/r = \Delta v/v$ for Δv gives: $\Delta v = v\Delta x/r$.

Earlier in this chapter we learned that the *instantaneous velocity* for any point along a path is found by taking the *limit* as the time interval becomes infinitesimal:

$$v = \lim_{\Delta t \to 0}(\Delta x/\Delta t)$$

Similarly, the *instantaneous acceleration* at a point along a path is the *limit* of $\Delta v/\Delta t$ or:

$$a = \lim_{\Delta t \to 0}(\Delta v/\Delta t)$$

Now substitute $\Delta v = v\Delta x/r$ for the centripetal acceleration:

$$a_c = \lim_{\Delta t \to 0}[(v\Delta x/r)/(\Delta t)] = \lim_{\Delta t \to 0}[(v/r)(\Delta x/\Delta t)]$$

Because v and r are constant and not affected by Δt getting small:

$$a_c = (v/r) \lim_{\Delta t \to 0}(\Delta x/\Delta t)$$

Because $v = \lim_{\Delta t \to 0}(\Delta x/\Delta t)$, the **centripetal acceleration** is:

$$\boxed{a_c = (v^2/r)}$$

Because $v = r\omega$:

$$a_c = (r\omega)^2/r$$

Therefore, **centripetal acceleration** is also written:

$$\boxed{a_c = r\omega^2}$$

We can also express **centripetal acceleration** in terms of the *period,* $\tau = 2\pi/\omega$:

$$a_c = r(2\pi/\tau)^2 = 4\pi^2 r/\tau^2$$

1.12. Key Concepts and Practice Problems

- Average speed is total distance traveled per total time: $v_{Ave} = \Delta x/\Delta t$.
- Instantaneous speed is the speed of an object at a specific time.
- Acceleration is the rate of change of speed or velocity with time.
- At constant acceleration $a = (v - v_0)/t$, $v = v_0 + at$, $x = v_0 t + (1/2)at^2$.
- Vectors describe both magnitude (or length) and direction.
- Speed is the rate of change of distance with time.
- Velocity is the rate of change of displacement with time.
- Angular velocity $\omega = \Delta\theta/\Delta t$.
- Centripetal acceleration $a_c = v^2/r$.

Practice Problems

1.1 A biplane pilot wants to win a U.S. Mail contract. To do so he needs to prove his capability by flying north from the airport 100 mi, turning, and flying back in no more than 4 h. **(a)** What speed does he need to average? **(b)** On his test day there is a steady 40 mph north wind. If his top airspeed is 80 mph, will he make it and if so by how many minutes? **(c)** At what wind speed must he cancel the trial and wait for a better day?

1.2 A runner must run 4 quarter-mile laps. She starts fast and completes the first lap in 1 min. **(a)** What speed did she average on the first lap? **(b)** She runs the second lap 3 mph slower, the third lap another 3 mph slower, and the last lap another 3 mph slower, finishing in about 6 min and 25 s. What was her average speed?

1.3 (a) Write an equation for the velocity as a function of time for an airplane that lands at 100 mph and stops in 5 s assuming constant deceleration. **(b)** At what time during the landing is the airplane's instantaneous speed equal to its average speed?

1.4 A rocket-powered dragster accelerates at a constant rate from standstill and does a quarter mile in 6 s. **(a)** What is the acceleration rate in ft/s^2 and top speed in mph? **(b)** Upon hitting the quarter mile mark, the driver pops the chute and brakes, slowing to a stop at a constant deceleration rate of one-third the magnitude of his acceleration. How far is he from the starting line when he stops?

1.5 A basketball player jumps with a hang time of 1 s (assume his legs remain straight and gravity is –32 ft/s). **(a)** How high did he jump? **(b)** What was his speed in ft/s when he came down? **(c)** With the same initial velocity, what would be his hang time and how high would he jump on the moon? (Assume $g_m = -5.3$ ft/s^2.)

1.6 If a skydiver falling at 125 mph pulls his ripcord and opens the parachute, what will happen? What will be the direction of acceleration until the rate of descent stabilizes?

1.7 Otis, a small white puppy, escapes his yard in a town made up of square blocks with 10 blocks per mile. He follows his nose 2 blocks west, chases a cat 3 blocks south, and then follows a boy on a bicycle 7 blocks east and then 7 blocks north. The boy finally tells Otis he should go home, so Otis sadly wanders 2 blocks west before realizing he is hopelessly lost. A blackbird offers to help. As the crow files, how far is Otis "displaced" from home?

1.8 At the beginning of the swim season, Donna could swim 100 m in 82 s (2 lengths of a 50 m pool). Practicing twice a day, she got her time down to 58 s. In m/s, how much had she increased her average velocity during a 100 m race?

1.9 A skydiver leaps from an airplane and reaches a terminal velocity of –50 m/s in 10 s. **(a)** What was her average acceleration? **(b)** When she reaches an altitude of 800 m, she releases her chute and within 4 s is descending at only 5 m/s. What was her average acceleration during those 4 s?

1.10 David slings a stone at 100 cubits/s toward a giant standing on a hill 25 cubits (cu) away whose eyes are 10 cu above the stone's release point. (Assume no air resistance, 1 cu = 18 in, and $g = -32$ ft/s^2). **(a)** If he aims at a point between the giant's eyes ($\theta \approx 21.8°$), where will the stone hit? **(b)** If he aims 14 in above the eyes ($\theta \approx 23.3214°$), will the stone hit the target? **(c)** If the giant ducks so the stone misses, how high and how far will the stone fly, and how long will it be airborne? (Assume release and landing points are ground level.)

1.11 (a) What is the angular velocity of the Earth around the Sun in rev/h? (Assume $r = 93{,}000{,}000$ mi, 1 year = 8,766 h, and circular orbit) **(b)** What is Earth's speed in its orbital path? **(c)** What is Earth's centripetal acceleration?

Answers to Chapter 1 Problems

1.1 (a) (100 mi + 100 mi)/4 h = 50 mph. **(b)** Speed north 80 – 40 = 40 mph, so 100 mi takes 2 h 30 min. Speed south 80 + 40 = 120 mph, so 100 mi takes 50 min. Total time is 3 h 20 min, beating the clock by 40 min. **(c)** Solve $100/(80 + s) + 100/(80 - s) \leq 4$ for max wind speed: $s = 49$ mph.

1.2 (a) (0.25 mi)/(1/60)h = 15 mph. **(b)** v_{Ave} = total distance/total time; time is 6 min 25 s ≈ 0.107 h, so 1 mi/0.107 h ≈ 9.35 mph.

1.3 (a) $v = v_0 + at$ or $v = 100 - 20t$. **(b)** average speed $(100 + 0)/2 =$ 50 mph, so 50 mph $= 100 - 20t$ or $t = 2.5$ s.

1.4 (a) $a = 2x/t^2 = 2(5280/4)/6.0^2 \approx 73.33$ ft/s^2. $(v - v_0) = at = 73.33 \times 6 =$ 440 ft/s $= 300$ mph. **(b)** $a = 73.33/3 = 24$ ft/s^2. Takes 3 times as long to stop, or 18 s. Average speed is 150 mph. Deceleration $x = vt =$ $150 \times 18s(1/3600) \approx 0.75$mi. Total distance $0.25 + 0.75 = 1$ mi.

1.5 (a) Time to top is 0.5 s; $x = (1/2)at^2$; $x = (1/2)(32)(0.5)^2 = 4$ ft. **(b)** $v = gt = (-32)(0.5) = 16$ ft/s downward. **(c)** $t = v/g = 16/5.3 = 3.02$ s. Time up and down is 2t or 6.04 s. $x = (1/2)at^2 = (1/2)(5.3)(3.02)^2 = 24.17$ ft.

1.6 The rate of descent decreases to a new lower terminal velocity. The acceleration is upward, opposite gravity.

1.7 Otis is 3 blocks east and 4 blocks north. $d^2 = 3^2 + 4^2 = 25$, $d = 5$ blocks $= 0.5$ mi.

1.8 Her velocity remained 0 since she finished where she started, but her speed increased by $(100$ m/s$) / (58$ m/s$) - (100$ m/s$) / (82$ m/s$) \approx 0.5$ m/s.

1.9 (a) $a_{Ave} = (v_1 - v_0)/t = (-50 - 0)/10 = -5$ m/s^2. **(b)** $a_{Ave} = (-5 - (-50))/4 = 11.25$ m/s^2.

1.10 (a) $v_{0x} = v_0 \cos \theta = 100 \cos 21.8 = 92.8$ cu/s; $t = x/v_{0x} = 25/92.8 =$ 0.269 s; $y = (v_0 \sin \theta)t - (1/2)gt^2 = 9.99$ cu $- 0.772$ cu $= 9.218$ cu; 10 cu $- 9.218$ cu $= 0.782 = 14$ in below eyes. Note: OK to use just: $y = (1/2)gt^2 = 0.772$ cu ≈ 14 in below eyes. **(b)** $v_{0x} = v_0 \cos \theta = 100 \cos 23.3214 = 91.83$ cu/s; $t = x/v_{0x} = 25/91.83 =$ 0.27224 s; $y = (v_0 \sin \theta)t - (1/2)gt^2 =$ $(100 \sin 23.3214)0.27224 - (1/2)(21.33)(0.27224)^2 = 9.9872$ cu; $(10$ cu $- 9.9872$ cu$) = 0.01278 = 0.23$ in. Very close, only 0.23 in below target. Alternatively, $y = (1/2)gt^2 = (1/2)(-21.33)(0.27224)^2 = -0.79045$ cu $= 14.23$ in, or 0.23 in below target. **(c)** Time to high point $T = (v_0 \sin \theta)/g = (100 \sin 23.3214)/(21.33) =$ 1.856 s, so total time $2T = 3.712$ s. Max height $h = (1/(2g))(v_0 \sin \theta)^2 =$ $(1/(2 \times 21.33))(100 \sin 23.3214)^2 = 36.739$ cu $= 55.11$ ft. Distance $X =$ $(v_0^2/g) \sin 2\theta = (100^2/21.33) \sin 46.6428° = 340.88$ cu $= 511.3$ ft. Alternate equation: $X = (v_0 \cos \theta)2T = 511.3$ ft.

1.11 (a) Using 8,766 h/yr, $\omega = \Delta\theta/\Delta t = (2\pi$ rev/yr$)(1/8760$ yr/h$) =$ 0.0007173 rev/h. **(b)** $v = r\omega = 93,000,000$ mi $\times 0.0007173$ rev/h $\approx$ 66,709 mph. **(c)** $a_c = v^2/r = 66,709^2/93,000,000 \approx 47.8$ mi/h^2. (This is 0.019 ft/s^2, or about 0.06% of the force of gravity on Earth.)

Chapter 2

FORCE, MOMENTUM, AND NEWTON'S LAWS

"I do not know what I may appear to the world, but to myself I seem to have been only like a boy playing on the sea-shore, and diverting myself in now and then finding a smoother pebble or a prettier shell than ordinary, whilst the great ocean of truth lay all undiscovered before me." Attributed to Sir Isaac Newton

"The more precisely the position is determined, the less precisely the momentum is known in this instant, and vice versa."
Werner Heisenberg, Uncertainty paper, 1927

2.1. Force

• Motion is influenced by the presence of forces. **Force** is required to produce a change in the state of motion of an object. To set an object in motion, change its direction, or stop its motion requires a push or pull force. A force is an influence that is capable of producing a change in the state of an object's motion. Force is a vector quantity and has both magnitude and direction. An object at rest will begin to move in the direction it is pushed by a sufficient net force.

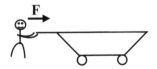
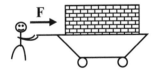

The amount of force required to move an object of great mass, such as a cart full of bricks, is much larger than the force required to move an object of smaller mass, such as an empty cart. Applying the same push for the same time period to the brick-loaded cart and the empty cart will result in the state of motion of the heavy cart being changed much less than the state of motion of the light cart.

Inertia is the property of matter that causes it to resist a change in its state of motion and is related to its mass. The brick-loaded cart has more inertia than the empty cart.

• The laws of physics are valid under different reference frames. For example, you can perform a set of physics experiments in a stationary lab on the ground and then repeat those experiments in a train moving at a *constant velocity,* and the results will be the same. Let's examine this by measuring acceleration on the ground:

$$\mathbf{a} = \Delta\mathbf{v}/\Delta t = (\mathbf{v}_1 - \mathbf{v}_0) / (t_1 - t_0)$$

Then get on the train and re-measure, but add a constant velocity $\mathbf{v}_T$ to account for the moving train:

$$\mathbf{a} = [(\mathbf{v}_1 + \mathbf{v}_T) - (\mathbf{v}_0 + \mathbf{v}_T)] / (t_1 - t_0) = (\mathbf{v}_1 - \mathbf{v}_0) / (t_1 - t_0)$$

The $\mathbf{v}_T$'s cancel! Newton's Laws are valid whether you are at rest or moving at a constant velocity (not undergoing acceleration).

• **Friction** is an important part of our world. Its effects keep cars on the road and buildings standing. We will mostly ignore friction in this chapter so we can focus on learning the basic principles of force, but will discuss friction in detail in Chapter 3. When you are modeling a situation or solving a problem, the effects of friction can only be ignored when they are negligible compared to other forces. Friction can often be reduced but never completely eliminated. Frictional forces generally exist and act *opposite to the direction of motion.* Friction is a *retarding force* that acts to slow down a moving object.

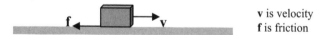

f ← ■ → v   v is velocity
                   f is friction

Units of Force

• Force is measured in units of (mass) × (length) / (time)^2, which is force times acceleration. Units of force in the SI, CGS, and ft-lb systems are as follows (see *Important Definitions* page xi):

SI or MKS system: A **Newton** N (named after Sir Isaac Newton 1642–1727) is the force required to accelerate a 1 kilogram mass 1 meter per second over a second's time.

$$1 \text{ N} = 1 \text{ kg·m/s}^2$$

CGS system: A **dyne** is the force required to accelerate a 1 gram mass 1 centimeter per second over a second's time.

$$1 \text{ dyne} = 1 \text{ g·cm/s}^2 \quad \text{and} \quad 1 \text{ N} = 100,000 \text{ dynes}$$

Foot-pound-second system: A **pound-force** (lbf) is the force required to accelerate a 1 **slug** mass 1 foot per second over a second's time.

$$1 \text{ pound-force} = \text{slug·ft/s}^2$$

The conversion factors between SI and foot-pound-second systems for force and mass are:

Force: 1 Newton = 0.2248 pounds, 1 pound = 4.448 Newtons.

Mass: 1 kilogram = 0.06852 slugs, 1 slug = 14.59 kilograms.

Note: The conversion between kilograms and pounds is often given as 1 kilogram = 2.2 lb, which only applies when the acceleration due to gravity is 9.8 m/s^2 near Earth's surface.

Note: Rather than just writing "pound", it is more accurate to write "pound-force" or "lbf" since there is often confusion with the term "pound" and whether it applies to mass or weight (force). Pounds and Newtons are units that measure force or weight. Kilograms and slugs are units of mass. Just to make this more confusing, while "slugs" are the units of mass in the foot-pound-second system, "pound-mass" (lbm) is also used. The relationship between them is: 1 slug = 32.17 lbm. When you are on the Earth's surface, where gravity is about 32 ft/s^2, one pound-mass weighs one pound-force and the mass of 1 slug weighs 32 lbf. We can also see that a pound-force is the force required to accelerate a 1 pound-mass 32.17 feet per second over 1 second's time.

2.2. Newton's Laws of Motion

• **Newton's Laws of motion** (developed by Sir Isaac Newton) relate force, mass, and the change in the state of motion. While modern physics has discovered that for infinitesimally small distances or extremely high velocities Newton's Laws require adjustments, for most circumstances the Newtonian Laws describe the dynamics of physical systems. The more extreme situations led to the development of quantum mechanics and relativity theories.

• **Newton's First Law**: *Every object persists in its state of rest or uniform motion in a straight line unless it is compelled to change that state by forces impressed on it.* In other words, an object will remain motionless or continue moving uniformly in a straight-line unless it is compelled to change by the application of an external force. This is also referred to as the *law of inertia.*

If no net force is applied to an object at rest (with zero velocity), it will remain at rest. If no net force is applied to an object in uniform motion, it will maintain that constant velocity. That is, if the net force $\mathbf{F}_{net}$ on an object is zero, so that the vector sum of all forces acting on it is zero, then the acceleration $\mathbf{a}$ of the object is zero, and it either remains at rest or continues to move with a constant velocity. If a non-zero force is applied, the velocity will change in the direction of the net force.

Newton's First Law suggests if $\mathbf{F}_{net} = 0$, then $\mathbf{a} = 0$ and $\mathbf{v} = $ zero or a constant. This means if we slide a smooth object across a *frictionless* icy surface, it will move forever without any change in speed or direction. Newton's First Law addresses zero force, zero acceleration, and constant velocity.

• **Newton's Second Law**: *For an object with constant mass, the force $\mathbf{F}$ is the product of the object's mass and its acceleration $\mathbf{a}$:*

$$\mathbf{F} = \mathbf{ma}$$
$$\mathbf{a} = \mathbf{F}/m$$

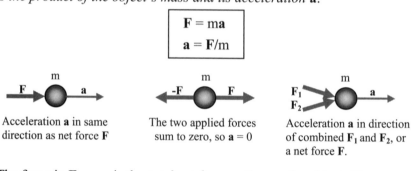

Acceleration **a** in same direction as net force **F**

The two applied forces sum to zero, so **a** = 0

Acceleration **a** in direction of combined $\mathbf{F}_1$ and $\mathbf{F}_2$, or a net force **F**.

The force in $\mathbf{F} = \mathbf{ma}$ is the *total net force acting on* the object. The accelerated motion of the object can only be produced by the application of a force to that object. The *direction* of the acceleration is the same as the *direction* of the net force, and the magnitude of the acceleration is proportional to the magnitude of the force.

A force will cause an acceleration and an acceleration will generate a force. The equation works both ways. When an external force is applied, the change in velocity depends on the mass or inertia of the object. Newton's Second Law suggests that an unbalanced force causes a mass to accelerate. When the net force on a system is zero, there is no acceleration.

Newton's Second Law specifies the dimensions of force as mass times acceleration. Using the system of kilograms, meters, and seconds, this gives the units as:

Force units = (mass units) × (acceleration units) = kg × m/s^2, which is often written kg·m/s^2. In the MKS system, a net force **F** acting on a 1-kg mass causes it to accelerate 1 m/s^2. The magnitude of the force is defined as 1 **Newton (N)** or:

$$1 \text{ N} = 1 \text{ kg·m/s}^2$$

Example: If 5 N of force is applied northward to a 5-kg object at rest for 5 seconds, how far will it move and what is its final velocity?

$$a = F/m = (5 \text{ N})/(5 \text{ kg}) = 1 \text{ N/kg} = (1 \text{ kg·m/s}^2)/(\text{kg}) = 1 \text{ m/s}^2$$

$$\text{Velocity } v = at = (1 \text{ m/s}^2)(5 \text{ s}) = 5 \text{ m/s} \text{ in the northward direction}$$

$$\text{Distance } x = (1/2)at^2 = (1/2)(1 \text{ m/s}^2)(5 \text{ s})^2 = 12.5 \text{ m}$$

• Calculus note: Newton's Second Law also defines a **force** to be equal to the differential change in **momentum** per unit time. The momentum is defined to be the product of the object's mass m and its velocity **v**. The differential equation for force **F** is: $F = d(mv)/dt$.

The law defines a **force** to be equal to the change in **momentum** (mass times velocity) per change in time.

$$F = \Delta p/\Delta t = \Delta(mv)/\Delta t \quad \text{or} \quad F = \lim_{\Delta t \to 0}(\Delta p/\Delta t)$$

(These "changes" are expressed most accurately using the derivative.) If the mass remains constant, we can use the definition of acceleration as the change in velocity with time ($a = \Delta v/\Delta t$) to reduce the Second Law to the product of mass and acceleration: $F = ma$.

Since **acceleration** is a change in velocity with a change in time **t**, we can also write this equation for the average force:

$$F = m(v_1 - v_0) / (t_1 - t_0)$$

• **Newton's Third Law**: *For every action (force), there is an equal and opposite reaction. If object A exerts a force on object B, then object B exerts an equal, oppositely-directed force on object A.*

$$F_{AB} = -F_{BA}$$

Note: Different notation may be used for the subscripts such as $F_A = -F_B$.

This Law suggests that single forces cannot occur and that forces always act in action/reaction pairs. When one object exerts a force on a second, the second exerts a force on the first that is equal in magnitude and opposite in direction.

The Third Law can be used to explain the production of thrust by a jet or rocket engine. Hot exhaust gases are produced which flow out the back of the engine, and, in reaction, a thrusting force is produced in the opposite direction. Another example is leaping off the end of a small boat toward a dock. If you don't have the boat tied up, as you leap the boat can lurch away from the dock. As you leap out of the boat toward the dock, your legs apply an equal but opposite force to the boat.

If you have two objects that interact only with each other and nothing else, you can use Newton's Third Law, $\mathbf{F}_A = -\mathbf{F}_B$, combined with the Second Law, $\mathbf{F} = m\mathbf{a}$, to give:

$$\mathbf{F}_A = -\mathbf{F}_B \quad \text{or} \quad m_A\mathbf{a}_A = -m_B\mathbf{a}_B$$

In an isolated system the equation $m_A\mathbf{a}_A = -m_B\mathbf{a}_B$ can be used to find one of the masses if the other mass and the ratio of $\mathbf{a}$'s are known:

$$m_B = m_A(-\mathbf{a}_A/\mathbf{a}_B)$$

The negative sign reflects that the accelerations have opposite directions.

2.3. Mass vs. Weight, Normal Force, and Tension

Mass vs. Weight

• **Mass** is an intrinsic property of an object and is independent of location and gravitational forces. An object's mass is the same on the Earth's surface, at the bottom of the ocean, on Mars, or in a weightless orbit. Mass depends on the number and type of atoms an object possesses. Newtonian physics defines mass as a measure of an object's inertia. **Inertia** measures an object's ability to resist a change in its state of motion when it is acted upon by a force. Remember from our discussion of Newton's Second Law, a net force of 1 Newton (N) acting on a 1-kg mass causes it to accelerate 1 m/s^2 (regardless of the object's location). An object's mass also determines how much gravitational force it will exert when acted on by gravity: $F = mg$.

Weight depends on, and will vary with, an object's location, such as its altitude above the Earth or what planet it is on. Weight reveals how Earth interacts with the object. Weight is the force a body *exerts on its support* (such as Earth). When an object is *not* accelerating, its weight **w** equals the gravitational force $\mathbf{F}_G$ that is *exerted on it*: $\mathbf{w} = \mathbf{F}_G$. In other words, weight is the amount of gravitational force acting on an object.

The **mass** and **weight** of an object are related by the acceleration due to gravity. On Earth's surface you can convert between **weight** and **mass** using: $w = mg$. The value of g is about g = 32 ft/s^2 = 9.8 m/s^2 on Earth's surface (though it varies). For example, a 60-kg woman on the Earth's surface weighs: $w = mg = (60 \text{ kg})(9.8 \text{ m/s}^2) = 588 \text{ N}$.

If you were to leave the Earth and go to the Moon, your mass would remain constant, but your weight would change. This is because the gravitational force exerted by the Earth on you as you stand on its surface is greater (by about 6 times) than the gravitational force exerted by the Moon as you stand on its surface.

Acting Forces

• As you stand on Earth, it is exerting a gravitational force $\mathbf{F}_G$ on you and you are exerting an opposite gravitational force $-\mathbf{F}_G$ on the Earth. If you stepped off a cliff you would accelerate toward the center of the Earth from the force of gravity (while the Earth accelerates toward you at a negligible rate). As long as you are resting on the surface, there is a balance of forces that keep you from falling.

The force you are exerting on Earth's surface is your weight $\mathbf{w} = \mathbf{mg}$, which is the same as the gravitational force exerted by the Earth on you: $\mathbf{w} = \mathbf{F}_G$. The reaction to force $\mathbf{w}$ is the "**normal**" force $\mathbf{N} = -\mathbf{w}$, which is the oppositely directed force the Earth's surface exerts on you. (The word "normal" comes from the fact that this force acts *perpendicular or normal* to the surface.)

While acting forces $\mathbf{F}_G$, $-\mathbf{F}_G$, $\mathbf{w}$, and $\mathbf{N}$ all have the same magnitude, the equal and oppositely directed *action-reaction pairs* are pair $\mathbf{F}_G$ and $-\mathbf{F}_G$ and pair $\mathbf{w}$ and $\mathbf{N}$. The forces acting on you are $\mathbf{F}_G$ downward and normal force $\mathbf{N}$ upward. The forces acting on the Earth are $\mathbf{w}$ and $-\mathbf{F}_G$. All the forces balance and there is no net force acting on you or the Earth and therefore no acceleration is experienced.

Normal Force

• The **normal force** is the force that the surface of one body exerts on the surface of another body. It is by definition *always perpendicular to the surfaces* in contact. The normal force is an *action-reaction force* that *reacts* to an external force pushing the object into the surface. Units are the same units as any force: kg·m/s^2, or Newtons N.

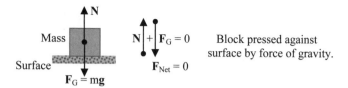

$$F_G = mg$$

$$N + F_G = 0$$
$$F_{Net} = 0$$

Block pressed against surface by force of gravity.

For an object sitting on a horizontal surface, the normal force is equal and opposite to the object's weight. The two forces acting on the object are the downward gravitational force $\mathbf{F}_G$ and the upward normal force $\mathbf{N}$. The net force is zero so the object is not accelerating.

$$F_G = mg$$

Block pressed against surface by an applied force.

For an object held against a vertical or slanted surface, the normal force acts perpendicular to the surface whether the surface points sideways or at an angle. The normal force is equal but opposite to the applied perpendicular force pressing the surfaces together.

• Suppose you decide to take your bathroom scale into an elevator. When the elevator is at rest the scale shows your weight: $\mathbf{w} = \mathbf{mg}$.

Scale measures your weight

Up Elevator: You weigh $\mathbf{w} = m(\mathbf{g}+\mathbf{a})$ during acceleration upward.

When the elevator accelerates upward, you feel downward pressure as the scale exerts an upward force on you of $\mathbf{w} = m(\mathbf{g} + \mathbf{a})$. During acceleration you are exerting a downward force on the scale reflecting the weight: $\mathbf{w} = m(\mathbf{g} + \mathbf{a})$.

After a few seconds the elevator stops accelerating and moves up as a constant speed, and scale reflects your normal weight: $\mathbf{w} = m(\mathbf{g} + \mathbf{0})$.

Scale measures your weight

Down Elevator: You weigh $\mathbf{w} = m(\mathbf{g}-\mathbf{a})$ during acceleration downward.

Uh oh. You hear a funny noise and the elevator begins to drop. If the elevator accelerates downward, the force you would exert on the scale would be: $\mathbf{w} = m(\mathbf{g} - \mathbf{a})$.

If the elevator goes into free fall, the downward acceleration would be **g**, and your weight would be: $w = m(g - g) = 0$. You would be weightless and therefore could not exert a force on another object, even though gravitational force $F_G = mg$ is acting on you. Fortunately, something grabs and the elevator decelerates and resumes a smooth descent. *Next time I'll take the stairs*, you say to yourself.

Inclined surface

• If you see an object on a flat, angled surface, you may be able to model it as an **inclined plane problem**. In the diagram the box is set on a *frictionless inclined surface* at an angle θ. It is acted on by F_G and **N**. The net force F_{Net} is the vector sum of F_G and **N** and takes into account the inclined angle: $F_{Net} = mg \sin θ$.

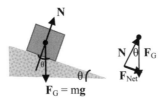

• **Example**: If a 100-pound block is sitting on a frictionless incline of 30°, what force is required to hold the block in place and what is the force of the block perpendicular to the incline plane?

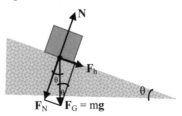

The object is acted on by the force of gravity F_G and normal force **N**. The net force F_{Net} produces acceleration and is the vector sum of F_G and **N**. The incline angle θ must be considered in the net force: $F_{Net} = mg \sin θ$. There is a force vector horizontal to the plane F_h and a force vector normal to the plane F_N, which are components of the object's weight mg. The component of mg in the "normal" N direction must be equal to **N** because these forces balance, and there is no net force or acceleration normal (perpendicular) to the incline surface. From the diagram we can use trigonometry to find the values of the vectors F_h and F_N. (Remember SohCahToa: sine = opposite/hypotenuse, cosine = adjacent/hypotenuse, tangent = opposite/adjacent.)

$$\sin \theta = |\mathbf{F_h}| / |mg| \quad \text{or} \quad \mathbf{F_h} = |mg| \sin \theta$$
$$\cos \theta = |\mathbf{F_N}| / |mg| \quad \text{or} \quad \mathbf{F_N} = |mg| \cos \theta$$

$\mathbf{F_h}$ is the force along the plane, therefore a force at least equal to and opposite to $\mathbf{F_h}$ is required to hold the block in place. Notice that we were given 100 lb, which is a force or weight that already accounts for the gravity on Earth's surface. To find the force required to hold up the block calculate:

$$\mathbf{F_h} = |mg| \sin \theta = (100 \text{ lb}) \sin 30° = 50 \text{ lb}$$

The block exerts a force $\mathbf{F_N}$ normal to the incline plane, which is equal and opposite to the normal $\mathbf{N}$ force. Its magnitude is:

$$\mathbf{F_N} = |mg| \cos \theta = (100 \text{ lb}) \cos 30° \approx 86.6 \text{ lb}$$

• An inclined plane can be helpful in everyday life. If $F_{horizontal}$ is much less than $F_{Gravity}$, the incline can act like a simple machine. For example, a person may not be able to lift a full oil drum up onto a truck, but she may be able to roll it up a ramp if the angle that the ramp makes with the ground is small.

Tension

• **Tension** is a force transmitted by ropes, strings, cables, chains, cords, etc., as part of an action-reaction pair. The direction of the force of tension is parallel to the rope or string. Because tension is a type of force, it has the same units as any force: kg·m/s² or Newtons N.

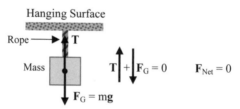

Hanging object (above): There are two forces acting on the hanging object: the downward gravitational force $\mathbf{F_G}$ and the upward force $\mathbf{T}$ due to tension in the rope. The net force is zero so the object is not accelerating. For an object hanging from a non-stretchable rope, the tension force will be equal and opposite to the object's weight.

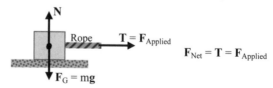

Pulled object (above): If an object is connected to a non-stretchable rope and pulled along a *frictionless surface* by a horizontally applied force $F_{Applied}$, the force will produce a tension T in the rope which will act on the object. The net forces acting on the object are:

$$F_{Net} = F_G + N + T = T = F_{Applied}$$

where F_G and N cancel since there is no net force or acceleration normal to the surface. Therefore, $F_{Net} = F_{Applied} = ma$ and the object is accelerating: $a = F_{Applied}/m$.

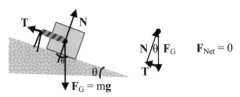

Object on incline (above): If an object is held motionless by a non-stretchable rope on a frictionless inclined surface at an angle θ, it will be acted on by tension T, F_G, and N. The net force F_{Net} is the vector sum of T, F_G, and N: $F_{Net} = 0$.

• You are still intrigued by the concept of weighing more and less depending on acceleration. You get a new idea and bravely venture back to the elevator. What if you suspend a rope from the ceiling of the elevator and then the elevator accelerates upward. You wonder what would be the tension T on the rope during the upward acceleration.

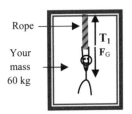

Elevator at Rest.
You weigh: $w = mg$
$= (60 \text{ kg})(9.8 \text{ m/s}^2) = 588 \text{ N}$
$F_{Net} = 0$

When the elevator (above) is *not moving* the gravitational force on you is $F_G = mg = (60 \text{ kg})(9.8 \text{ m/s}^2) = 588 \text{ N}$. You are exerting a force of the same magnitude on the rope, or $mg = 588 \text{ N}$. The tension T_1 on the rope is the reaction to your weight and is also $mg = 588 \text{ N}$. The net force on you is zero: T_1 up plus F_G down, which are equal and opposite, and result in zero acceleration: $F_{Net} = 0$.

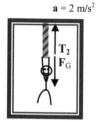

$a = 2$ m/s^2

T_2
F_G

Elevator accelerates upward.
You weigh: $\mathbf{w} = m(\mathbf{g} + \mathbf{a})$
$= (60$ kg$)(9.8$ m/s$^2 + 2$ m/s$^2) = 708$ N
$\mathbf{F}_{Net} = m\mathbf{a}$

The elevator (above) *accelerates upward* at 2 m/s^2. This increases the tension on the rope to $\mathbf{T}_2$. The net force on you is $\mathbf{T}_2$ up plus $\mathbf{F}_G$ down, which is equal to your mass times the acceleration: $\mathbf{T}_2 - \mathbf{F}_G = m\mathbf{a}$. The tension in the rope is found by solving $\mathbf{T}_2 - \mathbf{F}_G = m\mathbf{a}$ for $\mathbf{T}_2$:

$$\mathbf{T}_2 = \mathbf{F}_G + m\mathbf{a} = m\mathbf{g} + m\mathbf{a} = m(\mathbf{g} + \mathbf{a})$$
$$= (60 \text{ kg})(9.8 \text{ m/s}^2 + 2 \text{ m/s}^2) = 708 \text{ N}$$

Notice that the rope tension is equal to your weight with the added acceleration.

Tension and Pulleys

• **Tension** forces can be illustrated using non-stretchable ropes or strings whose mass and friction are ignored. The ropes or strings are held by *frictionless* **pulleys** which allow for changing the direction of forces and motion. These ropes or strings transmit tension forces in action-reaction pairs. For two objects connected by the rope or string threaded through a pulley, a force directed by the first object occurs concurrently with a force directed along the rope or string to the second object.

• **Example**: Suppose two objects are suspended on a frictionless, non-stretchable rope hanging over a frictionless pulley so that the magnitude of the tension T is the same on either side of the pulley. Object m_1 has a mass of 4 kg and object m_2 a mass of 6 kg. Calculate the magnitude of the acceleration and the magnitude of the rope tension.

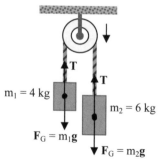

$m_1 = 4$ kg
T
T
$m_2 = 6$ kg
$F_G = m_1 g$
$F_G = m_2 g$

The net force $\mathbf{F}_{Net} = m_1\mathbf{a}$ acting on object m_1 is the sum of vectors $\mathbf{T}$ and $\mathbf{F}_G = m_1\mathbf{g}$. The net force $\mathbf{F}_{Net} = m_2\mathbf{a}$ acting on object m_2 is the sum of vectors $\mathbf{T}$ and $\mathbf{F}_G = m_2\mathbf{g}$. Choose the positive direction as the direction of motion, which will be toward m_2 falling, and write equations for the magnitudes of the acting forces. The magnitude of the acceleration will be the same throughout the rope. Let's calculate the magnitude of the acceleration by first looking at the forces on the two masses.

Object m_1: $T - m_1g = m_1a$ or $T = m_1g + m_1a$

Object m_2: $m_2g - T = m_2a$ or $T = m_2g - m_2a$

Rope tension is the same for each mass so setting $T = T$:

$$m_1g + m_1a = m_2g - m_2a$$
$$m_1a + m_2a = m_2g - m_1g$$
$$a(m_1 + m_2) = g(m_2 - m_1)$$
$$a(4 \text{ kg} + 6 \text{ kg}) = (9.8 \text{ m/s}^2)(6 \text{ kg} - 4 \text{ kg})$$
$$a(10 \text{ kg}) = (9.8 \text{ m/s}^2)(2 \text{ kg})$$
$$a = (9.8 \text{ m/s}^2)(2 \text{ kg}) / (10 \text{ kg}) = 1.96 \text{ m/s}^2 \text{ is the acceleration}$$

Calculate the magnitude of the tension T using either equation for T:

$$T = m_1g + m_1a \quad \text{or} \quad T = m_2g - m_2a$$
$$T = (4 \text{ kg})(9.8 \text{ m/s}^2) + (4 \text{ kg})(1.96 \text{ m/s}^2) = 47.04 \text{ N}$$
$$T = (6 \text{ kg})(9.8 \text{ m/s}^2) - (6 \text{ kg})(1.96 \text{ m/s}^2) = 47.04 \text{ N}$$

Therefore, acceleration $a = 1.96 \text{ m/s}^2$ and tension $T = 47.04 \text{ N}$.

• **Example**: Suppose two objects are connected by a frictionless, non-stretchable rope through a frictionless pulley. The tension T on either side of the pulley is the same. In this system object m_1 is resting on a support and object m_2 is hanging. How would you calculate the acceleration and rope tension?

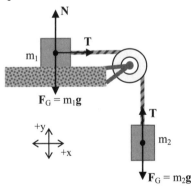

This is a 2-dimensional x-y system: m_1 moves in the x-direction and m_2 in the y-direction. If you choose the positive direction as the direction of motion, m_1 slides in the +x direction and m_2 falls in the –y direction. First write equations for the magnitudes of the acting forces. The magnitude of the acceleration will be the same throughout the rope.

Object m_1:

The net force $\mathbf{F}_{Net}$ acting on object m_1 is the sum of vectors $\mathbf{T}$, $\mathbf{N}$, and $\mathbf{F}_G$. Because there is no net force or acceleration of object m_1 in the y-direction, $\mathbf{F}_G$ and $\mathbf{N}$ having opposite directions cancel each other. The horizontal applied force $\mathbf{F}_{Applied} = m_1\mathbf{a}$ produces the tension $\mathbf{T}$ in the rope, which acts on object m_1. The net force on object m_1 is:

$$\mathbf{F}_{Net} = \mathbf{T} = \mathbf{F}_{Applied} = m_1\mathbf{a} \quad \text{or} \quad \mathbf{T} = m_1\mathbf{a}$$

Therefore, in the positive x-direction of motion: $T = m_1a$.

Object m_2:

The net force $\mathbf{F}_{Net}$ acting on object m_2 is the sum of vectors $\mathbf{T}$ and $\mathbf{F}_G$. The net force on object m_2 is the vector sum:

$$\mathbf{F}_{Net} = \mathbf{F}_G + \mathbf{T} \quad \text{or} \quad m_2\mathbf{a} = m_2\mathbf{g} + \mathbf{T}$$

In the direction of motion the magnitude of m_2g is positive and T is negative giving:

$$m_2a = m_2g - T$$

To solve for the magnitude of the acceleration a we can combine the m_1 and m_2 equations:

$$T = m_1a \quad \text{and} \quad m_2g - T = m_2a$$

Adding these equations cancels T and gives: $m_2g = m_1a + m_2a$.

Alternatively, solve the m_2 equation for T and set both equations equal:

$$T = m_2g - m_2a = m_1a$$
$$m_2g = m_1a + m_2a$$
$$m_2g = a(m_1 + m_2)$$
$$a = m_2g \, / \, (m_1 + m_2)$$

Therefore, we can calculate acceleration in this system using:

$$a = m_2g \, / \, (m_1 + m_2)$$

Note that in this example an equation for acceleration could also be found by modeling the two objects and the extended (mass-less) rope between them as one mass, $M = m_1 + m_2$. In this model the net force is $F_{Net} = Ma$, with the net force acting on the system being equal to the force of gravity acting on m_2, or m_2g. Therefore:

$$F_{Net} = Ma = m_2g$$

We can solve $Ma = m_2g$ for acceleration, a, (remember $M = m_1 + m_2$):

$$Ma = m_2g$$
$$(m_1 + m_2)a = m_2g$$
$$a = m_2g \, / \, (m_1 + m_2)$$

This is the same equation for calculating acceleration of this system.

• **Free pulleys** can reduce the amount of force required to lift or move an object by increasing the distance over which the force acts. A free pulley wheel divides the weight over two ropes—one on each side. To lift a weight, a pulley will exert a force on each rope equal to half the weight.

• **Example**: Pulley systems with free pulleys allow a weight to be raised with less force than lifting it directly. If m = 500 lb, what force **F** is required to raise the weight, and how far must the rope be pulled down to raise the weight 10 ft?

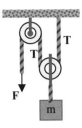

The weight is suspended on 2 ropes, each with T = 250 lb. Since F = T = 250 lb, any downward force over 250 lb will raise the weight. To raise the weight 10 ft, the rope segments on both sides of the free pulley must shorten by 10 ft. One end of the rope is fixed, so the free end must be pulled 20 ft. (You can exert twice the force, but the rope must be pulled twice as far.)

2.4. Linear Momentum and Impulse

• **Linear momentum p** is the product of an object's mass m and its velocity **v**. Momentum and velocity are both vectors, while mass is a scalar.

$$\mathbf{p} = m\mathbf{v}$$

• Newton's Second Law, **F** = m**a**, was originally written as the time rate of change of momentum:

$$\mathbf{F} = \Delta\mathbf{p}/\Delta t \quad \text{or} \quad \mathbf{F} = \lim_{\Delta t \to 0}(\Delta\mathbf{p}/\Delta t)$$

We usually think of mass as remaining constant, but it can change. For example, an accelerating rocket loses mass as it ejects gas. (Although the total mass of the rocket plus the ejected gas does not change, which is required in a Newtonian world by the law of conservation of mass.)

Because **p** = m**v**, force **F** can also be written:

$$\mathbf{F} = \lim_{\Delta t \to 0}(\Delta m\mathbf{v}/\Delta t)$$

When mass is constant:

$$\mathbf{F} = m \lim_{\Delta t \to 0}(\Delta\mathbf{v}/\Delta t)$$

Because acceleration is:
$$\mathbf{a} = \lim_{\Delta t \to 0}(\Delta \mathbf{v}/\Delta t)$$
We can again write Force **F** in terms of m and **a**:
$$\mathbf{F} = m\mathbf{a}$$

The relationship between applied force and the change of momentum reveals something we have all experienced—that stopping an object depends not only on its mass but also on its velocity. Think of someone walking into you vs. someone running into you.

• **Units of momentum**: The units for momentum are consistent with the equation for momentum, **p** = mv: mass times velocity or mass times length per time. MKS and CGS units are: kg·m/s and g·cm/s.

• **Example**: Find the momentum of a 1,000 kg car driving south at 20 m/s. What if the car's velocity was doubled? Its mass was doubled? Both velocity and mass were doubled?

Momentum is:
$$\mathbf{p} = m\mathbf{v} = (1{,}000 \text{ kg})(20 \text{ m/s}) = 20{,}000 \text{ kg·m/s, south}$$
Doubling velocity will double the momentum:
$$\mathbf{p} = m\mathbf{v} = (1{,}000 \text{ kg})(40 \text{ m/s}) = 40{,}000 \text{ kg·m/s, south}$$
Doubling mass will double the momentum:
$$\mathbf{p} = m\mathbf{v} = (2{,}000 \text{ kg})(20 \text{ m/s}) = 40{,}000 \text{ kg·m/s, south}$$
Doubling both velocity and mass quadruples momentum:
$$\mathbf{p} = m\mathbf{v} = (2{,}000 \text{ kg})(40 \text{ m/s}) = 80{,}000 \text{ kg·m/s, south}$$

• **Example**: You are looking into the side of a large fish tank when you see your friend, who is climbing the stairs next to the tank, drop his new camera. Oops. Splash! Within a tenth of a second of hitting the water, the camera's velocity goes to zero! If the camera weighs one pound and your friend dropped it from about 10 ft, what average force in Newtons did the water exert on the camera to stop its fall?

First convert to SI units: Since 1 m ≈ 3.28 ft, the camera dropped:
$$10 \text{ ft} \times 1 \text{ m} / 3.28 \text{ ft} \approx 3.05 \text{ m}$$
The camera weighs 1 pound and 1 lb ≈ 4.448 Newtons (N), or 4.448 kg·m/s^2. To find the mass in kilograms divide by gravity 9.8 m/s^2:
$$(4.448 \text{ kg·m/s}^2) / (9.8 \text{ m/s}^2) \approx 0.45 \text{ kg}$$
Note that because we are on the surface of the Earth we could have used the conversion 1 kg ≈ 2.2 lb: (1 lb)/(2.2 lb/kg) ≈ 0.45 kg.

We want to find the value for the average force imposed by the water, which is a retarding force on the camera. We can use Newton's Second Law in terms of momentum:

$$F = \Delta p / \Delta t = (p_2 - p_1) / \Delta t$$

where in this case the final momentum p_2 is zero.

Next we need the value of momentum p_1 when the camera hits the water. We can use the equation for linear momentum:

$$p_1 = mv_1$$

We find the value for velocity v_1 when the camera hits the water using $v^2 = 2gx$, which is the equation for an object dropped from rest under the constant acceleration of gravity. The value for x is the distance the camera dropped:

$$(v_1)^2 = 2gx = 2(9.8 \text{ m/s}^2)(3.05 \text{ m}) = 59.78 \text{ m}^2/\text{s}^2$$

Take the square root: $v_1 \approx 7.7$ m/s.

Knowing v_1 we can find the momentum $p_1 = mv_1$:

$$p_1 = (0.45 \text{ kg})(7.7 \text{ m/s}) \approx 3.5 \text{ kg·m/s}$$

Finally, to calculate the force the water exerted on the camera use Newton's Second Law in terms of momentum $F = \Delta p / \Delta t = (p_2 - p_1) / \Delta t$. The time we observed was "Within a tenth of a second" so we will use 0.1 s for Δt.

$$F = (p_2 - p_1) / \Delta t = (0 - 3.5 \text{ kg·m/s}) / (0.1 \text{ s}) \approx -35 \text{ kg·m/s}^2$$

Therefore, the retarding force of the water on the camera was about 35 N. (The negative sign denotes that the retarding force was in the opposite direction of the velocity of the camera.)

Impulse

• **Impulse** reflects how much momentum changes when a force is applied for a certain time period. Impulse is a vector since it has both magnitude and direction. If you tap a ball with force **F** for time period Δt, you have provided an impulse. Note that the force is the average force applied during the time period.

$$\boxed{\textbf{Impulse} = \textbf{F} \times \Delta t}$$

You can visualize an impulse from a tap by plotting force vs. time:

 This line represents the average force applied over the time period.

• **Units of impulse**: Impulse has units of force times time or: Newton·seconds (N·s), dyne·seconds (dyne·s), and pound·seconds (lb·s).

• **Impulse (FΔt)** and **momentum (p = mv)** are related. We can see this using Newton's Second Law:

$$F = ma$$

Because **a** = Δv/Δt:

$$F = m(\Delta v/\Delta t)$$

To develop an impulse, which is **F** × Δt, multiply both sides by Δt:

$$F\Delta t = m\Delta t(\Delta v/\Delta t)$$

Cancelling Δt's on right side:

$$F\Delta t = m\Delta v = mv_{final} - mv_{initial}$$

Because mΔv represents the change in momentum, or Δp = mΔv, impulse FΔt equals the change in momentum:

$$\boxed{\textbf{Impulse} = \textbf{F}\Delta t = \Delta \textbf{p}}$$

We can say that the impulse imparted to an object causes a change in its momentum. Also the total change in momentum that is caused by a force depends on how long that force is applied. The preceding equation shows that the same change in momentum can be caused by a stronger force acting over a short time or a weaker force acting over a longer time.

We can determine the impulse when force varies. The previous graph showed an impulse from a tap by plotting force vs. time and drawing a line to reflect the average force applied during Δt. A graph reflecting a constant force, shown below, depicts the impulse as the product of force and time, which is the area of the shaded rectangle. The graph showing variable force has the time axis divided into small intervals with the impulse during an interval approximated by the area of that interval's rectangle. The total impulse, which is the net change in momentum, is the sum of the areas under the curve.

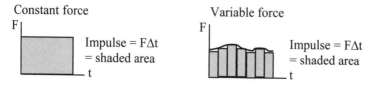

Constant force — Impulse = FΔt = shaded area

Variable force — Impulse = FΔt = shaded area

• **Example**: Suppose you decide to try your hand at hockey. You whack a 0.2-kg puck to the goal at the north end of the rink with an average force of 70 N. Your stick contacts the puck for about a tenth of a second. What was the impulse experienced by the puck?

$$\text{Impulse} = F\Delta t = (70\ N)(0.1\ s) = 7\ N{\cdot}s,\ \text{north}$$

Note that you did not need the mass of the puck in your calculation.

• **Example**: A 0.2-kg hockey puck experiences a 60-N force for a tenth of a second toward the south goal. What is the puck's momentum change?

The change in momentum is $\Delta\mathbf{p} = m\Delta\mathbf{v}$. But since we don't know the velocity change, we can use the equation that relates impulse, force, and momentum change:

$$\text{Impulse} = \mathbf{F}\Delta t = \Delta\mathbf{p} = (60 \text{ N})(0.1 \text{ s}) = 6 \text{ N·s, south}$$

2.5. Conservation of Linear Momentum

• In an isolated system the total **momentum** is constant:

$$\mathbf{p} = \text{constant}$$

This is called the law or principle of **conservation of linear momentum**, and says **momentum is conserved in a closed system**. Achieving a perfectly isolated system in reality is not possible, but some systems can be modeled as closed or isolated when *external forces are negligible compared to the interactions within the system*.

In a system of two objects, momentum is:

$$\mathbf{p} = \mathbf{p}_1 + \mathbf{p}_2 = m_1\mathbf{v}_1 + m_2\mathbf{v}_2 = \text{constant}$$

An example is to think of two identical air hockey pucks—one is white and the other gray. Suppose you hit the white puck and it strikes the gray puck perfectly centered. After the collision the white puck stops and the gray puck continues the motion of the white puck. If the mass of the gray puck is equal to that of the white puck and the white puck stops after hitting the gray puck squarely, the velocity of the gray puck will be the same as the white puck's original velocity. The momentum was transferred from the white puck to the gray puck. The momentum of this two-puck system was conserved.

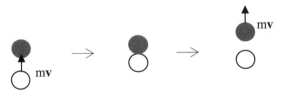

If instead, the white puck strikes the side of the gray puck and they both are in motion after the impact, their combined momentum after the collision will be equal to the momentum of the white puck before impact.

In the isolated system of two air hockey pucks, the total momentum of the system did not change even though they exerted force on each other. More generally, in a system of objects in which no external forces are acting on that system ($F_{net} = 0$), the total momentum of the objects is not changed by any forces the objects exert on each other (p = constant). Momentum is conserved.

If two colliding pucks have masses m_1 and m_2, initial velocities v_{i1} and v_{i2}, and final velocities v_{f1} and v_{f2}, then using conservation of momentum:

$$\boxed{m_1 v_{i1} + m_2 v_{i2} = m_1 v_{f1} + m_2 v_{f2}}$$

For an isolated system with no external forces, this holds whether the objects are air hockey pucks or asteroids.

• Conservation of linear momentum can be proven using Newton's Third Law that two bodies exert an equal and opposite force on each other:

$$F_1 = -F_2$$

with F_1 as the force exerted on object 1 by object 2 and F_2 as the force exerted on object 2 by object 1. If F_1 and F_2 are the only forces acting on the two objects, then using Newton's Second Law $F = ma$:

$$m_1 a_1 = -m_2 a_2$$

Since $a = \Delta v / \Delta t$:

$$m_1 \Delta v_1 / \Delta t = -m_2 \Delta v_2 / \Delta t$$

Since the time period is the same, the Δt's cancel:

$$m_1 \Delta v_1 = -m_2 \Delta v_2$$

Since $p = mv$:

$$\Delta p_1 = -\Delta p_2 \quad \text{or} \quad \Delta p_1 + \Delta p_2 = 0$$

The change in momentum is zero and momentum is conserved.

• We can also quickly see that momentum is conserved using the relationship between *force* and momentum, $F = \Delta p / \Delta t$. From this relationship we can see that if force $F = 0$, then momentum p does not change with time, which is conservation of momentum.

• The relationship, $p = p_1 + p_2 = m_1 v_1 + m_2 v_2$ = constant, is helpful when solving problems in which initial velocities and masses are known and final velocities are needed.

• **Example**: A 1.5-kg toy train car is sitting on a straight, *frictionless* track when it is hit by a second 1.0-kg toy train car traveling 0.1 m/s. If the two cars link and move forward together, what is their velocity?

Let m_1v_1 represent the 1.5-kg car's initial momentum: (1.5 kg)(0 m/s).
Let m_2v_2 represent the 1.0-kg car's initial momentum: (1.0 kg)(0.1 m/s).
Let v_f represent the final velocity of the combined cars: (? m/s).

Because we are ignoring friction and other external forces, we can use conservation of momentum:

$$m_1v_1 + m_2v_2 = m_1v_f + m_2v_f$$

$$m_1v_1 + m_2v_2 = (m_1 + m_2)v_f$$

$$v_f = (m_1v_1 + m_2v_2)/(m_1 + m_2)$$

$$v_f = ((1.5\,\text{kg})(0\,\text{m/s}) + (1.0\,\text{kg})(0.1\,\text{m/s})) / (1.5\,\text{kg} + 1.0\,\text{kg}) = 0.04\ \text{m/s}$$

$$v_f = 0.04\ \text{m/s} = 4\ \text{cm/s (or about 0.09 mph) in the forward direction}$$

• **Example**: You are in your canoe, which is not moving, with your mini-cannon. Directly to the north you imagine an enemy battleship and fire your cannon at it horizontally. The projectile's mass is $m_p = 1$ kg, and it is launched at $v_p = 250$ m/s. You, the canoe, and your gear have a combined mass of $m_c = 100$ kg. What change in velocity do you experience?

Momentum is conserved, so $\Delta p = 0$, and $m_cv_c + m_pv_p = 0$, or
$$v_c = -m_pv_p/m_c = -1 \times 250/100 = -2.5\ \text{m/s or 2.5 m/s, south (quite a kick!)}$$

2.6. Torque

• If a force is applied to an object at a point other than its central axis of rotation, that force may create a **torque**. Torque is a measure of how effective a force is in producing rotation about an axis.

Torque is the product of an applied force **F** times the length of a lever arm, which is the shortest radial *perpendicular* distance from the axis of rotation to a line drawn along the direction of force. Torque can be described in terms of the distance between the point of application of the force and the axis of rotation multiplied by the component of the force perpendicular to this distance line.

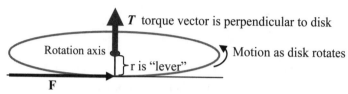

T torque vector is perpendicular to disk

Rotation axis   Motion as disk rotates

r is "lever"

F

Force **F** acts perpendicular to lever r along side of disk

When the **applied force F is perpendicular** to the radial line **r** from the axis, the entire force is effective in producing **torque *T***, and:

Torque *T* = **r** × **F**   providing **F** is perpendicular to **r**

The length of **r** and the applied force **F** perpendicular to **r** determine the torque. If you push on a swinging gate far from the hinge (long **r**), the gate will open more easily than if you push next to the hinge (short **r**).

If force **F** is applied perpendicular to the gate near its hinge (rotation axis) and **r** is very short, the turning force is minimal (above).

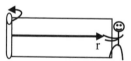

If force **F** is applied perpendicular to the gate far from the hinge and **r** is long, the gate turns, or opens, more easily (above).

• What if the *force is applied at an angle* other than 90° perpendicular?

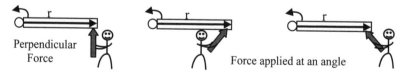

Perpendicular
Force   Force applied at an angle

If the force is applied at an angle, **the perpendicular component of that force contributes to the torque. Torque** is found using:

Torque *T* = r**F** sin θ

where r is the radial distance from the axis of rotation to the point at which the force is applied, and θ is the angle between **F** and the radial line **r** that connects the axis of rotation to the point at which the force is applied. Angle θ is also defined as the acute angle between the lines of vector **r** and the force vector **F**.

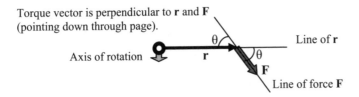

Torque vector is perpendicular to **r** and **F** (pointing down through page).

Axis of rotation   **r**   θ   θ   Line of r

F

Line of force F

- Mathematically, torque is usually expressed using the **vector cross product**, or *vector product*:

$$T = \mathbf{F} \times \mathbf{r} = |\mathbf{F}|\,|\mathbf{r}|\sin\theta$$

A force **F** can be applied to a lever arm, or radius vector **r**, which has its initial point located at the center of rotation. The torque is a vector having a magnitude that measures the force of the rotation and a direction along the axis of rotation. More generally the *vector product* or *cross product* of two vectors is:

$$\mathbf{A} \times \mathbf{B} = |\mathbf{A}|\,|\mathbf{B}|\sin\theta$$

where $|\mathbf{A}|$ and $|\mathbf{B}|$ represent the *magnitudes* (or *lengths*) of vectors **A** and **B** and θ is the angle between vectors **A** and **B**. The product exists in three dimensions with **A** and **B** in a plane and **A** **X** **B** normal (perpendicular) to the plane. The *cross product* of two vectors produces a third vector with length $|\mathbf{A}|\,|\mathbf{B}|\sin\theta$ and direction perpendicular to **A** and **B**. The length of **A** **X** **B** depends on sin θ and is greatest when θ = 90° or sin θ = 1.

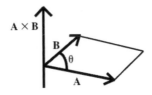

A × B

B

θ

A

The cross product of two vectors occurs geometrically according to what is referred to as the *right-hand screw rule* such that for **A** **X** **B**, if you move from vector **A** to vector **B** through angle θ, the result is vector **A** **X** **B**, which is perpendicular to both **A** and **B**. The right-hand screw rule can be visualized by curling the fingers of the right hand from **A** to **B**, where **A** **X** **B** points in the direction of the right thumb.

- Note that applying torque does not require that the object possess a physical "lever arm," just the application of some off-center force. The **r** vector can be a construct that represents the displacement between an object's rotation axis and the point where the force **F** is applied. A force can, for example, be applied to a free particle of mass such as a meteor in space.

• **Units of torque** are force times length. In SI this is Newton-meters (N·m) or kg·m^2/s^2, in the CGS system this is dyne-centimeters (dyn·cm), in the foot-pound-seconds system this is foot-pounds (ft·lb). The direction of a torque vector can be positive or negative such that a torque causing a counterclockwise rotation about the axis is positive and a torque causing a clockwise rotation is negative.

• **Example**: Suppose you sit on a seesaw 1.5 m from the pivot axis. What maximum torque will you cause assuming your mass is 70 kg?

The force you apply to the seesaw is your weight or mass times gravity (70 kg)(9.8 m/s^2). Remember, when the applied force **F** is perpendicular to the radial line **r** from the axis, the entire force is effective in producing torque. This is when the angle $\theta = 90°$, or $\sin \theta = 1$. In this example maximum torque will occur when the seesaw is horizontal because the full weight acts perpendicular to the lever arm. At horizontal the torque is:

$$T = r \times F = (1.5 \text{ m})(70 \text{ kg})(9.8 \text{ m/s}^2) = 1{,}029 \text{ kg·m}^2/\text{s}^2 \text{ or } 1{,}029 \text{ N·m}$$

2.7. Angular Momentum and Conservation of Angular Momentum

• First remember that as an object moves along a circle, angle θ changes. The rate at which angle θ changes is the *angular velocity* ω. In other words, the angular velocity ω is a measure of the change in angle θ as an object moves at a constant speed around a circle.

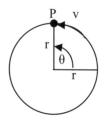

The speed of an object moving in a circular path of radius r is: $v = r\omega$.
The **angular velocity** is: $\omega = \Delta\theta/\Delta t$, where angle θ is in radians and is the
measure of where object P is at time t. The velocity vector **v** of the object
changes direction continually even though its speed may be constant.

• Now let's consider angular momentum: **Angular momentum (L)** is a
vector quantity having magnitude and direction along the axis of rotation. The direction of the angular momentum vector can be visualized
using the right-hand screw rule (discussed above for torque). When you
curl the fingers of the right hand along the path of motion, the direction
in which your right thumb points is the direction of the angular
momentum vector **L**.

For a mass rotating around a central axis:

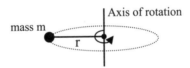

Angular momentum L can be defined as:
$$L = \text{(distance } \mathbf{r} \text{ from mass m to axis of rotation)}$$
$$\times \text{(perpendicular component of linear momentum, } \mathbf{p}_\perp)$$

$$\boxed{L = r \times p_\perp}$$

where **L** is perpendicular to both the radial vector **r** and the momentum
vector **p** and points in its direction according to the right-hand screw
rule. (Subscript $_\perp$ denotes perpendicular.)

Substitute linear momentum $p = mv$ to get **angular momentum**:

$$\boxed{L = rmv_\perp}$$

The perpendicular velocity component is $v_\perp = v \sin\theta$. The angle θ is
between vector **p** (or **v**) and the **r** line connecting mass m and the axis of
rotation. Therefore, **angular momentum** is also written:

$$\boxed{L = rmv \sin\theta}$$

Angular momentum depends on the perpendicular component of velocity even though the motion may not always be directly perpendicular. We use the relationship $v_\perp = v \sin \theta$ and can visualize it as:

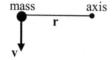
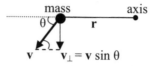

v is perpendicular to r
$L = rmv \sin 90° = rmv$

$v_\perp$ is perpendicular component of v
$L = rmv_\perp = rmv \sin \theta$

Because the speed of an object moving in a circular path of radius r is $v = r\omega$, where v is perpendicular to r, we can substitute this into the angular momentum expression $L = rmv \sin \theta$ to describe a **mass rotating around a central axis**:

$$L = mr^2\omega \sin \theta$$

• You will find that **angular momentum** is often described more generally using **moment of inertia** which allows you to model different geometries other than just a mass rotating around a central axis. The moment of inertia I for a single mass or particle rotating around a central axis is $I = mr^2$. Following are sample moment of inertia values for different geometries:

A particle revolving around a central axis: $I = mr^2$.
Thin hoop rotating about its center: $I = mr^2$.
Disk rotating about its center: $I = (1/2) mr^2$.
Solid sphere rotating about its center: $I = (2/5) mr^2$.
Thin rod rotating lengthwise through its center about an
     axis perpendicular to it: $I = (1/12) mr^2$.

We can express **angular momentum** using moment of inertia and angular velocity as:

$$L = (\text{moment of inertia I}) \times (\text{angular velocity } \omega) = I\omega$$

Remember the definition of linear momentum **p**: $p = mv$. Angular momentum can be thought of as a rotational counterpart of linear momentum with inertia related to mass and angular velocity related to liner velocity.

Conservation of Angular Momentum

• A few examples of conservation of angular momentum include: tops spinning without falling over, a figure skater spinning faster as arms and legs drawn in, and gyroscopic compasses giving direction. Earlier in this chapter we learned that linear momentum is conserved so that when the net *force* on an object is zero, the *linear momentum* remains constant. Similarly, angular momentum is also conserved, providing no external *torque* is applied and the *net torque on the object is zero*.

We can state **conservation of angular momentum** as: *When the net torque on an object or system of objects is zero, its angular momentum will remain unchanged in both magnitude and direction.* Therefore:

Initial angular momentum = final angular momentum

$$I_i\omega_i = I_f\omega_f \quad \text{providing net torque is zero}$$

Since $L = I\omega = rmv_\perp$:

$$r_imv_i = r_fmv_f$$

Remember that the force on an object is equal to the time rate of change of linear momentum, $F = \Delta p/\Delta t$. Similarly for angular momentum, the **torque** T is equal to the time rate of change of angular momentum:

$$T = \Delta L/\Delta t \quad \text{or} \quad T = \lim_{\Delta t \to 0}(\Delta L/\Delta t)$$

From this relationship we can see that if torque T equals zero, then angular momentum L does not change with time, which conserves angular momentum.

• **Example**: One way to visualize conservation of angular momentum is to imagine yourself spinning on a rotating platform. First you have your arms outstretched while holding weights, then you bring your arms close to your body. What will happen? Why? What is the ratio of moments of inertia if you are initially spinning 0.5 revolution per second, then after bringing in your arms you spin at 1 revolution per second?

Of course, you will spin faster when you bring your arms in, but why? Because initial angular momentum equals final angular momentum, you will not lose angular momentum. As your arms come in, r gets shorter and velocity must increase. You can determine the ratio of moments of inertia using $I_i\omega_i = I_f\omega_f$ and solve for I_i/I_f:

$$I_i/I_f = \omega_f/\omega_i = (1 \text{ rev/s}) / (0.5 \text{ rev/s}) = 2$$

• **Example**: The Earth moves in an elliptical orbit around the Sun. It is closest to the Sun, a distance of 1.47×10^8 kilometers (perihelion radius), in December and furthest from the Sun, a distance of about 1.52×10^8 km (aphelion radius), in June. Given its velocity at the far, aphelion position is 29.3 km/s, what is its velocity at its close, perihelion position?

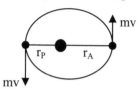

Because any external forces are negligible, the angular momentum is the same at every point on the orbit. Gravity is *not* an external force (and exerts no torque). Gravity is an internal force directed along the line connecting the bodies and provides the center-seeking force behind the Earth's centripetal acceleration. Earth's angular momentum vector points in a fixed direction so Earth always moves in the same plane and the product of r and v remains constant throughout the orbit (so v varies inversely to r).

Using subscript P for perihelion and subscript A for aphelion, by conservation of angular momentum:

$$I_A \omega_A = I_P \omega_P = r_A m v_A = r_P m v_P$$

$$r_A m v_A = r_P m v_P \quad \text{or} \quad r_A v_A = r_P v_P$$

Solve for v_P:

$$v_P = r_A v_A / r_P = (1.52 \times 10^8 \text{ km})(29.3 \text{ km/s}) / (1.47 \times 10^8 \text{ km}) = 30.3 \text{ km/s}$$

Therefore, the orbiting velocity at its close, perihelion position is 30.3 km/s or about 67,779 mph. (This compares to the far, aphelion orbiting velocity: 29.3 km/s or about 65,542 mph.)

• The **units for angular momentum** are: $kg \cdot m^2/s$.
In the study of rotational motion, physicists generally measure angles in radians. Therefore angular momentum units may be thought of in terms of **radians** as $(kg \cdot m^2)(radians/s)$, although it is simply written $kg \cdot m^2/s$. Remember, one complete rotation is 2π radians, which is equivalent to 360 degrees, or 2π radians = 360 degrees. Therefore: 1 **radian** = (360 degrees) / $(2\pi) \approx 57.3$ degrees.

2.8. Center of Mass

• We have discussed the principles of conservation of linear and angular momentum in the context of a point mass or individual masses, but how do you apply such principles to systems or aggregates? We need to find a single point within a system that behaves according to these principles of conservation. In a system this point is referred to as its **center of mass**.

The **center of gravity** is the point in a object at which gravity can be considered to act. In the presence of a *uniform gravitational field* the center of gravity and center of mass are the same. The **center of mass** is the point around which the object's mass is concentrated so that there is equal mass on either side of a plane that includes that point. It is the point that moves as if it were a point mass when subjected to external forces.

For example, if a net force **F** is acting on an object or system of objects of mass M, the acceleration of the center of mass of the object or system is $a_{cm} = F/M$. Newton's Second Law can be written in terms of the motion of the center of mass as $F_{net} = Ma_{cm}$. The net force F_{net} acting on a system is equal to the product of the total mass M of the system and the acceleration a_{cm} of the center of mass. Note that if the net force acting on a system is zero, then the center of mass does not accelerate.

Similarly, linear momentum is written in terms of the velocity of the center of mass as $p = Mv_{cm}$. If there are no external forces, the center of mass of a system of objects moves with a constant velocity.

If a rigid object is projected through the air, its center of mass will follow a natural parabolic arc even if it spins or rotates. For example, if you haphazardly toss a baseball bat into the air, while its motion may appear complicated as it rotates, its center of mass will trace out a parabolic path similar to a thrown ball.

You can determine an object's center of mass whether it is uniform or irregular in shape and density. When an object is supported at its center of mass, there is no net torque acting on it, and it will remain in static equilibrium. If an object has a uniform distribution of mass, so that its density is the same throughout, its center of mass is at its geometrical center. If an object has a point, line, or plane of symmetry, the center of mass will lie on that point, line, or plane. Note that the center of mass does not need to be located within the object itself, such as the center of mass of a donut being in the hole. In objects such as spheres, cubes, and rectangular solids, the center of mass is at the central point. If you hold a rigid pole, measuring stick, broom, or similar object in your hand or place it on a narrow support, you can physically determine the location of its center of mass by sliding it along that support until it balances.

If an object is uniform, such as a meter stick, the center of mass will be at its **geometric center**. If an object is irregular in shape, such as a broom, its center of mass will be closer to the end with more mass.

You can also visualize the center of mass by considering a system of two particle masses connected to each other by a ridged massless pole and balanced on a pivot at the system's center of mass.

Because the system is balanced and there is no rotation, there is no net torque. Assuming a uniform gravity field, the **torque** around the center of mass produced by the gravitational force on m_1 equal to but opposite the torque produced by the gravitational force on m_2. Remember that torque is $T = r \times F$, providing F is perpendicular to r. Using the balance of torques for a two object system:

$$r_1 m_1 g = r_2 m_2 g \quad \text{or} \quad r_1 m_1 = r_2 m_2$$

You can calculate the center of mass for a two-particle system separated by distance d. If you select the reference frame as the x-axis of a coordinate system, you can define m_1 and m_2 as being located at x_1 and x_2 respectively, and the center of mass to be located at x_{cm}.

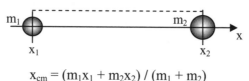

$$x_{cm} = (m_1 x_1 + m_2 x_2) / (m_1 + m_2)$$

The equation of center of mass x_{cm}, can be expanded to a number n masses in one dimension, $M = m_1 + m_2 + m_3 + \ldots m_n$, along the x-axis at n positions $x_1 + x_2 + x_3 + \ldots x_n$.

$$x_{cm} = (m_1 x_1 + m_2 x_2 + \ldots + m_n x_n) / (M)$$

If you expand this to two or three dimensions, each individual mass is described in terms of its x-, y-, and z-components, and each dimension is represented by an equation similar to the center of mass x_{cm} equation so you also have a y_{cm} equation and a z_{cm} equation.

• **Example**: Given that the mean distance from the Sun to the Earth is $r = 1.49 \times 10^{11}$ m, the mean radius of the Sun is $r_s = 6.96 \times 10^8$ m, the mean radius of the Earth is $r_e = 6.37 \times 10^6$ m, the mass of the Sun is $m_s = 1.99 \times 10^{30}$ kg, and the mass of the Earth is $m_e = 5.98 \times 10^{24}$ kg, can you find the location of the center of mass of the Sun-Earth system?

You can model this as a two-particle system where the center of mass lies on the line connecting the particles (or, more precisely, their individual centers of mass). Draw a line (x-axis) through the Sun and Earth and set the Sun at zero on the x-axis.

$$\text{Center of mass } x_{cm} = (m_s x_s + m_e x_e) / (m_s + m_e)$$

If we include the radii of the Sun and Earth, the center to center distance is: 1.49×10^{11} m $+ 6.96 \times 10^{8}$ m $+ 6.37 \times 10^{6}$ m, or writing it out:

149,000,000,000 m is the mean distance from the Sun to the Earth
    696,000,000 m is the mean radius of the Sun
+   6,370,000 m is the mean radius of the Earth
149,702,370,000 m is the distance from center to center

We can see that the distance between the two masses is very large compared to their radii. Rounding, the Sun to the Earth from center to center gives about: 1.50×10^{11} m. (Note that the way we modeled the Sun and Earth with Sun at $x = 0$, the distance between them is x_e.)

$$x_{cm} = \frac{(1.99 \times 10^{30}\,\text{kg})(0) + (5.98 \times 10^{24}\,\text{kg})(1.50 \times 10^{11}\,\text{m})}{(1.99 \times 10^{30}\,\text{kg}) + (5.98 \times 10^{24}\,\text{kg})}$$

We can calculate the non-zero term in the numerator by multiplying the two numbers (and adding the exponents):

$$(5.98 \times 10^{24}\,\text{kg})(1.50 \times 10^{11}\,\text{m}) = 8.97 \times 10^{35}\,\text{kg·m}$$

For the denominator we add the two numbers. We see that the mass of the Sun dominates the mass of the Earth:

1,990,000,000,000,000,000,000,000,000,000 kg
+   5,980,000,000,000,000,000,000,000 kg
1,990,005,980,000,000,000,000,000,000,000 or about 1.99×10^{30} kg

$$x_{cm} = (8.97 \times 10^{35}\,\text{kg·m}) / (1.99 \times 10^{30}\,\text{kg}) = 4.5 \times 10^{5}\,\text{m}$$

(For division we divide the numbers and subtract the exponents.) Therefore the center of mass between Sun and Earth is 4.5×10^{5} m from the Sun's center (since we set up the problem with the Sun located at zero). Notice that the center of mass between Sun and Earth, 4.5×10^{5} m, is much less than the radius of the Sun, 6.96×10^{8} m. The center of mass in the Sun-Earth system is deep inside the Sun.

2.9. Key Concepts and Practice Problems

- Newton's Laws of Motion. First Law: Objects maintain at rest or straight-constant velocity unless acted on by force. Second Law: For an object with constant mass, $F = ma$. Third Law: Any action (force) has an equal and opposite reaction, $F_{AB} = -F_{BA}$.
- Mass and weight are related by the acceleration due to gravity: $w = mg$.
- Normal force: perpendicular reaction force of one surface onto another.
- Linear momentum: $p = mv$.
- Impulse $= F\Delta t = \Delta p$.
- Conservation of linear momentum: $F_{net} = 0$, $m_1 v_{i1} + m_2 v_{i2} = m_1 v_{f1} + m_2 v_{f2}$.
- Torque: $T = rF \sin \theta$.
- Angular momentum: $L = rmv \sin \theta$.
- Conservation of angular momentum: $T_{net} = 0$, $r_i m v_i = r_f m v_f$.
- Center of gravity: point where gravity effectively acts.
- Center of mass: point where mass is concentrated.

Practice Problems

2.1 (a) A tugboat turned a giant oil tanker 1,000 times its weight, but could not free itself from a sand bar at low tide. Why? **(b)** A slingshot can launch a 1-lb rock at 50 mph horizontally on Earth. What would be the launch speed be on the Moon where the rock weighs 1/6 as much?

2.2 (a) A truck cruises along a straight highway at a steady 60 mph in a 50 mph crosswind. What forces are acting on the truck and with what net force? **(b)** E.T.'s 10,000-kg spacecraft is falling straight toward Earth at 1,000 m/s. Its deceleration thrusters produce a maximum force of 198,000 N. What is the minimum altitude he can fire the thrusters and still stop before impacting the ground? (Ignore air resistance and assume uniform gravity of -9.8 m/s^2. **(c)** The gravitational attraction of a star on a small planet is 3.537×10^{22} N. What additional information do you need to compute the gravitational attraction the planet has on the star?

2.3 (a) E.T.'s mass is 40 kg. What would his apparent weight be during the deceleration phase in Problem 2.2(b)? **(b)** What are the normal forces exerted on a gecko weighing 100 g as he (i) walks across the floor, (ii) up the wall, and (iii) along the underside of a 45° ceiling? **(c)** Young Ken is pulling his friend Barbie up a 20° hill in his red wagon at 2 mph. Wagon and passenger weigh 70 lb. What is the tension on the wagon handle? (Assume no rolling resistance.) **(d)** Pulley systems allow a weight to be raised with less force than lifting directly. If m = 500 lb, what force F is required to raise the weight?

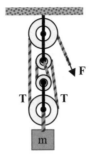

2.4 (a) A crossbow can launch a 0.1-kg arrow at 200 m/s. The arrow accelerates from 0 m/s to 200 m/s over distance d = 0.5 m as it is launched. What is its momentum at launch and what average force was exerted on the arrow? **(b)** What impulse did the crossbow impart to the arrow? **(c)** The crossbow is used to shoot 2 arrows. One hits a tree and is embedded into the trunk 0.1 m, and the other hits an iron shield and embeds only 0.01 m. Which arrow imparts the greater impulse?

2.5 Joe is unable to turn a screw with a thin screwdriver. Jody notices this and hands him a screwdriver with a much thicker handle. Joe is now able to turn the screw. Why?

2.6 (a) Will and Bill visit a funhouse at the fair and stand on a large rotating disk with a 4 m radius. It starts to rotate at an angular velocity $\omega = 1$ counterclockwise as Will stands at the outer edge and Bill stands halfway between Will and the center. They each have a mass of 60 kg. What is each boy's linear velocity and angular momentum? **(b)** Will thinks he's about to fly off and steps quickly inward toward Bill, but he feels his feet go out from under him and falls to his right. What happened?

2.7 If a sky rocket at a fireworks display explodes, what in general happens to its center of mass and to its aggregate linear and angular momentums (ignore air resistance)?

Answers to Chapter 2 Problems

2.1 (a) Despite its high inertial mass, the floating tanker presented very low opposing frictional forces, so it slowly accelerated in the direction it was pushed. The weight of the tug on the sand bar created an opposing frictional force the tug could not overcome despite its much lower inertial mass. **(b)** The launch speeds would be the same on the Moon since neither the force of the slingshot nor the rock's mass and inertia change in differing gravitational fields.

2.2 (a) Gravity pulling down and pavement pushing up; force of engine through wheels pushing forward against resistance of wind, rolling tires, and drive train; and cross wind pushing sideways against tire friction that is resisting sideways movement. Net force is zero, since velocity is constant. **(b)** $F_g = ma_g = (10,000)(-9.8) = -98,000$ N. $F_{net} = F_{thrust} + F_g = 198,000 + -98,000 = 100,000$ N. $F_{net}/m = a_{net} = 100,000/10,000 = 10$ m/s². $v_1 - v_0 = at$, or $(0 - (-1000)) = 10t$, so $t = 100$ s; $d = (1/2)at^2 = (1/2)(10)(100^2) = 50,000$ m or 50 km above the surface. **(c)** None. The planet exerts an equal and opposite force on the star.

2.3 (a) $w_{e.t.} = (m_{e.t.})(g + a) = (40)(9.8 + 10) = 792$ N. **(b)** (i) 100 g. (ii) 0 g (no horizontal component of gravity to produce opposing normal force). (iii) $n = (-100)(\sin \theta) \approx -70.71$ g. (Normal force is negative because it is opposite to the downward force of his sticky toes on the ceiling!) **(c)** Their velocity is constant so the only force is gravity. $F_w = mg \sin \theta \approx (70)(-9.8)(0.342) = -234.6$ N. $T = -F_w = 234.6$ N. **(d)** The weight is suspended (using 2 free pulleys) on 4 ropes, so each rope has a tension of 125 lb. At steady state, $F = T$, so any downward pull over 125 lb will raise 500 lb. (The rope must be pulled 4 ft to raise the weight 1 ft).

2.4 (a) $p = mv = (0.1$ kg$)(200$ m/s$) = 20$ kg·m/s. During launch, the average velocity is 100 m/s. Time to reach 200 m/s $= t = d/v_{Ave} = 0.5/100 = 0.005$ s. $a = (v_2 - v_1)/t = (200)/0.005 = 40,000$ m/s². $F_{Ave} = ma = (0.1)(40,000) = 4,000$ kg·m/s² $= 4,000$ N. **(b)** Impulse $= F\Delta t = (4,000)(0.005) = 20$ N·s. **(c)** Impulses are the same because Δp is the same (both arrows go to zero momentum). The shield stops the arrow with 10 times the force applied over one-tenth the time interval.

2.5 He probably is exerting the same force on the handle, but at a greater radius from the axis of rotation (torque $T = r \times F$). If the thick handle is 3 times the diameter of the thin handle, the torque is 3 times greater.

2.6 (a) $v_W = r_W \omega = (4)(1) = 4$ m/s. $v_B = r_B \omega = (2)(1) = 2$ m/s. $L_W = r_W m_W v_W = (4)(60)(4) = 960$ kg·m²/s. $L_B = r_B m_B v_B = (2)(60)(2) = 240$ kg·m²/s. **(b)** To conserve his angular momentum, Will would need to revolve at 4 times the rate if he moves halfway to the center, but the disk does not spin faster to accommodate him. Alternatively, Will's linear velocity needs to slow to 2 m/s. The disk's surface is moving slower than Will's center of mass as he moves inward, creating a torque that knocks him over.

2.7 Since no external forces are acting, no changes to center of mass or to linear or angular momentums occur.

Chapter 3

EQUILIBRIUM AND FRICTION

3.1. Static Equilibrium
3.2. Torques in Equilibrium
3.3. Friction
3.4. Key Concepts and Practice Problems

"This most beautiful system of the sun, planets, and comets could only proceed from the counsel and dominion of an intelligent and powerful Being."
Attributed to Sir Isaac Newton

3.1. Static Equilibrium

• When the net force acting on an object is zero, there is no acceleration, and the object is either moving at a constant velocity or it is at rest. When an object is in motion at *constant velocity* it is in **dynamic equilibrium**. When an object is at rest it is in **static equilibrium**. *For an object to be in a state of equilibrium there must be no net forces or torques acting on it.* Unless an object is in deep space far away from all other objects so gravitational forces are negligible, it will have some force acting on it, though the net force may be zero.

• **Static equilibrium** exists when *forces are balanced so that the net force on the object is zero* and it is at rest. A building is in static equilibrium relative to Earth.

• The condition of net force being zero can be written as the *sum of the forces equals zero*:

$$\sum \mathbf{F} = 0$$

where $\sum$ (the Greek letter sigma) denotes the "sum", and $\sum \mathbf{F} = 0$ signifies that the vector sum of the forces is zero. The sum $\sum \mathbf{F}$ is a shorthand way to write out all of the forces $\mathbf{F}_1 + \mathbf{F}_2 + \mathbf{F}_3 + \dots$ that are acting on an object:

$$\sum \mathbf{F} = F_1 + F_2 + F_3 + \dots = 0$$

When the forces on an object sum to zero, the components in each direction or dimension sum to zero. In three dimensions $\sum F = 0$ is:

$$\sum F_x = 0 \quad \sum F_y = 0 \quad \sum F_z = 0$$

Note that there may be situations where $\sum F_x = 0$ and $\sum F_y = 0$, but $\sum F_z \neq 0$, such as a rocket launching straight up.

If you know the forces acting on an object, you can use trigonometric functions to find the horizontal and vertical components of each force.

• We can also visualize balanced and unbalanced forces using their force vectors. By doing a graphical sum of the vectors, we can see if they sum to equilibrium.

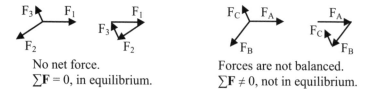

No net force.   Forces are not balanced.
$\sum F = 0$, in equilibrium. $\sum F \neq 0$, not in equilibrium.

• **Example**: You and your friend are about to hike across an ice field. With ice ax in hand, you survey the breathtaking azure and crystal-white panorama. Uh oh! The ice suddenly shifts beneath you. You slip off a precipice, and your friend slides in the opposite direction down a 45° ice incline. Fortunately, you already roped up and your falls are arrested. If your mass is 60.1 kg and your friend's is 85 kg, what will become of the two of you? (Assume no friction and a mass-less rope.)

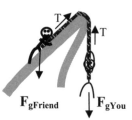

What you need to know is whether you and your friend are in equilibrium and $\sum F = 0$ for both of you together, or if the mass of one of you will overwhelm, and acceleration toward that person will occur. Assume the tension T is uniform along the rope. The tension your friend creates on the rope is:

$$T = mg \sin 45° = (85 \text{ kg})(9.8 \text{ m/s}^2) \sin 45° = 589 \text{ N}$$

The tension you create is:

$$T = (60.1 \text{ kg})(9.8 \text{ m/s}^2) = 589 \text{ N}$$

What a relief! You and your friend are in equilibrium. Now what?
Fortunately, your friend can reach his ice ax. He braces himself using his
ax as you begin climbing up the rope. Good thing you've been doing
your pull-ups lately!

• **Example**: After recovering from your last mishap, you lean against
your backpack and fall asleep. You are awakened by your friend's voice
shrilly calling to you for help. You don your backpack and crampons and
head toward his voice. You stop suddenly before stepping into a wide
crevasse and are shocked to see him hanging from a suspended rope. His
only question is, "Will this rope hold me? It's not the thick one we used
earlier." Can you find the maximum rope tension?

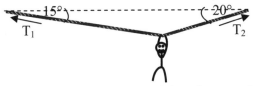

First, you help him out of the crevasse and then the two of you sit down
and calculate. With his gear your friend's mass is about 85 kg. You
estimate the rope angles at 15° and 20°. To calculate the tensions in each
segment you use equilibrium.

x-components:

$$T_{1x} + T_{2x} = 0$$

or:

$$-T_1 \cos 15° + T_2 \cos 20° = 0$$

$$T_1 = (T_2 \cos 20°) / (\cos 15°) = 0.973(T_2)$$

y-components:

$$T_{1y} + mg + T_{2y} = 0$$

or:

$$T_1 \sin 15° + (85 \text{ kg})(-9.8 \text{ m/s}^2) + T_2 \sin 20° = 0$$

You have two equations and two unknowns:

$$T_1 = (0.973)T_2 \quad \text{and} \quad T_1 \sin 15° - (833 \text{ N}) + T_2 \sin 20° = 0$$

Substitute:

$$(0.973)T_2 \sin 15° - (833 \text{ N}) + T_2 \sin 20° = 0$$

$$T_2 [(0.973) \sin 15° + \sin 20°] = 833$$

$$T_2 = 1{,}403 \text{ N}$$

Substitute:

$$T_1 = (0.973)T_2 = (0.973)(1{,}403) = 1{,}365 \text{ N}$$

Therefore, the two tensions were $T_1 = 1,365$ N and $T_2 = 1,403$ N, with the greater tension on the steeper side. When you get home you can look up the specs for your rope to see if it was close to snapping.

3.2. Torques in Equilibrium

• *For an object to be in a state of* **equilibrium,** *there must be no net forces or net torques acting on it.*

$$\sum F = 0 \quad \text{and} \quad \sum T = 0$$

In rotational equilibrium there is no angular acceleration, so an *object is either rotating at a constant angular velocity or it is not rotating at all. In* **rotational equilibrium, net torque is zero** *and the sum of all acting torques is zero*:

$$\sum T = 0$$

Remember: Torque can be created when a force is applied to an object at a point other than its central axis of rotation. Torque is a measure of how effective a force is in producing rotation about an axis.

$$\text{Torque} = T = rF \sin \theta$$

$$\text{When} \quad F \perp r \quad \text{then} \quad T = r \times F$$

Angle θ is the acute angle between vector **r** and force vector **F**.

The concept of rotational equilibrium can be used to determine the torque necessary to prevent rotation. In problems involving rotational equilibrium, the sum of the torques acting in one rotational direction must equal the sum of the torques acting in the opposite rotational direction. When performing a rotational equilibrium analysis, you need to choose a pivot point in a convenient location and be careful when assigning the plus and minus signs designating the rotational direction for each term.

Examples of torques in equilibrium include weights on a seesaw, a person on a ladder, a tall building in a wind, a diving board and diver, and the tension in the cables of a drawbridge bearing the torque of its weight.

• **Example**: You and your friend find a flat board in the woods and set up a seesaw. Given that your mass m_Y is 55 kg and your friend's mass

m_F is 70 kg, if you sit 2 m from the pivot how far must your friend sit on the other side to keep you off the ground? (Ignore the board's mass.)

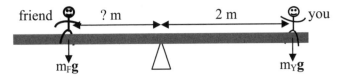

To hold you off the ground, the torques will be balanced. Set your torque axis in the center and make counterclockwise the positive direction.

$$\sum T = 0 = + (m_F g \times ?\ m) - (m_Y g \times 2\ m)$$
$$m_F g \times ?\ m = m_Y g \times 2\ m$$
$$m_F \times ?\ m = m_Y \times 2\ m$$
$$?\ m = (55\ kg \times 2\ m) / (70\ kg) = 1.57\ m$$

Therefore, if your friend sits at least 1.57 m from the pivot you will stay in the air.

• **Example**: You and your friend need some R&R after your latest ice-hiking experience and stop at a resort with a swimming pool. While waiting your turn to jump off the diving board, you notice it looks a bit rickety and wonder how much force is on the two diving board supports. Can you calculate the forces F_1 and F_2 for each support? Use two different axes or pivots for your torque calculations. Assume in your swimsuits your mass m_Y is 55 kg, your friend's mass m_F is 70 kg, and the board's mass m_B is 50 kg. (Designate board's mass at its center.)

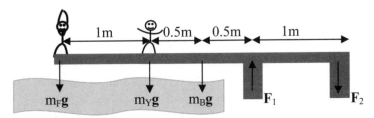

To determine F_1 and F_2 (two unknowns) you will need two equations. This is an equilibrium problem so $\sum F = 0$ and $\sum T = 0$. You can choose any pivot or axis, but let's first use the F_1 support.

$$\sum F = 0 = F_1 - F_2 - m_F g - m_Y g - m_B g$$
$$= F_1 - F_2 - (70\ kg)(9.8\ m/s^2) - (55\ kg)(9.8\ m/s^2) - (50\ kg)(9.8\ m/s^2)$$
$$F_1 = F_2 + 1{,}715\ N$$

Now sum the torques around the axis with counter-clockwise positive and remember $T = r \times F$, if $F \perp r$:

$$\sum T = 0$$
$$= (0\,\text{m} \times F_1) - (1\,\text{m} \times F_2) + (2\,\text{m} \times m_F g) + (1\,\text{m} \times m_Y g) + (0.5\,\text{m} \times m_B g)$$
$$0 = -(1\,\text{m} \times F_2) + 1{,}372\,\text{N} + 539\,\text{N} + 245\,\text{N}$$
$$F_2 = 2{,}156\,\text{N}$$

Substitute into $F_1 = F_2 + 1{,}715\,\text{N}$:
$$F_1 = 2{,}156\,\text{N} + 1{,}715\,\text{N} = 3{,}871\,\text{N}$$

Therefore, $F_1 = 3{,}871\,\text{N}$ and $F_2 = 2{,}156\,\text{N}$.

To check, we can redo torque using the board's center as axis:
$$\sum T = (0.5\,\text{m} \times F_1) - (1.5\,\text{m} \times F_2) + (1.5\,\text{m} \times m_F g) + (0.5\,\text{m} \times m_Y g) + (0 \times m_B g)$$
$$0 = 0.5 \times (F_2 + 1{,}715) - (1.5 \times F_2) + (1{,}029) + (269.5)$$
$$0 = 0.5\,F_2 + 857.5 - 1.5\,F_2 + 1{,}298.5$$
$$0 = -1F_2 + 2{,}156$$
$$F_2 = 2{,}156\,\text{N}$$
$$F_1 = F_2 + 1{,}715\,\text{N} = 3{,}871\,\text{N}$$

Same answer using different axis: $F_1 = 3{,}871\,\text{N}$ and $F_2 = 2{,}156\,\text{N}$.

3.3. Friction

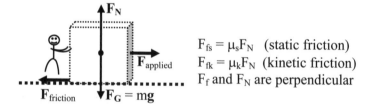

$F_{fs} = \mu_s F_N$ (static friction)
$F_{fk} = \mu_k F_N$ (kinetic friction)
F_f and F_N are perpendicular

• In a *frictionless* environment, when a net force is applied to an object, it accelerates: $F = ma$. In the *presence of friction*, acceleration is slowed or stopped due to the retarding force of friction: $F_{applied} - F_{friction} = ma$. Friction is a retarding force that reduces acceleration. If the friction force equals the applied force and the net forces are zero, the object can move at *constant velocity*: $F_{applied} - F_{friction} = ma = 0$.

• When two surfaces come into contact and exert forces on each other, the microscopic irregularities on each surface interlock, causing the surfaces to adhere and resist sliding. This resistance to moving across each other is called **friction**. There are different types of frictional forces: *static friction, kinetic friction*, and *rolling friction*.

Static friction occurs when there is no motion. It is the frictional force that resists motion or sliding. **Kinetic friction** occurs when there is a relative sliding motion. It is the frictional force that retards motion. **Rolling friction** occurs when a curved surface rolls along another, sticking and un-sticking as it rolls. Static friction is stronger than kinetic friction because motion depresses the microscopic irregularities in the surface and, when moving, surfaces do not have a chance to interlock so the adhering is reduced. You may have noticed that it takes a lot of effort to push a heavy box when it is at rest, but once you get it moving it takes less effort to keep it sliding.

Note: When we discuss friction we are assuming that it is the only retarding force on the surfaces and that there is no "digging in" of one surface or edge into the other surface.

- Properties of the **force of friction**, $F_f = \mu F_N$:

Its *magnitude* depends on the properties of the surfaces, which are reflected in the coefficient of friction μ.

It is proportional to the normal force F_N between the surfaces. Remember, the normal force equals the component of weight perpendicular to the surface and is $F_N = mg$ for level surfaces and $F_N = mg \cos \theta$ for inclines.

Its *direction* is opposite motion along, or parallel to, the surface.

For a given weight, friction does *not* depend on the area of contact between the surfaces. (A 3-kg cube and a 3-kg thin rectangle have the same friction.)

- The general mathematical relationships for friction are written:

$$F_{fs} = \mu_s F_N \text{ static friction } \quad \text{and} \quad F_{fk} = \mu_k F_N \text{ kinetic friction}$$

where the normal force can be written F_N or N. The Greek letter mu, or μ, with subscripts s and k represents the **coefficients of static and kinetic friction**. The values depend on the contacting surfaces. Coefficient values range from 0.0 to greater than 1.0 and can be found in scientific and engineering handbooks. A value of 0.0 would correspond to zero friction. Examples of μ for different surfaces include: wood on wood μ_s 0.25–0.5, μ_k 0.2; glass on glass μ_s 0.94, μ_k 0.4; Teflon on Teflon μ_s 0.04, μ_k 0.04; ice on ice μ_s 0.1, μ_k 0.03; waxed wood on wet snow μ_s 0.14, μ_k 0.1; and waxed ski on snow μ_s 0.1, μ_k 0.05.

- The **units** for the frictional forces and normal force, F_{fs}, F_{fk}, and F_N, are those of force whereas μ_k and μ_s are dimensionless constants having no units. Remember, force is measured in units of (mass)(length)/(time)^2, or 1 N = 1 kg·m/s^2 or 1 dyne = 1 g·cm/s^2. 1 N = 100,000 dynes.

• **Kinetic friction** is a constant motion-opposing force that depends on the properties of the surfaces and the normal force exerted. For an object to move at a **constant velocity**, the net force on the object must be zero. This means that the magnitude of the applied force must be balanced by the magnitude of the retarding force due to friction. For example, the frictional force between a box and the floor depends on the normal force exerted by the floor on the box and the coefficient of friction.

The **normal force** F_N, which is always *perpendicular to the surface*, is a reaction force to the weight of the box. For an object sitting on a flat level surface, the normal force is: $F_N = mg$. If the surface is at an angle θ, the normal force is: $F_N = mg \cos \theta$. Because the normal force is equal to the component of weight perpendicular to the surface, if the surface is at an incline, you will need to consider the angle of incline. The normal force and the force of friction are perpendicular to each other because the normal force is perpendicular to the surface and the frictional force is parallel to the surface.

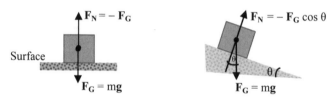

Static friction occurs in reaction to a force and depends on the magnitude of the applied horizontal force. If a box is at rest on the floor and is not being pushed, there is no static friction. Once a horizontal pushing or pulling force is applied to the box, the *static friction increases* until the applied force is greater than the friction, and the box begins to move. *Static friction must be overcome to get an object to begin to slide* across a surface.

• To demonstrate that static friction increases until the applied force exceeds the force of friction, you can push on a heavy box. Suppose you push with 5 N of force and nothing happens. You push harder with 10 N of force—still nothing. Finally you push with 12 N and the box is in motion!

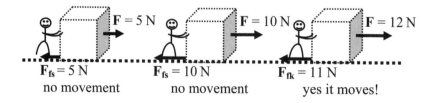

As long as your pushing force is inadequate to move the box, the force of static friction is equal to your applied force. As you push harder on the box, the applied force is equal to the force of static friction until a maximum static friction F_{fs} exists to hold the box in equilibrium. This maximum static friction occurs just before motion. The **coefficient of static friction** μ_s *corresponds to the maximum static friction just before movement*. Once the box moves it is operating under kinetic friction F_{fk}.

• **Example**: A large 40-kg wood box is sitting on the floor. If the coefficients of static and kinetic friction are $\mu_s = 0.4$ and $\mu_k = 0.2$, what are the forces required to get the box to slide and then to keep it sliding at a constant velocity?

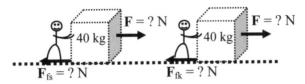

We need to determine the force required to overcome the maximum static friction. That force is equal to the maximum static friction:

$$F = F_{fs} = \mu_s F_N = \mu_s mg = (0.4)(40 \text{ kg})(9.8 \text{ m/s}^2) = 156.8 \text{ N}$$

The force required to keep the box moving at a constant velocity must equal the force of kinetic friction:

$$F = F_{fk} = \mu_k F_N = \mu_k mg = (0.2)(40 \text{ kg})(9.8 \text{ m/s}^2) = 78.4 \text{ N}$$

• **Example**: Adjustable incline planes are used to determine the coefficient of static friction by placing an object of known weight on the incline surface and adjusting the angle until it begins to slide. What is the equation you could use to calculate the coefficient of friction?

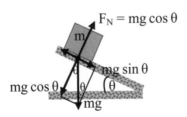

The normal force is: $F_N = mg \cos \theta$
The component of the object's weight parallel to the incline is: $mg \sin \theta$
The mathematical relationship for static friction is: $F_{fs} = \mu_s F_N$

Just before the object begins to slide, the component of the object's weight parallel to the incline surface equals the maximum force of static friction so that:

$$F_{fs} = mg \sin \theta \quad \text{and} \quad F_{fs} = \mu_s F_N = \mu_s mg \cos \theta$$

Solving for the coefficient of static friction:

$$\mu_s = F_{fs}/F_N = (mg \sin \theta)/(mg \cos \theta) = (\sin \theta)/(\cos \theta) = \tan \theta$$

• When your surface is at an incline you need to consider the angle of incline, the coefficient of friction, the net normal force, the parallel component of an object's weight, and the force of friction.

• **Example**: Suppose you are wondering at what angle you would begin to slide down a slope of hard ice in case you slip while hiking. If the coefficient of static friction between you and the ice is about 0.14 and you weigh 100 lb, will you slide if the slope is a 10° incline? What about a 5° incline? At what angle will you slide?

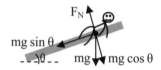

Angle 10°:

$$F_N = mg \cos \theta = (100 \text{ lb}) \cos 10° \approx 98.5 \text{ lb}$$
$$F_{parallel} = mg \sin \theta = (100 \text{ lb}) \sin 10° \approx 17.36 \text{ lb}$$

The maximum static friction is:

$F_{fs} = \mu_s F_N = (0.14)(98.5 \text{ lb}) = 13.8 \text{ lb}$ which is less than 17.36 lb

The static friction value is less than the component of your weight that is parallel to the slope so *you will slide at 10°.*

Angle 5°:

$$F_N = mg \cos \theta = (100 \text{ lb}) \cos 5° \approx 99.6 \text{ lb}$$
$$F_{parallel} = mg \sin \theta = (100 \text{ lb}) \sin 5° \approx 8.7 \text{ lb}$$

The maximum static friction is:

$F_{fs} = \mu_s F_N = (0.14)(99.6 \text{ lb}) = 13.9 \text{ lb}$ which is greater than 8.7 lb

Static friction value is greater than the parallel component of weight, so *you will NOT slide at 5°.*

The angle just before you begin to slide is where $F_{fs} = F_{parallel}$. Or:

$$F_{fs} = \mu_s F_N = \mu_s \, mg \cos \theta = F_{parallel} = mg \sin \theta$$
$$\mu_s \, mg \cos \theta = mg \sin \theta$$
$$\mu_s \cos \theta = \sin \theta$$
$$\mu_s = \sin \theta / \cos \theta = \tan \theta$$
$$\theta = \arctan \mu_s = \tan^{-1}(0.14) = 7.97 \text{ degrees}$$

When the angle exceeds 7.97°, or about 8°, you will begin to slide.

• **Example**: You and your friend have been hiking all day on an ice field and find a flat place to pitch a tent. Unfortunately there are a few bears around so you don't want any food in your tent. You put your food in a steel box and set it away from your tent (there are no trees from which to suspend it). Your friend says the slope where you are about to place the food is too steep and the food would slide if it were bumped. He said that once it got going, the food would surely accelerate down the hill. What equation would you use to calculate the acceleration of the food assuming you can estimate the coefficient of friction for metal on ice?

First remember that the components of gravity acting on the food are mg cos θ perpendicular and mg sin θ parallel to the surface. Write down the following equations:

F_N = mg cos θ, the normal force.

$F_{parallel}$ = mg sin θ, parallel component of gravity.

$F_{fk} = \mu_k F_N = \mu_k$ mg cos θ, the kinetic friction force.

Net force along the slope is: $F_{acceleration}$ = ma = mg sin θ – μ_k mg cos θ.

Net acceleration along the slope is: a = g sin θ – μ_k g cos θ.

• **Example**: How high can your 70.0 kg friend climb up a 4.0 m, 15.0 kg ladder leaning at 40° before the ladder begins to slip? Assume the coefficient of static friction between the ladder and the wall is μ_{sw} = 0.2 and between the ladder and the ground is μ_{sg} = 0.4.

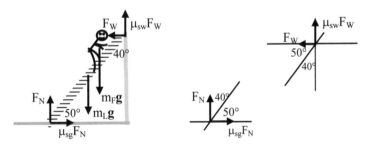

When he reaches maximum height and the ladder is about to slip, $\mu_{sg}F_N$ is at its maximum. As long as the ladder is not moving the forces and torques balance: $\sum F = 0$ and $\sum T = 0$.

$$\sum F_{y\text{-direction}} = 0 = \mu_{sw}F_W + F_N - m_Lg - m_Fg$$

$$0 = (0.2)F_W + F_N - (15 \text{ kg})(9.8 \text{ m/s}^2) - (70 \text{ kg})(9.8 \text{ m/s}^2)$$

$$(0.2)F_W + F_N = 833 \text{ N}$$

$$\sum F_{x\text{-direction}} = 0 = \mu_{sg}F_N - F_W = (0.4)F_N - F_W$$
$$(0.4)F_N = F_W \quad \text{or} \quad F_N = F_W/(0.4)$$

Combine and solve for F_W and F_N:

$$(0.2)F_W + F_N = 833 \text{ N}$$
$$(0.2)F_W + F_W/(0.4) = 833 \text{ N}$$
$$F_W(0.2 + 1/0.4) = 833 \text{ N}$$
$$F_W = 833 \text{ N} / (2.7) = 308.5 \text{ N}$$
$$F_N = F_W/(0.4) = 308.5 \text{ N} / 0.4 \approx 771.3 \text{ N}$$

Now we need to use the torque balance since it considers lengths and distances. Use clockwise as direction of torque and choose where the *ladder rests on the ground as the pivot* (forces at pivot multiply by zero length thus are zero). Also, use the center of the ladder as its length when considering its weight on the ground, since it has an equal distribution of mass along its length. Also remember the angle θ in $T = rF \sin \theta$ used when calculating torque is the acute angle between extended lines of vector $\mathbf{r}$ and the force vector $\mathbf{F}$.

$$\sum T = \mathbf{r}\mathbf{F} \sin \theta = 0 = -(4 \text{ m}) F_W \sin 50° - (4 \text{ m}) \mu_{sw}F_W \sin 40°$$
$$+ (2 \text{ m}) m_L g \sin 40° + (x) m_F g \sin 40°$$
$$0 = -(4 \text{ m})(308.5 \text{ N})(\sin 50°) - (4 \text{ m})(0.2)(308.5 \text{ N})(\sin 40°)$$
$$+ (2 \text{ m})(15 \text{ kg})(9.8 \text{ m/s}^2)(\sin 40°) + (x)(70 \text{ kg})(9.8 \text{ m/s}^2)(\sin 40°)$$
$$0 = -945 \text{ N} - 159 \text{ N} + 189 \text{ N} + (x)441 \text{ N}$$
$$915 \text{ N} = (x)441 \text{ N}$$
$$x = 915 / 441 \approx 2.07 \text{ m}$$

As your friend reaches about 2.07 m up, the ladder will begin to slip.

Alternatively, choose where the *ladder rests on the wall as the pivot*:

$$\sum T = \mathbf{r}\mathbf{F} \sin \theta = 0 = +(4 \text{ m}) F_N \sin 40° - (4 \text{ m}) \mu_{sg}F_N \sin 50°$$
$$- (2 \text{ m}) m_L g \sin 40° - (4 - x) m_F g \sin 40°$$
$$0 = (4 \text{ m})(771.3 \text{ N})(\sin 40°) - (4 \text{ m})(0.4)(771.3 \text{ N})(\sin 50°)$$
$$- (2 \text{ m})(15 \text{ kg})(9.8 \text{ m/s}^2)(\sin 40°) - (4 - x)(70 \text{ kg})(9.8 \text{ m/s}^2)(\sin 40°)$$
$$0 = 1983 \text{ N} - 945 \text{ N} - 189 \text{ N} - (4 - x)441 \text{ N}$$
$$849 \text{ N} = (4 - x)441 \text{ N}$$
$$1.93 = 4 - x$$
$$x = 2.07 \text{ m}$$

Same answer using different pivot. The ladder slips as he reaches 2.07 m.

3.4. Key Concepts and Practice Problems

- For an object to be in equilibrium: $\sum F = 0$ and $\sum T = 0$.
- Dynamic equilibrium: object in motion at constant velocity.
- Static equilibrium: object at rest.
- Force of Friction: $F_{friction} = \mu F_N$.
- Static friction: $F_{fs} = \mu_s F_N$ is overcome for an object to begin to slide.
- Kinetic friction: $F_{fk} = \mu_k F_N$ a constant motion-opposing force.

Practice Problems

3.1 The heater in your hot air balloon malfunctions, and you begin to lose altitude. You estimate your downward acceleration to be -1 m/s². The balloon and its payload weigh 500 kg. How many 5-kg sandbags do you need to jettison immediately to stop the downward acceleration?

3.2 Mario is walking a tightrope across Niagara Falls, holding a 30-ft, 20-lb balance beam at its midpoint. Suddenly a large parrot descends and perches 2 ft from the left end of his pole and starts to admonish Mario to be very careful. To keep his balance, Mario must shift the pole 3 ft to the right. What was the weight of the parrot?

3.3 If your 70-kg friend in the ladder example in Section 3.3 wants to try again to reach the top of the ladder, at what minimum angle from the ground must he lean the ladder? (Note: $\sin \theta = \cos(90 - \theta)$. The ladder, wall, and ground form a right triangle with 2 non-90° angles summing to 90°.)

Answers to Chapter 3 Problems

3.1 $F_g = mg = (500 \text{ kg})(-9.8 \text{ m/s}^2) = -4{,}900$ N.
$F_{net} = ma = (500 \text{ kg})(-1.0 \text{ m/s}^2) = -500$ N. To stop downward acceleration, reduce F_g by 500 N to $-4{,}400$ N. Let m_1 be mass after jettisoning sandbags, so $F_g = (m_1)(-9.8) = -4{,}400$ N and $m_1 = 449$ kg. You must reduce the weight by $(500 \text{ kg} - 449 \text{ kg}) = 51$ kg. Jettisoning 11 of the 5-kg sandbags will stop the downward acceleration and begin slowing your rate of descent.

3.2 After shifting the pole, the left side is 12 ft long (left-midpoint at 6 ft), with the parrot perched 10 ft out, and the right side is 18 ft (right-midpoint at 9 ft). Each foot of pole weighs 20 lb / 30 ft = 2/3 lb/ft. The 2 sides balance:
$(W_{\text{left-pole}})(r_{\text{left}}) + (W_{\text{parrot}})(r_{\text{parrot}}) = (W_{\text{right-pole}})(r_{\text{right}})$
$(12 \times 2/3)(6) + (W_{\text{parrot}})(10) = (18 \times 2/3)(9)$
$48 + (W_{\text{parrot}})(10) = 108$, and $(W_{\text{parrot}}) = (108 - 48)/10 = 6$ lb

3.3 Using the base of the ladder as the pivot point:

$\Sigma T = \mathbf{rF} = 0 = -(4\text{ m})(F_w \sin \theta) - (4\text{ m})(\mu_{sw} F_w \cos \theta)$

$\qquad + (2\text{ m})(m_L\, g \cos \theta) + (4\text{ m})(m_F\, g \cos \theta)$

$\qquad = -(4\text{ m})(308.5\text{ N})(\sin \theta) - (4\text{ m})(0.2)(308.5\text{ N})(\cos \theta)$

$\qquad + (2\text{ m})(15\text{ kg})(9.8\text{ m/s}^2)(\cos \theta) + (4\text{ m})(70\text{ kg})(9.8\text{ m/s}^2)(\cos \theta)$

$\qquad = -1{,}234(\sin \theta) - 246.8(\cos \theta) + 294(\cos \theta) + 2{,}744(\cos \theta)$

$1{,}234(\sin \theta) = 2{,}791.2(\cos \theta)$

$\tan \theta \approx 2.262$ or $\theta \approx 66.15°$ is the minimum angle from ground.

Chapter 4

NATURAL FORCES

"Your theory is crazy, but it's not crazy enough to be true."
Attributed to Niels Bohr

"The chief aim of all investigations of the external world should be to discover the rational order and harmony which has been imposed on it by God and which He revealed to us in the language of mathematics."
Attributed to Johannes Kepler

4.1. Gravitation

• *A force acts between any two objects or masses and depends on the inverse square of their distance from each other. This attracting force is described by the **Law of Universal Gravitation**.*

• Johannes Kepler (1571–1630) and Isaac Newton (1642–1727) were both intrigued by gravity and the orbits of planets and moons. Early on, Kepler developed an explanation of planetary motion and described it in what are called **Kepler's Laws**, which are summarized as follows.

• **Kepler's First Law**: Planets move in ellipses, with the Sun at one focus.

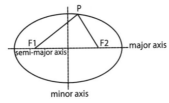

The figure shows an ellipse and its two *focus* points, F1 and F2. (A circle has one focus point at its center.) The long axis is the major axis, and half of this length is the semi-major axis. The short axis is the minor axis. All points on an ellipse have the property for point P that the:

(distance between P and F1) + (distance between P and F2) = constant

• **Kepler's Second Law**: As a planet travels in its orbit, an imaginary line connecting the planet to the Sun sweeps equal areas in equal intervals of time.

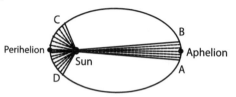

Equal areas (shaded) in equal time periods

As the planet moves from point A to point B, a line drawn from the Sun to the planet sweeps across and creates an area as shown. If the planet moves for a set period of time anywhere in its orbit, the area swept out by a line extending from the Sun to the planet during that time period will always be the same. In other words, if it takes the same time to go from A to B as from C to D, the areas swept will be equal. By this law we see that a planet moves slowest at its furthest, or aphelion, position and moves fastest at its closest, or perihelion, position. This is consistent with conservation of angular momentum.

• **Kepler's Third Law**: The square of a planet's orbital period is proportional to the cube of its mean distance, or semi-major axis, from the Sun:

$$\tau^2 \propto R^3$$

If the period τ of the orbit is in years and the semi-major axis R of the orbit is in units of Sun-Earth distance or *Astronomical Units* (which is 1.496×10^{11} m or 93,000,000 mi), then the proportionality becomes an equality and:

$$\tau^2 = R^3$$

This law applies to planets revolving around the Sun.

• **Newton** further explained the motions of the planets through his **Laws of Motion** and the development of the **Universal Law of Gravitation**. In fact, these laws describe the motion of everything from falling objects to planetary motion.

Remember **Newton's Laws of Motion: First Law of Motion**: An object does not accelerate, but rather remains at rest or moves in straight-line motion at constant velocity, unless it is acted on by an external force. **Second Law of Motion**: For an object with constant mass, the relationship between force $\mathbf{F}$ and acceleration $\mathbf{a}$ is $\mathbf{F} = \mathbf{ma}$. **Third Law of Motion**: For any action (force), there is an equal and opposite re-action, so that if object A exerts a force on object B, then B exerts an equal, oppositely-directed force on A, $\mathbf{F}_{AB} = -\mathbf{F}_{BA}$.

Kepler's Laws are consistent with Newton's Laws, and Newton's work revealed that Kepler's Laws were a natural result of his laws. Newton reasoned that some force must be exerted by the Sun on the planets and by the Earth on the Moon. Using centripetal acceleration and Kepler's Third Law relating a planet's distance R and period τ ($\tau^2 = R^3$), it could be shown that the centripetal acceleration of a planet varies as the inverse square of its distance R from the Sun. Remember from Section 1.11 for an object in circular motion, **centripetal acceleration** equals $a_c = v^2/r = (r\omega)^2/r = r\omega^2$. If you substitute the *period* $\tau = 2\pi/\omega$, and rearrange to $\omega = 2\pi/\tau$, centripetal acceleration can be expressed as:

$$a_c = r\omega^2 = r(2\pi/\tau)^2 = 4\pi^2 r/\tau^2$$

Using $a_c = 4\pi^2 r/\tau^2$ and assuming a nearly circular planetary orbit:

$$a_c \propto R/\tau^2 \quad \text{where } \propto \text{ means "is proportional to"}$$

Substituting Kepler's Third Law, $\tau^2 = R^3$, gives:

$$a_c \propto R/R^3 \quad \text{or} \quad a_c \propto 1/R^2$$

Newton found that the *centripetal acceleration* of a planet around the Sun, or of the Moon around the Earth, *depends on the inverse square of distance*.

Because $F_c = ma_c$:

$$F_c \propto 1/R^2$$

Since gravity provides the centripetal force for an orbiting body:

$$F_G \propto 1/R^2$$

From his work Newton concluded that *a universal force acts between any two objects or masses and depends on the inverse square of their distance from each other*. This is apparent by the Sun attracting Earth or the Earth attracting the Moon, but by Newton's Third Law (for any force, there is an equal and oppositely-directed force), he realized that a planet must also exert a force on the Sun, and the Moon must exert a force on the Earth. Therefore the attracting force between two objects or masses is called the **Law of Universal Gravitation** and described as:

$$\boxed{F_G = Gm_1m_2/r^2}$$

where F_G, the force of gravitation, a vector, represents the force acting on either of the masses m_1 or m_2, and depends on the inverse square of distance r between their centers. $G = 6.67 \times 10^{-11}$ m^3/kg·s^2 or N·m^2/kg^2, derived experimentally, is the constant of proportionality, known as the **universal gravitational constant,** which characterizes the intrinsic strength of the gravitational force. This law shows that any two masses gravitationally attract each other with forces of equal magnitude.

• Note: The **universal gravitational constant** G was first determined in 1798 by **Henry Cavendish**, who measured the minuscule force between two lead balls by using a highly sensitive torsion balance. Since his experiment there have been slightly more accurate measurements. A recent measurement using a new torsion balance method obtained: $G = 6.674215 \pm 0.000092 \times 10^{-11} \, \text{m}^3/\text{kg·s}^2$. From this an accurate value for the Earth's mass was also determined as $5.972245 \pm 0.000082 \times 10^{24} \, \text{kg}$ and the Sun's mass as $1.988435 \pm 0.000027 \times 10^{30} \, \text{kg}$. (*Physical Review Letters v. 85, pp. 2869–2872, 2000.*)

• **Example**: Imagine your friend asks you to help him correlate Newton's three Laws of Motion with planetary motion. How could you answer?

First, you suggest approximating the orbits of the planets as circles so the equations for uniform circular motion apply. You explain that when an object or planet is in uniform circular motion, in order for it to stay in that orbit and not fly out, it must be experiencing a constant acceleration toward the center of its orbit. You explain that apparently Newton noted that the Moon continuously *falls* in its path around the Earth because of the acceleration due to gravity, thereby creating its orbit. You point out that because a planet in uniform circular motion is constantly accelerating toward the center of its orbit, by Newton's First Law of Motion, there must be a center-directed force acting on it. By Newton's Second Law, $F = ma$, you suggest that he can find the magnitude of that force. Using the equation for acceleration of an object in uniform circular motion, $a_c = v^2/r$, you further suggest he can calculate the centripetal acceleration. You finally add that from Newton's Third Law, if there is a force attracting a planet to the Sun, there must be an equal and oppositely-directed force attracting the Sun toward the planet. In fact, you exclaim that not only is a planet accelerating toward the Sun, but the Sun accelerates minutely toward a planet—though we can generally approximate the Sun as fixed compared to a relatively small planet.

• **Example**: Calculate the mass of the Earth using Newton's Universal Law of Gravitation. (Use Earth's radius as $6.37 \times 10^6 \, \text{m}$.)

Begin with $F_G = Gm_E m_O/r_E^2$, where m_E is Earth's mass, m_O is the mass of some object on the Earth's surface, and r_E is Earth's radius. We know G and r_E but will need another relationship to determine F_G and m_E. We know that the gravitational force F_G on any object m_O on Earth's surface is given by: $F_G = m_O g$. We can combine these two equations for F_G and solve for the mass of the Earth m_E:

$$F_G = Gm_E m_O / r_E^2 = m_O g \quad \text{or} \quad m_E = gr_E^2 / G$$

Substitute values for g, G, and r_E:

$$m_E = (9.8 \text{ m/s})(6.37 \times 10^6 \text{ m})^2 / (6.67 \times 10^{-11} \text{ m}^3/\text{kg·s}^2) \approx 5.96 \times 10^{24} \text{ kg}$$

This value for Earth's mass, 5.96×10^{24} kg, differs slightly from the more accurate value we reported above, $5.972245 \pm 0.000082 \times 10^{24}$ kg, due to rounding errors.

• **Example**: Imagine a teacher asks you and your friend to calculate the force of gravity on the Moon by the Earth. The two of you make the calculations using two different approaches. What might those two different methods be?

You like reasoning through things methodically, so you decide to approach the problem using the Moon's centripetal acceleration in the formula $F_G = m_M a_c$. You know that to hold its orbit, the Moon feels a centripetal acceleration from the Earth of $a_c = v^2/r_{EM}$. To determine v you remember that velocity is distance per time. In the case of the Moon's orbit, it is the distance of one revolution around the Earth divided by the time it takes to travel around once:

$$v = (\text{orbit circumference}) / (\text{orbit period}) = 2\pi r_{EM}/\tau$$

You look up the Earth-Moon distance r_{EM} to be 384,400 km and the period of the Moon's orbit as 27.3217 days, and plug into $v = 2\pi r_{EM}/\tau$:

$$v = 2\pi(384{,}400 \text{ km})(1{,}000 \text{ m/km}) / (27.3217 \text{ days})(86{,}400 \text{ s/day}) = 1{,}023 \text{ m/s}$$

Next you plug v into the centripetal acceleration equation and calculate:

$$a_c = v^2/r_{EM} = (1{,}023 \text{ m/s})^2 / (3.844 \times 10^8 \text{ m}) = 2.7225 \times 10^{-3} \text{ m/s}^2$$

Looking up the Moon's mass as 7.3483×10^{22} kg, you finally find the force of gravity on the Moon using:

$$F_G = m_M a_c = (7.3483 \times 10^{22} \text{ kg})(2.7225 \times 10^{-3} \text{ m/s}^2) = 2.00 \times 10^{20} \text{ N}$$

Using an alternative approach, your friend looks up values for G, m_E, m_M, and r_{EM}, and calculates the force of gravity on the Moon by the Earth using Newton's Universal Law of Gravitation, $F_G = Gm_E m_M / r_{EM}^2$:

$$F_G = (6.6742 \times 10^{-11} \text{ m}^3/\text{kg·s}^2)(5.9722 \times 10^{24} \text{ kg})(7.3483 \times 10^{22} \text{ kg}) / (3.844 \times 10^8 \text{ m})^2$$
$$= 1.98 \times 10^{20} \text{ N}$$

You and your friend get essentially the same results—the two answers round to about 2×10^{20} N! You feel that you learned more by reasoning through your approach, but your friend brags that his way was faster. The teacher points out that there is often more than one approach to solving a problem if you reason through and use the applicable formulas.

• **Satellites**: Gravity is the only significant force acting on a satellite. It provides the centripetal force that keeps a satellite in orbit. For a satellite in orbit, *the gravitational force equals the centripetal force*:

$$F_G = F_c = m_s a_c = m_s v^2 / r_{SE} = G m_E m_s / r_{SE}^2$$

where subscript s represents satellite, subscript E Earth, and r_{SE} is center-to-center satellite-Earth distance. Satellites in circular orbits travel at a constant speed v, while force F_G and acceleration point toward the center of the Earth perpendicular to **v**. Can you guess how to calculate that **velocity for a satellite**? Simply solve the above equation for v:

$$m_s v^2 / r_{SE} = G m_E m_s / r_{SE}^2$$

$$\boxed{v^2 = G m_E / r_{SE} \quad \text{or} \quad v = [G m_E / r_{SE}]^{\frac{1}{2}}}$$

This equation not only calculates the **velocity of a satellite** orbiting the Earth, but applies to any object that is orbiting another due to their gravitational attraction. The mass used is that of the body being orbited. Therefore, if it is the Sun that is being orbited by Earth, then the mass of the Sun would be used.

• **Example**: How would you calculate a satellite's velocity and circular orbital period if it is 36,000 km above Earth, given the Earth's mass is 5.9722×10^{24} kg?

To calculate the satellite's velocity at 36,000 km above Earth you can use the equation we just developed for satellites:

$$v^2 = G m_E / r_{SE}$$

But first remember to add the radius of Earth, 6,370,000 m, to the 36,000 km given in the question, since r_{SE} in the velocity equation represents the center-to-center distance:

$$r_{SE} = 36{,}000{,}000 \text{ m} + 6{,}370{,}000 \text{ m} = 42{,}370{,}000 \text{ m}$$

Now you calculate v:

$$v^2 = G m_E / r_{SE}$$

$$= (6.6742 \times 10^{-11} \text{ m}^3/\text{kg·s}^2)(5.9722 \times 10^{24} \text{ kg})/(42{,}370{,}000 \text{ m}) \approx 9{,}407{,}519 \text{ m}^2/\text{s}^2$$

Take the square root resulting in the satellite's velocity: $v \approx 3{,}067$ m/s.

To find the **period τ of the orbit of a satellite**, you can first relate velocity to period. Knowing that velocity is distance per time, you can reason that the velocity of an orbiting object is the distance around the orbit divided by the time it takes to travel around once, or:

$$v = (\text{orbit circumference}) / (\text{orbit period}) = 2\pi r / \tau$$

Set this equal to the *velocity of a satellite*, $[G m_E / r_{SE}]^{\frac{1}{2}}$:

$$v = 2\pi r_{SE} / \tau = [Gm_E/r_{SE}]^{\frac{1}{2}} \quad \text{or} \quad \tau = 2\pi r_{SE} / [Gm_E/r_{SE}]^{\frac{1}{2}}$$

$$\tau = 2\pi r_{SE} / [Gm_E]^{\frac{1}{2}}/[r_{SE}]^{\frac{1}{2}} = 2\pi r_{SE}[r_{SE}]^{\frac{1}{2}}/[Gm_E]^{\frac{1}{2}} = 2\pi [r_{SE}^{3}]^{\frac{1}{2}}/[Gm_E]^{\frac{1}{2}}$$

$$\boxed{\tau = 2\pi [r_{SE}^{3}/Gm_E]^{\frac{1}{2}}}$$

This equation calculates the **period for a satellite**.
Finally, substitute the values:

$$\tau = 2\pi[(42{,}370{,}000\,\text{m})^3 / (6.6742 \times 10^{-11}\,\text{m}^3/\text{kgs}^2)(5.9722 \times 10^{24}\,\text{kg})]^{\frac{1}{2}} \approx 86{,}796\,\text{s}$$

Then convert to get the satellite period:

$$86{,}796\text{ s} \times (1\text{ day} / 86{,}400\text{ s}) \approx 1\text{ day period}$$

• A satellite can be placed in a **geostationary orbit** at a certain height over Earth's equator which will hold the satellite in a position fixed over that point so that the period of its orbit will exactly equal the period of the rotation of the Earth. A satellite must be placed at a certain height above the Earth's equator in order to remain geosynchronous.

• **Example**: Develop an equation for calculating the altitude of a geostationary satellite. Then make the calculation.

You can begin with the equation just developed for the period τ of a satellite and solve for r_{SE} to determine the height. For a geostationary satellite the period is one day or about 86,400 s. Begin with:

$$\tau = 2\pi [r_{SE}^{3}/Gm_E]^{\frac{1}{2}} \quad \text{or} \quad \tau = 2\pi [r_{SE}^{3}]^{\frac{1}{2}}/[Gm_E]^{\frac{1}{2}}$$

$$[r_{SE}^{3}]^{\frac{1}{2}} = \tau[Gm_E]^{\frac{1}{2}}/2\pi$$

Square both sides to remove the square root:

$$\boxed{r_{SE}^{3} = \tau^2 Gm_E / 4\pi^2}$$

This equation calculates the **distance above Earth's center for a satellite**. Now substitute the values:

$$r_{SE} = [\tau^2 Gm_E /4\pi^2]^{\frac{1}{3}}$$

$$= [(86{,}400\,\text{s})^2(6.6742 \times 10^{-11}\,\text{m}^3/\text{kg·s}^2)(5.9722 \times 10^{24}\,\text{kg})/4\pi^2]^{\frac{1}{3}} \approx 42{,}241{,}000\,\text{m}$$

Since r_{SE} represents the center-to-center distance, we need to subtract the radius of the Earth, 6,370,000 m, for the height above Earth:

$$42{,}241{,}000\text{ m} - 6{,}370{,}000\text{ m} = 35{,}871{,}000\text{ m}$$

Therefore, the height above Earth for a geostationary satellite is just under 36,000 km (or about 22,289 mi).

• Note that in a *geosynchronous orbit*, the orbital period is one day, so that the satellite matches the rotation rate of the Earth. A *geostationary orbit* is a special case of a geosynchronous orbit in which the satellite is in a circular orbit and remains "stationary" over a certain point on the

Earth's surface above the equator. For a satellite to remain in a *geostationary orbit* above Earth, it must be about 36,000 km above the equator, have a velocity of about 3.07 km/s, and have a period that tracks Earth's rotation. You may notice the values for velocity and height differ slightly depending on the source. This is likely due to slightly differing values used in their calculations.

4.2. Electrostatic Force

Note: This section will discuss the electrostatic force. We will introduce the electromagnetic force and moving charges in Chapters 10 through 12.

• Every charged particle or object exerts a force on every other charged particle or object. **Electrostatic phenomena** arise from the forces that *stationary or slow moving electric charges* exert on each other. These forces can be described by the Electrostatic Force Law, also called Coulomb's Law.

• **Atoms** are made up of *protons, neutrons,* and *electrons*. The **protons** and **neutrons** reside in the atom's nucleus, and the **electrons** revolve in "clouds" or "shells" around the nucleus. Protons possess a positive charge and neutrons possess no charge. Electrons possess a negative charge that is *equal in magnitude* to the positive charge of a proton.

Electron

Neutron and Proton

In its natural, electrically neutral state, an atom has an equal number of protons and electrons. The number of protons specifies the **atomic number** of an atom or of an element. The element carbon, for example, has an atomic number of six reflecting its six protons. In its most prevalent natural state carbon also has six neutrons and six electrons.

The charge on an electron, which is sometimes called the basic unit of charge, is: Electron charge = -1.602×10^{-19} Coulombs.
The equal and opposite positive charge of a proton is:
Proton charge = $+1.602 \times 10^{-19}$ Coulombs.

Note: A **Coulomb** is based on the measurement of current, which is the flow of charge through a conducting medium. The Coulomb is defined in

MKS (meter-kilogram-second) units as the amount of charge that flows past a fixed point in one second in a current of one Ampere.

• The force that attracts an electron to protons in the nucleus and holds it in "orbit" around the nucleus is the **electrostatic force**. Without this electrostatic force, the electron, which is traveling at a high speed, would not remain in its orbit.

In nature, *unlike or opposite charges attract each other, and like charges repel each other.*
Opposite charges attract: $\Theta \rightarrow \leftarrow \oplus$
Like charges repel: $\Theta \leftarrow\!\longrightarrow \Theta$ and $\oplus \leftarrow\!\longrightarrow \oplus$

• *Certain atoms can gain or lose electrons, resulting in electrons being transferred from one atom or group of atoms (object) to another.* The protons, however, tend to be a fixed part of an atom's nucleus. The distribution of electric charge usually occurs because of the movement of electrons. Once electrons are transferred, an atom or object can be left with either an excess or deficiency of electrons. An excess of electrons results in a negative charge. Conversely, a deficiency of electrons results in a positive charge.

Conservation of Charge: Electric charge cannot be created or destroyed, only transferred or redistributed. The total amount of electric charge in an isolated system remains constant.

Electrostatic fields exist around charged particles or objects. These fields create the force that acts between charged particles causing them to either attract or repel one another. The *field lines point toward negative charges and away from positive charges.*

Electrostatics can involve the study of the buildup of charge on the surface of objects due to contact with other surfaces. When two surfaces contact each other, charges can transfer and electrostatic forces can develop. These charges can remain until they either bleed off to ground or are neutralized with a discharge—the familiar experience of *static shock*.

• In 1785 French physicist **Charles Augustin de Coulomb** studied electrostatic forces using a sensitive torsion balance similar to the balance Cavendish used to study the gravitational force. Coulomb found that the electrostatic force between two charged objects varies as the *inverse square of the distance between them*. Remember, the gravitational force also varies as the inverse square of the distance between two masses. The electrostatic force is proportional to the product of the charges involved. This is similar to the gravitational force with respect to masses. The electric force acting on a point charge q_1 as a result of the presence of a second point charge q_2 is described by the **Electrostatic Force Law** or **Coulomb's Law** given as:

$$F_E = Kq_1q_2/r^2$$

This law shows that **the force of electrostatic attraction or repulsion** between two electric charges *at rest* is directly proportional to the product of the two charges and **inversely proportional to the square of the distance** between them. In the equation: F_E is the *force of electrostatic attraction or repulsion* (in Newtons); q_1 and q_2 represent the charges (in Coulombs C) of the two particles or objects (note upper-case Qs may be used); r is their separation (in meters); and K is **Coulomb's Constant** $K \approx 8.98755 \times 10^9$ N·m^2/C$^2 \approx 9.0 \times 10^9$ N·m^2/C^2.

Coulombs Law of Electrostatic Force can also be written as:

$$F_E = q_1q_2/4\pi\varepsilon_0r^2 \quad \text{where } \varepsilon_0 = 1/4\pi K$$

and where $\varepsilon_0 \approx 8.8542 \times 10^{-12}$ C^2/N·m^2 and is called the *permittivity of free space* or *electric constant*, which measures the effect of a substance (or a vacuum) on the electric field.

You can quickly see the similarity between the **Coulomb's Electrostatic Force Law**, $F_E = Kq_1q_2/r^2$, and the **Newton's Law of Universal Gravitation**, $F_G = Gm_1m_2/r^2$. The electric force F_E is a **vector** quantity having magnitude and direction. The direction depends on the signs of the charges. *If the signs of q_1 and q_2 are different, the force is attractive and the direction of the force on each charge is toward the other. If the signs of q_1 and q_2 are the same, the force is repulsive and the direction of the force on each charge is away from the other.* Similar to gravitational force calculations, the distance of the charges is considered to be the center-to-center distance.

From Coulomb's Law of Electrostatic Force we can see that electrostatic force depends on (1) the type of charges—like charges repel and opposite charges attract; (2) the amount of charge—more charge equals more force; and (3) the distance between charged objects—the closer together, the greater the force (by the inverse square).

• **Example**: Hydrogen atoms contain a proton and an electron having charges of plus and minus 1.6×10^{-19} C. If the average distance between the proton and electron in the hydrogen atom is 5.3×10^{-11} m, what is the electrostatic force between them?

Use Coulomb's Law of Electrostatic Force with $K = 9.0 \times 10^9$ N·m²/C²:

$$F_E = Kq_Eq_P/r^2$$

$$= (9.0 \times 10^9 \text{ N·m}^2/\text{C}^2)(-1.6 \times 10^{-19} \text{ C})(1.6 \times 10^{-19} \text{ C}) / (5.3 \times 10^{-11} \text{ m})^2$$

$$= -2.3 \times 10^{-28}/(5.3 \times 10^{-11})^2 = -0.082 \times 10^{-6} = -8.2 \times 10^{-8} \text{ N}$$

The negative sign reflects the attractive force between the proton and electron.

• **Example**: Suppose your friend noticed you solve the previous example and thought he could trip you by asking you how fast the electron is traveling around the nucleus of the hydrogen atom.

You remind yourself that in order to maintain "orbit" there must a centripetal acceleration, $a_c = v^2/r$. You also know that, since the electrostatic force is the dominant force, the centripetal force, $F_c = ma_c = mv^2/r$, must be about equal to the electrostatic force, $F_E = Kq_Eq_P/r^2$. Therefore:

$$F_c = F_E = m_ea_c = m_ev^2/r = Kq_Eq_P/r^2$$

Since we calculated F_E in the previous example as $F_E = 8.2 \times 10^{-8}$ N, we can solve for v in terms of F_E:

$$F_E = m_ev^2/r$$
$$v^2 = F_Er/m_e$$

You look up the **mass of an electron** to be about 9.1×10^{-31} kg and r was given in the last example as 5.3×10^{-11} m, so v^2 becomes:

$$v^2 = F_Er/m_e$$
$$v^2 = (8.2 \times 10^{-8} \text{ N})(5.3 \times 10^{-11} \text{ m}) / (9.1 \times 10^{-31} \text{ kg}) = 4.78 \times 10^{12} \text{ m}^2/\text{s}^2$$

You take the square root to get velocity: $v = 2.19 \times 10^6$ m/s.

"Wow, that little whiz is moving pretty fast!" your friend muses. The he asks, "How does it keep from flying away from its nucleus?"

"At close range the electrostatic force between electrons and protons is one of the strongest forces in the Universe!" you respond. "It's way more powerful than gravity. Even though the proton and electron in the hydrogen atom possess the fundamental force of gravity, each proton and electron can also develop a much stronger electrostatic force. While the electron is rocketing around its nucleus trying to get away, the electrostatic force is pulling the proton and electron together. These effects balance and maintain the electron in orbit." You pause, "Hey, just for fun let's calculate the force of gravity between the proton and electron in the hydrogen atom." You set up the gravity equation:

$$F_G = Gm_e m_p / r^2$$
$$= (6.67 \times 10^{-11} \, m^3/kg \cdot s^2)(9.1 \times 10^{-31} \, kg)(1.67 \times 10^{-27} \, kg) / (5.3 \times 10^{-11} \, m)^2$$
$$= 3.6 \times 10^{-47} \, N$$

"Wow, it is certainly a lot less than the attractive electrostatic force of -8.2×10^{-8} N," your friend remarks.

• **Example**: Imagine there are three charges: $q_1 = -5 \, \mu C$, $q_2 = +10 \, \mu C$, and $q_3 = -5 \, \mu C$. What is the net force or charge on q_2? (Note that μ, pronounced "mu", designates micro or 10^{-6}.)

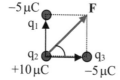

q_1 to q_2 distance is 0.2 m
q_2 to q_3 distance is 0.2 m

This is a vector problem, which means we calculate the forces along the x- and y-directions. From Section 1.7, the component vectors of **V** are $V_x = V \cos \theta$ and $V_y = V \sin \theta$, where **V** has magnitude $|V| = [V_x^2 + V_y^2]^{1/2}$ and the direction is the angle **V** makes with x-axis, or: $\theta = \tan^{-1}(V_y/V_x)$.

For x-direction: The force on q_2 is from q_3, with no x-component from q_1 because it is perpendicular and $\cos 90° = 0$.

$$F_{Ex} = Kq_2q_3/r^2 = (9.0 \times 10^9 \, N \cdot m^2/C^2)(10 \times 10^{-6} C)(-5 \times 10^{-6} C)/(0.2 m)^2 = -11.25 N$$

where the minus sign reflects that the forces are attractive.

For y-direction: The force on q_2 is from q_1, with no y-component from q_3 because it is perpendicular.

$$F_{Ey} = Kq_2q_1/r^2 = (9.0 \times 10^9 \, N \cdot m^2/C^2)(10 \times 10^{-6} C)(-5 \times 10^{-6} C)/(0.2 m)^2 = -11.25 N$$

The magnitude of the net force or charge is vector **F**:

$$|F| = [F_x^2 + F_y^2]^{1/2} = [11.25^2 + 11.25^2]^{1/2} \approx 15.91 \, N$$

The direction is the angle θ above the x-axis, which we can see by inspection, but let's calculate to show method:

$$\theta = \tan^{-1}(F_y/F_x) = \tan^{-1}(11.25/11.25) = 45°$$

• **Example**: You can charge an electrically neutral object by transferring electrons. This can occur by friction (rubbing it against another material), conduction, or induction. You can actually lift an object using electrostatic force. Imagine you take your socks out of the dryer on a cold, dry day. You notice a small half-ounce sock stuck to the top of the dryer, so you peel it off. If you immediately hold the sock 0.2 m below an object carrying a charge of 30 µC and the sock is pulled up to the object, how much charge must the sock have had?

In this example the electrostatic force between the charged object and the charged sock must be equal to the weight w of the sock.

$$F_E = w = m_{sock}g = Kq_{sock}q_{object}/r^2$$

We are given or can look up the values:

$F_E = m_{sock}g = w = (0.5 \text{ oz})g = (0.014 \text{ kg})(9.8 \text{ m/s}^2) \approx 0.14 \text{ N}$

$K = 9.0 \times 10^9 \text{ N·m}^2/\text{C}^2$

$q_{object} = 30 \times 10^{-6} \text{ C}$

We can rearrange to solve for q_{sock}:

$$q_{sock} = wr^2/Kq_{object} = (0.14 \text{ N})(0.2 \text{ m})^2/(9.0 \times 10^9 \text{ N·m}^2/\text{C}^2)(30 \times 10^{-6} \text{ C})$$
$$= 2 \times 10^{-8} \text{ C} = 0.02 \times 10^{-6} \text{ C} \quad \text{or } 0.02 \text{ µC}$$

So 0.02 µC is the minimum charge required to overcome the sock's weight.

• **Coulomb's Law of Electrostatic Force** models stationary charges or charged objects and is a good approximation for forces between slow moving charges or objects. When movement of charges occurs, magnetic fields are produced, which alters the force on the two charged objects. The magnetic interaction between moving charges is likened to having the force from an electrostatic field combined with relativistic effects caused by very high speeds. The electromagnetic force is discussed later in the book.

4.3. Strong Nuclear and Weak Forces

"Just when you thought you were beginning to understand reality, we take a brief detour into the ever-so-strange world of nuclear physics!"

The information presented in the following pages is meant to provide you with some foundational information on nuclear physics and introduce you to the jargon. In the future when you hear words like "quark" and "gluon", you will have an idea what they mean and how they fit into your greater universe.

• If you think for a moment, there are undoubtedly many questions you have about the subatomic universe that exists all around you. This section offers a brief glimpse into our strange subatomic world. This world is filled with fascinating concepts and yet-to-be-discovered answers to our reality. Perhaps you may be intrigued by this strange world and decide to seek answers to some of the many remaining questions surrounding nuclear physics.

One question you may be wondering about is: What holds protons in a nucleus of an atom together, considering the strong repulsive electrostatic forces they exert on each other and the miniscule space they occupy?

 Nucleus with protons and neutrons

Let's estimate the electrostatic repulsive force between two protons. If the typical distance between protons in a nucleus is about 2×10^{-15} m, the electrostatic force is:

$$F_E = Kq_pq_p/r^2$$
$$= (9.0 \times 10^9 \, \text{N·m}^2/\text{C}^2)(1.6 \times 10^{-19} \, \text{C})(1.6 \times 10^{-19} \, \text{C})/(2 \times 10^{-15} \, \text{m})^2 \approx 58 \, \text{N}$$

58 N is a huge force for such a tiny space. In fact, it is the same force as the weight of a 6-kg object! So why don't protons blow apart in the nuclei of atoms due to their strong repulsive force? The answer is the presence of an even stronger force called the **strong nuclear force**. But before we go into that, let us back up and look at some of the most fundamental particles that make up our Universe.

The Fundamental Building-Block Particles

• Based on current research, the Universe seems to be composed of twelve fundamental building-block particles and governed by four fundamental forces. The current thought on how the twelve particles and three of the four forces (gravity excluded) fit together is described by the "Standard Model." The **Standard Model** is a description of the elementary particles and forces in the Universe.

"Did you know you and everything around you is made up of unimaginably small and strange particles?"

You know about protons, neutrons, and electrons, but there are particle building blocks even more fundamental. In fact, the **matter** around (and inside) us is made of two basic categories of particles called **fermions**.

The two categories of **fermions** are **quarks** and **leptons**. Each category consists of six kinds of particles (**6 quarks and 6 leptons**), which are in related pairs. Quarks are the building blocks of protons and neutrons. Interestingly, the Standard Model of physics also posits that *everything has an exact opposite, or* **antiparticle**. The universe of particles around us includes not only the fermions (matter particles), but also four force "particles" called **bosons**, which we introduce in the next subsection.

FERMIONS
Matter particles

QUARKS	LEPTONS
Up +2/3	Electron −1
Down −1/3	Electron-neutrino 0
Charm +2/3	Muon −1
Strange −1/3	Muon-neutrino 0
Top +2/3	Tau −1
Bottom −1/3	Tau-neutrino 0
Also:	Also:
6 antiquarks	6 antileptons

Note: The above numbers reflect the charge.

BOSONS
Force Carriers

GLUONS—Strong Force

W & Z Bosons—Weak Force

PHOTONS—Electromagnetic Force

GRAVITONS—Gravitational Force
(Gravitons are as yet unverified)

• There are six types or "flavors" of **quarks**, together with their six associated antiparticles. The six quarks are paired in three subgroups— the Up and the Down quark, the Charm and the Strange quark, and the Top and the Bottom quark. It is the Up and Down quarks that are the constituents of neutrons and protons. Quarks carry a type of "**charge**" called "**color**", which can be red, blue, or green, with antiparticles anti-red, anti-blue, or anti-green. These are not actual colors but rather properties, and can add up to be colorless the way blue, green, and red light add up to white light. Quarks are believed to be held together by this so-called color charge or color force, which we discuss below.

Quarks also carry fractional electrical charges, as noted in the above diagram. Since protons and neutrons are made up of quarks we can get some insight into their positive and neutral charges. A proton is made up of two Up quarks and one Down quark, and has a total charge of $(+2/3) + (+2/3) + (−1/3) = +1$, which is the net charge on a proton. Similarly, a neutron is made up of two Down quarks and one Up quark, giving it a total charge of $(−1/3) + (−1/3) + (+2/3) = 0$, which is the net charge on a neutron.

Particles made up of quarks are called **Hadrons**. Particles made of *two quarks* are called **Mesons**, and particles made of *three quarks* are called **Baryons**. *Neutrons and protons are Baryons.* Only quarks interact through the *strong force*, but they can also interact through the weak and electric forces.

• There are six types or "flavors" of **leptons** together with their six associated antiparticles. The six leptons are paired in three subgroups— the electron and the electron-neutrino, the muon and the muon-neutrino, and the tau and the tau-neutrino. The electron, the muon, and the tau all have an electric charge and a mass, whereas the neutrinos are electrically neutral with very little or negligible mass. **Leptons** are susceptible to the electric and weak forces but not the strong force. **Neutrinos** are leptons and are produced in processes such as beta decay and reactions that involve the *weak force* and also in nuclear fusion reactions. Neutrinos interact through the *weak force* and the much weaker *gravitational force*, have no electric charge, very little or negligible mass, move at close to the speed of light, and are difficult to detect.

• There are also four **boson** "particles" that transmit or "carry" the fundamental forces. As we will discuss below, each fundamental force has its own corresponding boson particle that *transmits or carries* that force. The boson particles are the **gluon** (for the strong force), the **photon** (for the electromagnetic force), the **W and Z bosons** (for the weak force), and the not-yet-discovered "**graviton**" (for gravity).

The Four Fundamental Forces

"*Here's an introduction to the truly bizarre fundamental forces that act in our strange and astonishing Universe!*"

• There are **four fundamental forces** of nature at work in the Universe: the *strong force*, the *weak force*, the *electromagnetic force*, and the *gravitational force*. They have different strengths and ranges. The **Standard Model** states that strong, weak, and electromagnetic forces between interacting matter particles *occur by the exchange of boson force carrier particles*. As mentioned, each fundamental force has its own mediating or "carrying" boson particle such that the **gluon** carries the **strong force**, the **photon** carries the **electromagnetic force**, the **W and Z bosons** carry the **weak force**, and the yet-to-be-discovered *graviton* would carry the **gravitational force**. These four basic forces of nature can be summarized as:

(1) The **strong nuclear force**, or **strong force**, carried by the **gluon** is the strongest force, but acts only at very short range. Strong nuclear forces hold protons and neutrons together in atomic nuclei and help

power the Sun. Only quarks interact via the strong force, but quarks can also interact through the weak and electric forces.

(2) The **weak force**, carried by the **W and Z bosons**, acts at very short distances, and is stronger than gravity. Weak nuclear forces are responsible for decay of radioactive nuclei, specifically Beta decay, and the changing of types or "colors" such as changing a quark from one type to another. Weak nuclear forces exhibit some peculiar symmetry characteristics not seen with the other forces. Quarks and leptons can interact via the weak interaction.

(3) The **electromagnetic or electric force**, carried by **photons**, acts between electrically charged particles at up to very long distances, varies as the inverse square of distance, and causes electrical and magnetic effects. Electromagnetic forces and interactions are responsible for visible light, X-rays, microwaves, radio waves, etc., and are fundamental to the telecommunications and electronics industries. Quarks and leptons can interact electromagnetically. Transient virtual photons carry the electrostatic force.

(4) The **gravitational force**, carried by the still elusive **"graviton"**, is the weakest force, but acts between all masses at up to extremely long distances and varies as the inverse square of distance.

• The **Standard Model** does not include the gravitational force due to the lack of confirmation of the theoretical graviton and mathematical inconsistencies when modeling the macro and atomic worlds. Quantum Theory models the atomic world and general relativity models the larger world. We look forward to the day with these two models may be seamlessly joined. Fortunately, in the scale of particle physics, the effects of gravity are negligible so the Standard Model for describing the strong, weak, and electromagnetic forces works well.

If you study the Standard Model you will find that the weak and electro-magnetic forces, also called *interactions*, are alternatively described together as the **electroweak theory**. This theory presents the force-carrying particles for the weak interactions as behaving similarly to how photons behave when carrying the electromagnetic interactions. That is, W and Z bosons mediate the weak force just as photons mediate the electromagnetic force. The *electroweak theory* involves the four force-carrying boson particles: photons, W^-, W^+, and Z^0 bosons. The W and Z bosons have much greater mass that the photon.

"This is a fascinating, evolving, and dynamic field of study! The following subsections should give you a little insight into the strong and weak forces and the strange nature of the world in which you live."

The Strong Nuclear Force

• The **strong nuclear force**, also referred to as the "**strong force**," is the *strongest of the four basic forces in nature*. It has the shortest range and acts at extremely short distances so that particles must be very close to each other before its effects are felt—the distance of about the diameter of a proton or a neutron. This force is *not* an inverse-square force like the electromagnetic and gravitational forces. The strong nuclear force maintains the stability of nuclei even though strong repulsive forces are present. It acts between protons and neutrons, protons and protons, and neutrons and neutrons. Protons and neutrons are collectively referred to as **nucleons** since, except for protons having a charge, protons and neutrons share many properties.

Protons and neutrons are made up of **quarks**. The strong nuclear force that holds protons and neutrons together in the nucleus of an atom acts between the quarks. The strong force is called an **exchange force** and is mediated by the **exchange of gluons between the quarks**. The strong force between quarks is the primary force holding an atom's nucleus together. The force holding protons and neutrons together is referred to as a residual strong force since it is a second order effect of the primary strong force between quarks.

Interestingly, isolated quarks have not been observed, suggesting they are permanently bound together by strong nuclear forces. In spite of that, studies of high-energy collisions between electrons and protons suggest the quarks that make up a proton appear to be nearly free and somewhat unbound. How can nuclear forces be strong enough to overcome repulsive forces and hold quarks together while, in experiments, they appear weak enough for quarks to seem more free? It turns out that the strong nuclear force that holds quarks together actually increases as the quarks are pushed apart, but weakens as quarks move extremely close together.

This seeming dichotomy is explained *through Quantum Chromodynamics* and *asymptotic freedom*. **Quantum Chromodynamics** (QCD) is a model of the strong nuclear force in which quarks are held together by exchanging **gluons** when they are at a distance of about the diameter of a proton. **Asymptotic freedom** explains the discrepancy that high-energy collision experiments revealed when they showed characteristics of a much weaker force and freer quarks at extremely close distances. When asymptotic freedom was overlaid on QCD, it established QCD as the viable theory describing the strong nuclear force. Asymptotic freedom is a phenomenon allowing quarks to behave as free particles when they are extremely close together, while becoming very strongly attracted to each other as their distance from each other increases. Asymptotic freedom allows the interaction between particles to be weak at very short distances (asymptotically approaching zero) and provides some understanding of the behavior of matter under extreme conditions.

The **gluon exchange** that occurs between quarks in atomic nuclei in QCD is similar to the **photon exchange** that occurs as electrons interact. In **Quantum *Electro*dynamics** (QED), the force between two *electrically charged* particles is mediated by the *exchange of a photon* or particle of light (which has no charge). Similarly in **Quantum *Chromo*dynamics**, *the force between two quarks is mediated by the exchange of a gluon.*

As mentioned above, **quarks** carry a type of "**color charge.**" While electrically charged particles can be positive or negative, the "color charge" of the strong force has the properties of "blue", "green", or "red". These are properties, not actual colors, and can add up to be colorless the way blue, green, and red light add up to white light. The interactions of these color charges form the *strong force or interaction*, so that quarks are believed to be held together via this so-called color charge or color force. The color force is believed to be mediated by the exchange of gluons. The color charges also have an antiparticle nature, or anti-red, anti-blue, and anti-green. Even gluons carry color charges. When quarks exchange gluons, a change in color can occur. Even more interesting—quarks are also surrounded by "virtual" gluons which can affect the quarks' color.

To summarize, the **strong nuclear force** holds protons and neutrons together, acts between quarks at very short distances of about a proton-diameter distance, gets weaker as quarks get closer, gets stronger as quarks move further apart, manifests as gluon exchange, and overcomes the repulsion of positively charged protons. Further, remember that quarks make up both protons and neutrons, and the presence of neutrons supplies strong force through gluon exchange. By their neutral presence, neutrons also help mitigate some of the repulsive proton-proton electrostatic force.

The Weak Force

• The **weak force**, also called **weak nuclear force** and **weak interaction**, is one of the four fundamental forces of nature and the most recent to be discovered. Though it is called weak, it is stronger than gravity. Gravity, however, can act over extremely long distances while the weak force acts over extremely small distances.

The weak interaction was discovered while scientists were studying and classifying types of nuclear radioactive decay processes including alpha, beta, and gamma decays. Scientists could explain alpha and gamma decays through the electromagnetic and strong interactions between nucleons, but **beta radiation** had unusual characteristics. Further studies led to an understanding of the weak interaction and its role in governing **beta decay**.

As we discussed above, neutrons and protons are made up of quarks. During **beta decay** a neutron having two *Down* quarks and one *Up* quark disappears and is replaced by a proton having two *Up* quarks and one *Down* quark, an electron, and an anti-electron neutrino. An intermediary in this process is a W boson which decays to produce the electron and anti-electron neutrino. In fact, the weak interactions involve the exchange or production of W and Z bosons (or more specifically W^+, W^-, and Z^0 bosons). Like the strong force exchanging gluons, the weak force is mediated by an **exchange of W and Z boson particles**.

Weak interactions are also responsible for a quark changing from one type or **flavor** to another. This phenomenon of the weak force changing the flavor of quarks enables decays of certain nuclear particles which require a quark change. The weak interaction uniquely provides a process in which flavor can change, thereby allowing transformation of quarks and leptons and the particles that contain them such as protons and neutrons. The weak interaction is in fact the only process by which a quark can change to another quark, or a lepton to another lepton, which occur through so-called flavor changes. (This ability for these fundamental particles to change is an important part of our Universe even though it may be difficult to visualize.)

The weak interaction acts between both quarks and leptons, whereas the strong force does not act between leptons. In fact, all fundamental particles except gluons and photons are believed to be subject to weak interactions. These interactions occur at very short distances of approximately 10^{-18} m or about a tenth of a percent of the diameter of a proton. Weak interactions are also noted for having strange properties as they exhibit peculiar symmetry features not seen with the other forces and allow effects that seem to violate normal behavior.

• Weak interactions are important to the functioning of the Universe. For example, in the Sun the transformation of hydrogen into deuterium (an isotope of hydrogen with one neutron rather than zero neutrons) occurs via the weak force before helium is formed. Also, the phenomena of radioactive decay, including radioactive elements used in medicine and technology, and the **beta-decay** of carbon to nitrogen in carbon-14 dating, involve the weak interactions.

• The deeper you dig into the world of nuclear physics, the more mysterious it gets! This interesting world leads to some of the most fascinating quests and questions, including the ever-elusive attempts to unify the fundamental forces between all of the elementary particles into a single framework.

4.4. Key Concepts and Practice Problems

• Kepler's Laws tell us: planets move in ellipses with the Sun at one focus, the Sun sweeps equal areas in equal intervals of time, and the square of a planet's orbital period is proportional to the cube of its mean distance from the Sun.
• The Law of Universal Gravitation is: $F_G = Gm_1m_2/r^2$.
• Velocity of an object orbiting the Earth or a large body: $v = (Gm_E/r)^{1/2}$.
• The Electrostatic Force Law or Coulomb's Law is: $F_E = Kq_1q_2/r^2$.
• Matter is made of fermions, including 6 types of quarks and 6 types of leptons.
• The forces are transmitted or carried by boson "particles".
• Bosons include: gluons, photons, W and Z bosons, and undiscovered gravitons.
• There are four fundamental forces of nature at work in the Universe:
  Strong nuclear force, or strong force: carried by gluons, strongest force, acts at very short range, and holds protons and neutrons together.
  Weak force: carried by W and Z bosons, acts at very short range, is stronger than gravity, and is responsible for Beta decay.
  Electromagnetic force: carried by photons, acts at up to long distances, varies as inverse square of distance, is responsible for light, X-rays, etc.
  Gravitational force: may be carried by the still elusive "graviton", is the weakest force, varies as the inverse square of distance, and acts at up to long distances.

Practice Problems

4.1 At what distance from the center of the Earth in the direction of the Sun are the gravitational forces equal and opposite? Assume the distance between the centers of the Earth and Sun is 1.497×10^8 km, the Sun's mass is 1.99×10^{30} kg, and the Earth's mass is 5.98×10^{24} kg.

4.2 Two objects are rubbed together, causing a transfer of electrons from one to the other. Without losing the electrostatic charge, the objects are separated by 1 cm and found to attract one another with a force of 0.09 N. How many electrons were transferred?

4.3 In this chapter we have been discussing a lot of weird particles. Have you wondered how they were discovered? Subatomic particles are often identified in linear or circular accelerators. In a linear accelerator, groups of electrons can be accelerated through a copper structure of discs and cylinders using an electromagnetic wave. This wave is made up of magnetic and electric fields, which are created by high energy microwaves. These microwaves are guided into the accelerator and create a pattern of oscillating electric fields pointing down the accelerator and oscillating magnetic fields in a circular pattern, forming an electromagnetic wave which travels down the accelerator. Once you produce your unknown particles by colliding electrons and positrons (antimatter electron counterpart), what do you need to know to identify them—charge, mass, momentum, speed, energy?

Answers to Chapter 4 Problems

4.1 Begin with $F_g = Gm_1m_2/r^2$. The distance from Earth's center is r and the distance from the Sun's center is $(1.497 \times 10^{11}) - r$. $F_E = Gm_Em_2/r^2$ and $F_S = Gm_Sm_2/(1.497 \times 10^{11} - r)^2$. We want to find the value for R where $F_E = F_S$. $Gm_Em_2/r^2 = Gm_Sm_2/(1.497 \times 10^{11} - r)^2$.
$m_E/r^2 = m_S/(1.497 \times 10^{11} - r)^2$ or $(1.497 \times 10^{11} - r)^2 m_E = r^2 m_S$
$(2.241 \times 10^{22}) - (2.994 \times 10^{11})r + r^2 = r^2(1.99 \times 10^{30})/(5.98 \times 10^{24})$
$(3.33 \times 10^5)r^2 + (2.994 \times 10^{11})r - (2.241 \times 10^{22}) = 0$
Using the quadratic formula $-b \pm [b^2 - 4ac]^{1/2}/2a$ and taking the smaller root which represents the side closer to the Sun, we get $r \approx 2.58 \times 10^8$ m. Because the Earth's radius is 6.37×10^6 m, the altitude is about 2.52×10^8 m or about 157,000 mi. Don't worry about falling into the Sun if you fly too high—your rotational velocity will keep you and the Earth in orbit around the Sun!

4.2 $F_E = Kq_1q_2/r^2$ or $q_1q_2 = F_Er^2/K = 0.09(0.01)^2/(9.0 \times 10^9) = 1.0 \times 10^{-15}$ C. Since the 2 objects have equal and opposite charges, take the square root, yielding 3.16×10^{-8} C. One electron has a charge of 1.6×10^{-19} C, so the number of electrons transferred was $(3.16 \times 10^{-8})/(1.6 \times 10^{-19}) = 1.976 \times 10^{11}$ or about 200 billion electrons.

4.3 You can identify a particle by determining its *charge* and *mass*. While in theory, you would say we can calculate a particle's mass if we know its *momentum* and *either* its *speed* or its *energy*, it turns out that when a particle is traveling near the speed of light, uncertainties in momentum or energy require us to measure momentum, speed, and energy. Detectors within the accelerator are set up to measure and track as many resultant particles as possible. They have several layers that identify the charge (if any) and the track each particle takes, as well as its momentum, velocity, and energy. Isn't it interesting that we use these familiar basic physics concepts of mass, velocity, momentum, and charge to understand this very strange and somewhat elusive subatomic world!

Chapter 5

ENERGY AND WORK

> *"It is evident that an acquaintance with natural laws means no less*
> *than an acquaintance with the mind of God therein expressed."*
> Attributed to James Prescott Joule

5.1. Work

• **Work** involves an applied force and the distance or displacement through which the force acts. For example, if you lift a weight, you are exerting the force required to overcome gravity and raise the object a certain distance upward. If you push an object, you are applying a force which can overcome friction and move the object a certain distance in the direction of the force. Mathematically, work W is the product of the component of force in the direction of the displacement and the displacement.

Work = (component of force in the direction of displacement $\mathbf{F_d}$)

$\times$ (displacement produced by the force $\mathbf{d}$)

$$W = \mathbf{F d} \cos \theta$$

The angle θ is the angle between the applied force vector and the displacement of the object once it is moved. The formula for work is also written using the component of force in the direction of displacement F_d times displacement **d**:

$$W = F_d d = Fd \cos \theta$$

or as the dot product (discussed below):

$$W = F \cdot d = Fd \cos \theta$$

• When the **applied force and displacement are parallel** to each other, the angle θ between the applied force and displacement vectors is zero. Because the cosine of zero equals one, $\cos 0 = 1$, the work done by the force is simply the product of the magnitude of the force and the magnitude of the displacement:

$$W = Fd$$

• **Example**: Suppose you push a box 10 m using a force of 50 N in the direction you want it to move. What is the work done on the box?

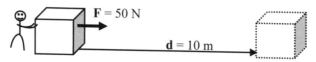

F = 50 N   d = 10 m

Because the applied force and displacement are parallel, so θ is zero, the work done on the box is force times displacement: W = Fd.

$$W = Fd = (50 \text{ N})(10 \text{ m}) = 500 \text{ N·m or 500 Joules}$$

• **Units of work** are the product of force and distance. In SI units this is Newton·meter (or N·m), which is kg·m/s² × m or kg·m²/s². A N·m is also defined as a **Joule (J)** in honor of **James Prescott Joule** (1818–1889), so work is measured in kg·m²/s² = N·m = J. In CGS (centimeter-gram-seconds) units, work is in dyne·cm, which is an erg. 1 J = 10^7 ergs. In English units, work is in foot-pounds (ft·lb).

• If you push (or lift) a box with a constant force and it moves with a constant velocity (due to opposing gravity or friction). The acceleration is zero, and there is no net external force. When the applied force is just enough to overcome the forces of friction or gravity, then $F_{net} = 0$. Whether an object moves at a constant speed or is accelerated, if it is displaced from one point to another due to an applied force, work is done.

• **Example**: Suppose you push a 20-kg box with a constant velocity in the direction you want it to move 10 m over a surface having a friction coefficient of 0.2. What is the work done on the box?

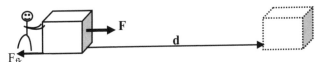

Because the applied force and displacement are parallel, angle $\theta = 0$ and $\cos 0 = 1$, so the work done on the box is: $W = Fd$. Also, because the box is sliding at a constant velocity, the net force acting on the box is zero and your pushing force is equal to the opposing force of friction. Remember from Chapter 3, the force required to keep the box moving at constant velocity must equal the **force of kinetic friction**, $F_{fk} = \mu_k F_N = \mu_k mg$, where μ_k is the *coefficient of kinetic friction* and F_N is the normal force, which is the weight of the box. We can calculate work using:

$$W = Fd = F_{fk}d = \mu_k F_N d = \mu_k mgd = (0.2)(20\,\text{kg})(9.8\,\text{m/s}^2)(10\,\text{m}) = 392\,\text{J}$$

• When the **applied force and displacement are NOT parallel** to each other, work is the displacement of an object multiplied by the component of the force in the direction of that displacement.

$$W = Fd \cos \theta$$

• **Example**: Suppose you push a 20-kg box with a constant velocity 10 m over a surface having a friction coefficient of 0.2. Instead of pushing in the direction of motion, you push downward at a 40° angle. What is the work done on the box?

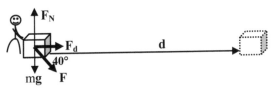

Because the applied force and displacement are NOT parallel, angle θ (which is between the applied force and the displacement) must be used to find the component of force in the direction of motion. The work done on the box is the component of force in the direction of motion that overcomes friction ($F \cos \theta$) times displacement d:

$$W = Fd \cos \theta = \mu_k F_N d$$

We are given d but need to find F. Because velocity is constant (no acceleration), the net force is zero and the horizontal force $F \cos \theta$ equals the frictional force:

$$F \cos \theta = \mu_k F_N$$

where the normal force F_N includes the weight of the box, mg, and the downward component of **F** from you pushing, which is $F \cos 50°$, or equivalently, $F \sin 40°$:

$$F_N = [(20 \text{ kg})(9.8 \text{ m/s}^2) + (F)(\sin 40°)]$$

Let's write the forces and then solve for F:

$$F \cos 40° = \mu_k F_N = \mu_k[(20 \text{ kg})(9.8 \text{ m/s}^2) + (F)(\sin 40°)]$$
$$F \cos 40° = (0.2)(196 \text{ kg·m/s}^2) + (F)(0.2)(\sin 40°)$$
$$F \cos 40° = (39.2 \text{ kg·m/s}^2) + (F)(0.129)$$
$$F \cos 40° - (F)(0.129) = (39.2 \text{ kg·m/s}^2)$$
$$F(\cos 40° - 0.129) = (39.2 \text{ kg·m/s}^2)$$
$$F \approx 61.5 \text{ N}$$

The work done on the box is:

$$W = Fd \cos \theta = (61.5 \text{ N})(10 \text{ m})(\cos 40) \approx 471 \text{ J}$$

Note that this is greater than the work needed to push the same box the same distance in the previous example (392 J), in which the force was applied parallel to motion, due to the greater opposing frictional force resulting from the additional downward force on the box.

• **Work W is a scalar not a vector** even though it is the product of two vectors, the force **F** vector and the displacement **d** vector. Even though work is a scalar and has magnitude but no direction, the direction of the vectors **F** and **d** matter during the calculation since work is the product of the component of force in the direction of displacement times displacement. As we discussed above, when applied force and displacement are parallel so that **F** acts along the direction of **d**, then $W = Fd$, since $\cos 0 = 1$. But when **F** and **d** are *not* parallel, the angle between them must be calculated.

The product $Fd \cos \theta$ is called a **scalar product** of vectors **F** and **d**. A scalar product is also called the dot product. The **dot product** of two vectors **A** and **B** is written $\mathbf{A \cdot B} = AB \cos \theta$, where θ is the angle between **A** and **B**. In words, the dot product of two vectors is the product of the magnitudes of the two vectors multiplied by the cosine of the angle in between those vectors.

• Work is not always done when a force is exerted. There are two primary cases when is work zero:

When displacement d = 0, then W = 0, because $W = \mathbf{F}d = (F)(0) = \mathbf{0}$. If you hold a huge rock, there is no displacement, so no work is done. (You do work to lift the rock, but not to hold it.) If you push against a 10-storey building, there is no displacement and no work done. If you lift your refrigerator 2 ft off the ground and set it back down in the same spot, you have done no work on the refrigerator. This is because the displacement of the refrigerator was zero. Also, as you lifted and lowered the refrigerator there was no frictional force that you were moving against.

(Remember, displacement is a vector representing the separation and direction between two points and is different from distance traveled.)

When the applied force is perpendicular to displacement, there is no work done. When $F \perp d$, $\theta = 90°$ and $\cos 90° = 0$. Therefore, $W = Fd \cos 90° = 0$ and no work is done. If you walk 10 m *horizontally at a constant speed* while carrying 10 large bricks, no work is done on the bricks. This is because your force holding the bricks is opposite gravity, which is perpendicular to the direction of motion. (If you carry the bricks up or down an incline there is a component of gravity acting in the direction of motion and work is done.) If you tie an object to a rope and twirl it in a circle horizontally over your head, there is no work done on the object by the rope since the tension T in the rope is always perpendicular to the motion.

• **Example**: Suppose you either drop a 10-kg rock off a 1,000-m cliff or shoot it horizontally off the same cliff using your rock launcher. Which scenario has greater work done on the rock by gravity after it is in motion?

When you drop the rock, gravity is the force acting on the rock, and the gravitational force **mg** is parallel to the rock. Work done on the rock is $W = mgd$. When you shoot the rock horizontally, gravity is still the acting force. In this case, **F** and **d** are not parallel and work is the force times the component of displacement in the direction of force, so that work done by gravity is still the same, **mgd**. The work done by the force of gravity is the same whether the rock drops directly downward or is shot horizontally and falls in a parabolic curve. This happens because the force acting on the rock is the gravitational force, which does no work in the horizontal direction since gravity and the horizontal component of displacement are perpendicular. Therefore the work done on the rock by gravity is the same.

• **Example**: Imagine you and your friend have been out snowshoeing and sledding all day. At the end of the day you pile all your stuff on one of your sleds and push it 200 m at a constant velocity up a 20° incline to the lodge. If the sled and contents have a mass of 30 kg and the snow has a kinetic friction coefficient of 0.05, how much work will you do pushing the sled? What if there is no friction?

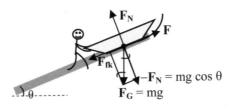

To push the sled you must overcome the forces of gravity and friction. (See Chapter 3 for discussion of friction.) The work done involves the component of gravity in the direction of displacement and the friction force that is opposite to motion. The force of gravity along the slope is:

$$F_G = mg \sin \theta = -(30 \text{ kg})(9.8 \text{ m/s}^2)(\sin 20°) \approx -100.6 \text{ N}$$

Remember $\sin \theta$ = opposite/hypotenuse and $\cos \theta$ = adjacent/hypotenuse. Also, the negative sign shows force opposite motion. Force of friction along the slope and opposite to motion is the product of the kinetic friction coefficient and the normal force, which is the component of the weight of the sled perpendicular to the slope:

$$F_{fk} = \mu_k F_N = \mu_k mg \cos \theta = -(0.05)(30 \text{ kg})(9.8 \text{ m/s}^2)(\cos 20°) \approx -13.8 \text{ N}$$

The force you need to exert to push the sled up the incline must overcome the combined parallel components of F_G and F_{fk} that the sled exerts on you, which are:

$$F_G + F_{fk} = -100.6 \text{ N} + -13.8 \text{ N} = -114.4 \text{ N}$$

The work is $W = Fd \cos \theta$, but $\mathbf{F}$ and $\mathbf{d}$ vectors are parallel and cos 0 is one, so the work you must do to overcome gravity and friction and push the sled at a constant velocity up the slope is:

$$W = Fd = (114.4 \text{ N})(200 \text{ m}) = 22{,}880 \text{ J}$$

If there is no friction (you put on your crampons), the work would only be against gravity, so:

$$W = F_G d = (mg \sin \theta)d = (30 \text{ kg})(9.8 \text{ m/s}^2)(\sin 20°)(200 \text{ m}) \approx 20{,}111 \text{ J}$$

As we will discuss later, this can also be calculated $W = mgh$, where h is the total height. We calculate height using trigonometry as $\sin 20°$ = opposite/hypotenuse, or $h = (\sin 20°)(200 \text{ m})$, so:

$$W = mgh = (30 \text{ kg})(9.8 \text{ m/s}^2)(\sin 20°)(200 \text{ m}) \approx 20{,}111 \text{ J}$$

(It's the same math, just a different formula.)

• **Calculus note—area and the integral**: Work done can be calculated using an integral. In fact, the value of the work done is the value of the integral, which can be represented as the area under the curve of a graph of force versus distance that is defined by that integral. We have learned that if an object is displaced distance x by a constant force F, then the *work* done is $W = \mathbf{F} \cdot \mathbf{x}$. When an object is displaced by a variable force acting in the direction of motion from point x_1 to point x_2, work can be represented by the integral: $_{x1}\int^{x2} F(x) \, dx$.

Whether there is constant or variable force, the motion can be represented along an axis of a coordinate system, which can be shown on a graph of force vs. displacement. The *work done is equal to the area under the curve or line* that runs between the beginning and ending points of displacement. In the graph below, the work done in lifting a 500-lb crate 10 ft from $x_1 = 0$ to $x_2 = 10$ ft is the shaded area. Work is:

$$\text{Shaded Area} = W = (500 \text{ lb})(10 \text{ ft}) = 5{,}000 \text{ J}$$

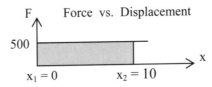

The integral describing work done to lift a 500-lb crate 10 ft is:

$$_0\!\int^{10} (500) \, dx = (500)(10) - (500)(0) = 5{,}000 \text{ ft·lb}$$

The more general definition of work using the integral considers that force can vary in magnitude and direction and the path can also vary. For example, if motion varies along curve ds from s_1 to s_2, then the total work done can be represented by the integral:

$$_{s1}\!\int^{s2} \mathbf{F \cdot ds} = \,_{s1}\!\int^{s2} (F \cos \theta) \, ds$$

• When *work* is done on or by a system, there is a *change in the total energy* of the system. We will discuss energy in the next few sections.

5.2. Kinetic Energy

• **Energy** is the ability to do **work** or the capacity to produce an effect. It is a property of matter, causing movement or a change in condition. Energy allows change to occur. Energy exists in many forms including mechanical, chemical, nuclear, thermal, electrical, electromagnetic, and electrostatic energy. In an *isolated system*, energy can be transformed from one form to another, but the total amount of energy is considered to be conserved. Solving problems using energy equations is often simpler than using Newton's Laws. Energy can be divided into two general categories: **kinetic energy** which involves *motion* and **potential energy** which involves *position*. Energy can be transferred by matter or by waves. In fact, much of the information scientists gather about the distant Universe comes from detecting and analyzing waves.

• **Kinetic energy** is the energy of motion. An object in motion carries kinetic energy and can perform work on another object that it collides with or pushes. An object in motion is able to use its motion to do work.

The faster an object is moving, the more energy it can transfer to another object it contacts. Kinetic energy exists in different forms including: rotational—the energy of rotational or turning motion; vibrational—the energy of vibrational or oscillating motion; and translational—the energy of moving from one location to another. Kinetic energy can involve waves, molecules, substances, and objects in the form of radiant, thermal, sound, and motion energy.

Kinetic energy ($KE = (1/2)mv^2$), momentum ($p = mv$), force ($F = ma = \Delta p/\Delta t$), and work ($W = Fd$) are interrelated. Kinetic energy is related to momentum because a moving object's momentum affects the force it will apply if it strikes another object. That force is also related to the amount of work performed. If you bump into your friend and move him a short distance, the work done on your friend is the force times the distance he was displaced. When you crash into your friend, you use your motion to do work on him. If you hit him and stop moving while he begins moving, you transfer kinetic energy to him.

• We can develop the equation for kinetic energy using relationships we already know. If an object is at rest with no force acting on it, and then a constant force is applied, the object will accelerate ($F = ma$). The object will move some distance, $x = (1/2)at^2$, at which time its velocity will be $v = at$. (Remember, when initial velocity v_0 is zero, then $v = v_0 + at$ and $x = v_0t + (1/2)at^2$ become $v = at$ and $x = (1/2)at^2$.) The work done on the object as it moves in the direction of the applied force is:

$$W = F \times d = ma \times (1/2)at^2 = (1/2)m(at)^2$$

Because $v = at$:

$$W = Fd = (1/2)mv^2$$

The work that has been done on this object is $W = Fd$ and, due to its motion, it has gained the equivalent energy of $(1/2)mv^2$. *The energy an object gains due to its motion is called kinetic energy.* Therefore, for an object of mass m moving at velocity v, its **kinetic energy** (KE) is defined as:

$$\boxed{KE = (1/2)mv^2}$$

• *Like work, energy is a* **scalar**. **Units** for **kinetic energy**, like work, are in Joules. Units in SI are $J = kg \cdot m^2/s^2 = N \cdot m$; in CGS are dyne·cm or erg; and in the English system are ft·lb.

• **Example**: If you are moving 2 m/s and have a mass of 50 kg, what is your kinetic energy?

$$KE = (1/2)mv^2 = (1/2)(50 \text{ kg})(2 \text{ m/s})^2 = 100 \text{ J}$$

• **Example**: If you are traveling at a certain speed and then double that speed, what is the change in kinetic energy? What if you halve the speed instead?

Always begin these types of problems by looking at the equation or relationship. In this case kinetic energy $KE = (1/2)mv^2$. You can see that KE is directly proportional to the square of velocity, $KE \propto v^2$. Therefore if velocity is doubled, KE will quadruple. To see this, choose a speed of $v = 4$ where $v^2 = 4^2 = 16$. If $v = 4$ is doubled to $v = 8$, you have $v^2 = 8^2 = 64$. Sixty-four is 4 times 16.

If the speed is halved, the kinetic energy is quartered. Again, we see this by selecting a speed of 4 so that $v^2 = 4^2 = 16$. When halved, $v = 4$ becomes $v = 2$ and $2^2 = 4$. Four is one quarter of sixteen.

• **Example**: Suppose a 10-kg ball at rest receives a 20-N force evenly for 5 s. What is the resulting kinetic energy and work on the ball?

When the applied force and displacement are parallel, the work is: $W = Fd$. To find d, we can use: $d = v_0t + (1/2)at^2$. Since initial velocity is zero:
$$d = (1/2)at^2 = (1/2)(F/m)(t)^2 = (1/2)[(20\,N)/(10\,kg)](5\,s)^2 = 25\,m$$
Work done is: $W = Fd = (20\,N)(25\,m) = 500\,J$.

The kinetic energy is: $KE = (1/2)mv^2$.
To find velocity we can use: $v = v_0 + at$.
Since initial velocity is zero:
$$v = at = (F/m)(t) = [(20\,N)/(10\,kg)](5\,s) = 10\,m/s$$
Remember a Newton, N, is kg·m/s². The kinetic energy is:
$$KE = (1/2)mv^2 = (1/2)(10\,kg)(10\,m/s)^2 = 500\,J$$
All the work done, 500 J, is transformed into kinetic energy, 500 J.

The Work-Energy Theorem

• **When work is done to or by a system, the energy of the system changes.** This relationship is referred to as the **work-energy theorem**. We said earlier that energy is the ability to do work. We can also think of work as a measure of the transfer of energy. If work is done on an object, then that amount of energy is transferred to it. When that energy is kinetic energy, and it moves the object some distance with a certain force, the work-energy theorem is between work W and kinetic energy KE. This means that the net work done on an object by an external net force equals the change in kinetic energy of the object:

$$\boxed{W = KE_f - KE_i = (1/2)(m)(v_f^2 - v_i^2) \quad \text{or} \quad W = \Delta KE}$$

If you apply a force to an object in the direction of displacement causing it to accelerate, and the object's speed increases ($v_f > v_i$), the force is doing positive work on the object ($W > 0$), and its KE increases. If you apply a force (opposite to direction of displacement) to decelerate an object, the force is doing negative work on the object ($W < 0$), and the object's KE decreases. (Since applied force is decelerating the object, the object is doing work on the source exerting the net force).

The relationship between work and kinetic energy, $W = \Delta KE$, can be used to find the velocity of an object.

• **Example:** (a) Suppose you do 500 J of work pushing a 20-kg crate across your garage, but friction between the crate and ground creates 400 J of heat. What is the final velocity of the box? (b) Suppose you pushed the 20-kg crate, and then it slides with an initial velocity of 2 m/s across a smooth section of your garage floor. If a 2-N frictional force is acting against the crate, what is the crate's final velocity after it slides 10 m?

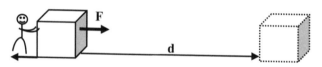

(a) The work done *on* the crate equals the kinetic energy it gains, and the work done *by* the crate equals the amount of kinetic energy it loses. In this case, the crate gains 500 J of kinetic energy from you pushing it and has a friction loss of 400 J from its kinetic energy to friction, resulting in a net gain of 100 J of kinetic energy. We can calculate the final velocity using the equation for kinetic energy, $KE = (1/2)mv^2$:

$$v^2 = 2KE/m = 2(100\ J)/(20\ kg) = 10\ kg{\cdot}m^2/s^2kg$$

$$v \approx 3.16\ m/s$$

(b) The initial kinetic energy of the crate is:

$$KE = (1/2)mv^2 = (1/2)(20\ kg)(2\ m/s)^2 = 40\ J$$

The crate decelerates due to friction as the floor does work on the crate opposite to the crate's motion:

$$W = F_f d = (-2\ N)(10\ m) = -20\ J$$

As work is done on the crate, its kinetic energy decreases and after sliding 10 m, the crate's kinetic energy becomes $40 - 20 = 20$ J. Therefore, the final kinetic energy of the crate is 20 J. Using the equation for kinetic energy, $KE = (1/2)mv^2$, we can determine final velocity:

$$v^2 = 2KE/m = 2(20\ J)/(20\ kg) = 2\ kg{\cdot}m^2/s^2kg \quad \text{or} \quad v \approx 1.41\ m/s$$

- **Example**: How will the final velocities differ in the two scenarios depicted if h is the same (ignore friction)?

By the work-energy theorem, $W = \mathbf{F} \cdot \mathbf{d} = KE_f - KE_i = (1/2)m(v_f^2 - v_i^2)$, where in this case, $\mathbf{F} \cdot \mathbf{d} = mgh$. Rearranging $mgh = (1/2)m(v_f^2 - v_i^2)$ gives $v^2 = 2gh$. Gravity is the only force doing work on you and your sled, and only the change from initial to final vertical position matters. Since the final and initial heights are equal in the two scenarios depicted, when you ignore friction, the final velocities are equal.

5.3. Potential Energy

- **Potential energy** involves the position of an object or the arrangement of its constituents or atoms. It can be thought of as *stored energy* or energy stored in an object. Potential energy comes in different forms including gravitational, mechanical, chemical, nuclear, and electrical. An object acquires potential energy when work is done on it that causes a change in its position (e.g. lifting), its configuration in relation to other objects, or the position of its constituent parts resulting in a change in its shape (e.g. compressing a spring).

- **Gravitational potential energy** involves position and reflects the potential for work to be done on an object by the gravitational force. If you lift a brick upward to height h above the ground, there is work done against the force of gravity as you lift: $W = F_g h = -mgh$. (The negative sign reflects the brick is lifted in the opposite direction of the force of gravity.) Doing $-mgh$ of work on the brick increases its gravitational potential energy by mgh. The force of gravity has the potential to do mgh of work on the brick at height h, so the brick has mgh gravitational potential energy. The **gravitational potential energy** is:

$$PE_G = mgh$$

where m is the mass of an object, g is the acceleration due to gravity, and h is the object's height above the ground or above a reference height. This equation is accurate when the distance **h is small relative to Earth's radius, h $\ll$ R$_E$**, allowing us to ignore any variation of g with height.

• **Units** for **potential energy**, like kinetic energy and work, are: Joules J $= kg \cdot m^2/s^2 = N \cdot m$ in SI; dyne·cm or erg in CGS; or ft·lb in English system.

• The higher an object is above the ground, the greater its gravitational potential energy. For example, if you hold a 1-kg ball 2 m above the ground, it will have a *gravitational potential energy* of $PE_G = mgh = (1 \text{ kg})(9.8 \text{ m/s}^2)(2 \text{ m}) = 19.6$ J. If you hold the same 1-kg ball 200 m above ground, it will have a gravitational potential energy of $PE_G = mgh = (1 \text{ kg})(9.8 \text{ m/s}^2)(200 \text{ m}) = 1{,}960$ J.

• While any object on Earth's surface has some gravitational potential energy (PE_G) reflecting its distance to Earth's center of gravity, PE_G is usually defined as zero at some low point of rest, and then it is acquired as the object is raised to a new higher point of rest.

If a brick having mass m is at rest, and you lift it above its location to height h and let it rest at its new higher location, you have done work W against gravity to move the brick higher to the distance h. Notice, however, that the brick would have no net change in velocity, so it would acquire no kinetic energy. Would the brick have energy after its move? Yes, the energy of position. If it were to fall to its original location, it would travel a distance h and acquire a velocity $v = [2gh]^{1/2}$, or $v^2 = 2gh$. Note that we developed this equation, $v^2 = 2gh$, in Section 1.5 for an object falling from rest using $x = (1/2)at^2 = (1/2)gt^2$, which we rearranged to $t^2 = 2x/g$ or $t = [2x/g]^{1/2}$. We also used the average acceleration $a_{Ave} = \Delta v/\Delta t = g$, and rearranged it to $v = gt$. By substituting $t = [2x/g]^{1/2}$ into $v = gt$, we got $v^2 = 2gx$. Since distance x is the height h, $v^2 = 2gh$. Therefore, the kinetic energy of the falling brick, if it had been lifted to height h, would be:

$$KE = (1/2)mv^2 = (1/2)(m)(2gh) = mgh$$

where mgh is equal to the work W that would need to be done to overcome the force of gravity g on the brick (or any mass m) and lift it a distance of height h:

$$KE = W = Fd = mgh$$

For any raised object, the **work done to overcome the gravitational force** to lift the object is stored in the object as potential energy. At the raised position the object has acquired the energy mgh that would become kinetic energy if it fell a distance of h. Therefore, an object lifted to a height h acquires a **gravitational potential energy** of mgh:

$$PE_G = mgh$$

So:

$$\boxed{KE = (1/2)mv^2 = W = Fd = PE_G = mgh}$$

• **Example**: You and your friend are on another expedition, this time to Ptarmigan Peak. (a) You climbed about 8 m up a high rocky cliff but forgot your camera. If your camera weighs 0.5 kg, what velocity will your friend need to toss it up so you can catch it? (b) You climbed further up the cliff and knocked a 30-kg rock down near to where your friend was standing (oops). If the rock hit the ground near him at 25 m/s, how high above him were you when you bumped it (ignore air resistance)?

(a) The *kinetic energy of the camera when your friend throws it must equal the camera's potential energy at the top* of the cliff where you are standing, so KE = PE or $(1/2)mv^2 = mgh$. Solve $(1/2)mv^2 = mgh$ for v: First cancel m's: $(1/2)v^2 = gh$.

$$v^2 = 2gh = (2)(9.8 \text{ m/s}^2)(8 \text{ m}) = 156.8 \text{ m}^2/\text{s}^2 \quad \text{or} \quad v \approx 12.5 \text{ m/s}$$

Your friend would need to throw the camera at a velocity of at least 12.5 m/s for it to reach you.

(b) The initial PE of the rock equals the final kinetic energy of the rock as it hits the ground. First find KE:

$$KE = (1/2)mv^2 = (1/2)(30 \text{ kg})(25 \text{ m/s})^2 = 9,375 \text{ J}$$
$$PE = KE = mgh = 9,375 \text{ kg·m}^2/\text{s}^2 = (30 \text{ kg})(9.8 \text{ m/s}^2)(h)$$
$$h = (9,375 \text{ kg·m}^2/\text{s}^2)/(30 \text{ kg})(9.8 \text{ m/s}^2) \approx 31.9 \text{ m}$$

We could also calculate this without using mass:

$$(1/2)mv^2 = mgh \text{ so } h = v^2/2g = (25 \text{ m/s})^2/(2)(9.8 \text{ m/s}^2) \approx 31.9 \text{ m}$$

You were about 32 m above your friend when you bumped the rock.

5.4. Conservative and Non-Conservative Forces

• In this chapter we learned that *work done* can be changed into *kinetic energy*. It is also clear that if work done on an object changes its potential energy, then the *change in potential energy can be equal to the work done*. If we do work to lift an object, the gravitational potential energy gained is:

$$\boxed{PE_G = W = Fh = mgh}$$

This equation is accurate when the distance **h is small relative to Earth's radius, h << R$_E$.** As you can see, energy can be changed from one form to another. Work is done on an object to lift it. When it falls from rest it begins with PE$_G$ and no KE. As it falls toward the Earth's center, it loses PE_G and gains KE. Just before it hits the ground its initial PE_G has been transformed into an equal amount of KE.

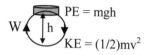

$$PE = mgh$$
$$KE = (1/2)mv^2$$

Just like lifting an object causes a conversion in energy, winding a watch spring does work that changes its shape, which is stored as potential energy. That potential energy is then converted to kinetic energy as the spring regains its original shape. The equation for potential energy is expressed in different forms depending on the type of potential energy described.

• We have learned that when an object is lifted or lowered (and no friction is present), the gravitational potential energy gained or lost only depends on the vertical displacement h, not on the path taken. Whether the path is straight up or along a diagonal or incline plane, *as long as friction is ignored, the **work** to lift an object depends only on vertical displacement h*. We saw this in our example near the end of Section 5.1 when you pushed a 30-kg sled up the 20°-incline hill to the lodge. When friction was ignored, you could calculate work using either $W = F_G d$ or $W = mgh$. The result was the same. When you pushed the 30-kg sled 200 ft up the incline, you did the same work as raising it straight up to the final height (h = (sin 20°)(200 m) in the sled problem). If you lift a mass from a low resting point up to a higher point, the work done is $W = mgh$ and does not depend on path taken, providing there is no friction.

W = mgh regardless of path taken providing friction is zero

• When you move an object and the work done against a force *depends only on the initial and final positions* of that object, not on the path taken, that force is a conservative force. The work a **conservative force** does on an object when moving it depends only on the end points of the motion—it is ***independent of the path taken***. When work is done by a conservative force, it is stored as some form of **potential energy**. The **gravitational force** fits this definition and is a **conservative force**. The **electrostatic force** and the **spring force** are also **conservative forces**. Conservative forces often act in a certain direction regardless of the direction of motion of the object the force is acting on. For example, the gravitational force acts in a downward direction regardless of the direction an object is moving.

• A **non-conservative force** is a force that *depends on the path* between the initial and final positions. The work a non-conservative force does on an object when moving it *depends on the path taken*. Non-conservative forces like friction are considered dissipative forces since energy can be dissipated, primarily as heat energy. The force of friction is a non-conservative force since the work done against friction to move an object depends on the path taken and its length, not just the initial and final positions. Pushing a box over rough ground in a circle does require work against friction even though the starting and ending points coincide. Friction and air resistance are non-conservative and act against motion regardless of the direction of the motion.

• We have learned about relationships between work and kinetic energy, $W = KE = (1/2)mv^2$, and work and potential energy, $W = PE_G = mgh$. In our discussion of gravitational potential energy we mentioned that $W = mgh$ does not depend on path taken, providing there is no friction. When friction, which is a non-conservative force, is present, work depends on path taken. Therefore, **potential energy is equal to work only when work is done against a conservative force**.

This is true because potential energy is a stored energy as an object is lifted, building potential energy, and then, as the object falls, its potential energy is transformed back into kinetic energy. The work put in to raise the object can be retrieved. Conversely, work done against a non-conservative force, like pushing against friction, is not stored for later use in a form that can be readily retrieved.

The **work-energy principle** states that *work done on an object by a net external force equals the total change in energy of that object*, providing the net external force is a **conservative force**. This is written:

$$W = \Delta KE + \Delta PE$$

where *PE includes any other acting forces including gravity*. Note that *mechanical energy* is defined as the sum of potential and kinetic energy.

5.5. Gravitational Potential Energy When h Is Large

• We have learned that when we are near Earth's surface where the gravitational force can be considered as nearly constant, and an object is raised a certain height that is small relative to the radius of the Earth R_E, the work done to lift is $W = mgh$ and the gravitational potential energy of the raised object is $PE_G = mgh$.

What if the distance that an object is lifted is significant compared to the Earth's radius R_E? Then the variation of the gravitational force must be considered. Remember from Section 4.1, the **Law of Universal Gravitation**: $F_G = GMm/r^2$, where F_G is the force of gravity acting on either of the masses M or m, r is the distance between their centers, and $G = 6.67 \times 10^{-11}$ $m^3/kg \cdot s^2$ is the experimentally derived universal gravitational constant. This formula tells us that F_G varies as $1/r^2$.

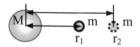

To calculate the work required to raise an object having mass m from an initial distance from Earth's center r_1 to a final further distance from Earth's center r_2, we write the force on the object at r_1 and r_2:

$$F_{1G} = GMm/r_1^2 \quad \text{and} \quad F_{2G} = GMm/r_2^2$$

The **work** done on the object is the average force F_G exerted times the distance that the object is raised, or $W = F_G d$, where $d = r_2 - r_1$. Force F, which varies as $1/r^2$, is normally found using calculus, but in this case also works out to be the *geometric mean* of $F_{1G} = GMm/r_1^2$ and $F_{2G} = GMm/r_2^2$. The *geometric mean* of two numbers is the square root of their product, which gives us the average force F_G as:

$$F_G = [(GMm/r_1^2)(GMm/r_2^2)]^{\frac{1}{2}} = GMm/r_1 r_2$$

This gives us the **work** to raise the object from r_1 to r_2:

$$W_{12} = F_G d = F_G(r_2 - r_1) = (GMm/r_1 r_2)(r_2 - r_1) = (GMm)(1/r_1 - 1/r_2)$$

Therefore: $W_{12} = (GMm)(1/r_1 - 1/r_2)$.

The *gravitational potential energy of the raised mass equals the work done to raise it*. Therefore when the distance that an object is lifted is significant compared to the Earth's radius R_E, the **gravitational potential energy** created by raising an object from r_1 to r_2 is:

$$\boxed{PE_G = (GMm)(1/r_1 - 1/r_2)}$$

Remember, the **gravitational force is conservative**, so PE_G depends on the initial and final positions, not the path taken. Also, when the distance the object is raised is small compared to Earth's radius R_E, this equation reduces back to $PE_G = mgh$. To show this, substitute R_E for r_1 and $R_E + h$ for r_2, then if $h << R_E$, $PE_G = GMmh/R_E^2$, where $F_G = GMm/R_E^2$. Since $GM/R_E^2 = g$, so $GMm/R_E^2 = mg$ or $GMm = R_E^2 mg$, combining this with $PE_G = GMmh/R_E^2$ reduces to $PE_G = mgh$.

The potential energy is generally set equal to zero at r = infinity since when two objects are extremely distant from each other, they have no influence on each other's movements. Taking the potential energy to be zero at infinity gives the simple form for potential energy:

$$PE_G = -GMm/r$$

Gravitational potential energy near a planet is negative, since gravity is doing positive work as the mass approaches. Once a mass is near a large body, it is held until sufficient energy is provided for it to escape.

• **Example**: What is the **escape velocity** an object needs to leave Earth's surface and move away an "infinite" distance (ignore air resistance)?

$$KE_{surface} + PE_{Gsurface} = KE_{r \to \infty} + PE_{Gr \to \infty}$$

We begin with all kinetic and end with all potential, leaving:

$$((1/2)mv^2)_s = (GMm)(1/R_E - 1/r_2)_{r \to \infty}$$

As $r \to \infty$, $1/r \to 0$, so: $(1/2)mv^2 = GMm/R_E$.

Solve for v^2: $v_{escape}^2 = 2GM/R_E$.

Substitute mean radius of the Earth, $R_E = 6.37 \times 10^6$ m, mass of Earth $M = m_e = 5.97 \times 10^{24}$ kg, and $G = 6.67 \times 10^{-11}$ m^3/kg·s^2:

$$v_{escape}^2 = 2(6.67 \times 10^{-11} \, m^3/kg \cdot s^2)(5.97 \times 10^{24} \, kg)/(6.37 \times 10^6 \, m)$$

Take square root of result to get v_{escape} for leaving Earth:

$$v_{escapeEarth} \approx 11{,}181 \text{ m/s is escape velocity (just over 25,000 mph)}$$

If you escaped Earth you would encounter the gravitational force from the Sun before escaping the solar system. To escape the Sun, you need $v_{escapeSun}$. This involves using the Sun's mass ($M = 1.99 \times 10^{30}$ kg) for M and the Sun-Earth distance for R_E (1.496×10^{11} m), so:

$$v_{escape}^2 = 2(6.67 \times 10^{-11} \, m^3/kg \cdot s^2)(1.99 \times 10^{30} \, kg)/(1.496 \times 10^{11} \, m)$$

$$v_{escapeSun} \approx 42{,}125 \text{ m/s is escape velocity (over 94,000 mi/h)}$$

Note: The energy balance can also estimate an incoming impact velocity.

5.6. Electrostatic Potential Energy

• Remember from Section 4.2 the **Electrostatic Force Law** or **Coulomb's Law** is: $F_E = Kq_1q_2/r^2$, where q_1 and q_2 are charges of two particles or objects, r is their separation, and K is **Coulomb's Constant**, $K \approx 9.0 \times 10^9$ N·m^2/C^2. F_E is the force of electrostatic attraction or repulsion between two electric charges *at rest* and is *inversely proportional to the square of the distance* between the charges.

Except for the fact that the sign on the charges can be positive or negative, the electrostatic force and gravitational force have similar forms. They both have a dependence that is *inversely proportional to the square of the distance* between their respective masses or charges. Correspondingly, the electrostatic potential energy has a form similar to the gravitational potential energy ($PE_G = -Gm_1m_2/r$). The **electrostatic potential energy** is the potential energy of a charge q_2 in the vicinity of source charge q_1:

$$PE_E = Kq_1q_2/r$$

Remember, if the signs of q_1 and q_2 are different, the force is attractive, $PE_E < 0$, and increases (becoming less negative) with separation. If the signs of q_1 and q_2 are the same, the force is repulsive, $PE_E > 0$, and it decreases with separation.

We learned the gravitational potential energy of a raised mass equals the work done to raise it from r_1 to r_2, or: $W_{12} = PE_G = (GMm)(1/r_1 - 1/r_2)$. Similarly, the **work done to move a charge** from r_1 to r_2 equals the **change in electrostatic potential energy**:

$$W_{12} = \Delta PE_E = (Kq_1q_2)(1/r_2 - 1/r_1)$$

Note that the term ($1/r_2 - 1/r_1$) is reversed (or a negative sign inserted) since the electrostatic force is attractive like gravity when the term q_1q_2 is negative. When work is done against the forces F_G or F_E, the potential energy increases and $W > 0$ and $\Delta PE_E > 0$. Conversely when work is done *by* F_G or F_E, potential energy decreases and $W < 0$ and $\Delta PE_E < 0$.

• **Example**: In Section 4.2 we calculated the force between the proton and electron in a hydrogen atom given that the charges are plus and minus 1.6×10^{-19} C. If the distance between the proton and electron is increased from 5.3×10^{-11} m to 53×10^{-11} m, what is ΔPE_E?

$$\Delta PE_E = (Kq_1q_2) \times (1/r_2 - 1/r_1)$$
$$\Delta PE_E = (9.0 \times 10^9 \text{ N·m}^2/\text{C}^2)(1.6 \times 10^{-19} \text{ C})(-1.6 \times 10^{-19} \text{ C})$$
$$\times [(1 / 53 \times 10^{-11} \text{ m}) - (1 / 5.3 \times 10^{-11} \text{ m})]$$
$$\Delta PE_E = -2.3 \times 10^{-28} \times -1.698 \times 10^{10} = 3.9 \times 10^{-18} \text{ J}$$

$\Delta PE_E > 0$ reflects work done against the attractive force to separate the proton and electron.

5.7. Introduction to the Electron Volt

• The **electronvolt** or **electron volt** (eV) is a unit of energy commonly used in atomic and nuclear physics. It is equal to the energy an electron gains when it travels or accelerates through a **potential difference** of one volt. One eV is equivalent to 1.602×10^{-19} J.

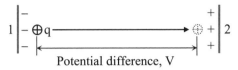

Potential difference, V

Work *is required to move positive charge* q from a negatively charged environment at plate 1 to a positively charged environment at plate 2. If 1 Joule (J) of work is needed to move a 1 Coulomb positive charge from point 1 to point 2, then the potential difference between the two points or plates must be 1 volt (V). The **potential difference** between two points is a measure of the work per unit charge required to move a charge from one of the points to the other. The potential difference is expressed:

> Potential difference: $\Delta V = W/q$
> In measured units: $1\ V = 1\ J/C$

The *potential difference* between two terminals or points in an electrical circuit *is the voltage* between the two points. Potential difference is measured in units of volts. For example, a 12-V **battery** has a potential difference of 12 V between the positive and negative terminals. A zero potential is considered *ground*, and potential differences are often measured with respect to ground.

• **Example**: If a charge *equal in magnitude* to the charge of one electron or one proton, which is $\pm 1.602 \times 10^{-19}$ Coulombs, is moved from one point to another point against a potential difference of 1 V which exists between the two points, how much work has been done on the charge?

$$W = qV = eV = (1.602 \times 10^{-19}\ \text{C})(1\ \text{V}) = 1.602 \times 10^{-19}\ \text{J}$$

This example shows that the work or energy to move the charge of an electron or proton across a 1 V potential difference is 1.602×10^{-19} J. This **unit of work or energy is called an eV or electronvolt**.

> $1\ \text{eV} = (1.602 \times 10^{-19}\ \text{C})(1\ \text{V}) = 1.602 \times 10^{-19}\ \text{J}$

$1{,}000\ \text{eV} = 10^{3}\ \text{eV}$ is a kiloelectronvolt (keV) and 1.602×10^{-16} J
$1{,}000{,}000\ \text{eV} = 10^{6}\ \text{eV}$ is a megaelectronvolt (MeV) and 1.602×10^{-13} J
$1{,}000{,}000{,}000\ \text{eV} = 10^{9}\ \text{eV}$ is a gigaelectronvolt (GeV) and 1.602×10^{-10} J

• **Example**: If a proton leaves a resting position and crosses a potential difference of 1 keV (from positive to negative), what is its final velocity?

The kinetic energy of the moving proton is: $KE = W = qV$.

$KE = (1.602 \times 10^{-19}\,C)(1\,keV) = (1.602 \times 10^{-19}\,J)(10^3\,eV) = 1.602 \times 10^{-16}\,J$

The KE of the proton in terms of velocity is: $KE = (1/2)m_{proton}v^2$.

Solve for v^2: $v^2 = 2KE/m_{proton} = 2(1.602 \times 10^{-16}\,J)/(1.67 \times 10^{-27}\,kg)$

The square root of the result is its velocity: $v \approx 4.38 \times 10^5$ m/s.

• Note that the *electronvolt* can be used as a unit of energy for particles that have not moved though a potential difference. A neutral particle such as a neutron, which has no charge and does not experience an electrostatic force or move through a potential difference, can nevertheless have its energy measured in electronvolts.

5.8. Rotational Kinetic Energy

• In this chapter we learned about kinetic energy for an object moving in a straight-line direction—up, down, sideways, etc.—known as translational motion. Suppose an object does not have translational motion, but is rotating about an axis through its center of mass. This object would have **rotational kinetic energy**.

The sphere is rotating at angular velocity ω. Each particle or atom within this sphere rotates around the central axis with the angular velocity ω. Remember from Section 2.7 we discussed that as an object moves along a circle, angle θ changes, and the rate at which angle θ changes is the angular velocity $\omega = \Delta\theta/\Delta t$, where θ is in radians. The speed of an object moving in a circular path of radius r is: $v = r\omega$, where vector **v** changes direction continually even though its speed may be constant. The kinetic energy of the particle is $(1/2)mv^2$, and velocity can be written $v = r\omega$. Therefore: $KE = (1/2)mv^2 = (1/2)m(r\omega)^2 = (1/2)mr^2\omega^2$. Since the sphere is full of particles, the rotational KE of the sphere is the sum of all its particles' rotational KEs: $KE_{rot} = (1/2)(\sum mr^2)\omega^2$, where $\sum mr^2$ is the **rotational inertia** $I = \sum mr^2$ of the particles in the sphere and represents the distribution of mass. (Remember, I is the **moment of inertia**.) Therefore, we can write the **rotational kinetic energy** as:

$$KE_{rot} = (1/2)I\omega^2$$

For an object with both translational and rotational KE, the total KE is:

$$KE_{total} = KE_{translational} + KE_{rotational} \quad \text{or} \quad KE_{total} = (1/2)mv^2 + (1/2)I\omega^2$$

Examples of **moment of inertia** I configurations were given in Section 2.7 including: Particle rotating $I = mr^2$; particle revolving $I = mr^2$; thin hoop rotating $I = mr^2$; disk rotating $I = (1/2)mr^2$; solid sphere rotating $I = (2/5)mr^2$; thin rod rotating lengthwise $I = (1/12)mr^2$.

• **Example**: If a solid sphere rolls from rest down an incline from 1 m height to the floor, what translational velocity will it have after it leaves the incline (ignoring friction)?

By conservation of energy, $KE_1 + PE_1 = KE_2 + PE_2$ or $PE_1 = KE_2$ where $KE_2 = (1/2)mv^2 + (1/2)I\omega^2$.

Therefore: $PE_1 = mgh = (1/2)mv^2 + (1/2)I\omega^2$.

Substitute $I = (2/5)mr^2$ for a sphere and $v = r\omega$:
$$mgh = (1/2)mv^2 + (1/2)((2/5)mr^2)(v/r)^2$$

Cancel and reduce:
$$gh = (1/2)v^2 + (1/5)v^2 = v^2(1/2 + 1/5) = (7/10)v^2$$
$$v^2 = (10/7)gh = (10/7)(9.8 \text{ m/s}^2)(1 \text{ m}) = 14 \text{ m}^2/\text{s}^2$$

Taking the square root gives velocity: $v \approx 3.74$ m/s.

5.9. Conservation of Energy

• The **law of conservation of energy** states that the total amount of energy in an isolated system remains constant, so that in an isolated system energy is neither created nor destroyed, but can change form. With the conservation of energy, the sum of the various forms of energy remains constant in a physical process. We saw this above when we discussed the transfer of work done W to lift an object into gravitational potential energy PE_G. When the object falls from rest it loses PE_G and gains KE until just before it hits the ground, when all the energy has been transformed into an equal amount of KE.

Conservation of energy allows us to design machines which can use energy to produce work such as burning fuel (thermal energy) to turn a turbine (mechanical energy) and generate electricity (electrical energy). The law of conservation of energy applies to any closed system and suggests that the energy in the Universe is constant. In light of relativity, conservation of energy may be thought of in a larger picture as the conservation of mass-energy, i.e., the total mass-energy of the Universe is constant even though energy can be transferred, transmitted, and transformed.

A *closed system* is defined as a system where no energy leaves the system and no energy from the outside enters the system. In reality, while a truly closed system does not exist since energy is often dissipated to some extent through heat, sound, or friction, there are situations where it is reasonable to ignore these losses and assume a closed system.

Conservation of mechanical energy is one form of the more general law of conservation of energy and is useful when solving problems. When frictional forces are present, energy is dissipated as heat and sound, and the mechanical energy is not conserved. Therefore, conservation of mechanical energy only applies when the system is closed. Consider a box sliding over a rough surface where the force of friction creates heat in the box and surface. In this case, mechanical energy is transformed into heat energy, but the total energy stays constant. When the forces acting on a system are **conservative**, energy is conserved. We learned earlier that if the net work done on an object is by conservative forces (and non-conservative forces such as friction are zero) the total energy of the object is conserved and does not change.

$$\Delta KE + \Delta PE = 0$$
$$KE + PE = \text{constant}$$

The law of conservation of energy is very useful when solving a variety of problems such as finding an object's velocity or height. We can use the equation for the law of conservation of energy:

$$\Delta KE + \Delta PE = 0$$

and write it for an object at two different points, 1 and 2:

$$KE_2 - KE_1 + PE_2 - PE_1 = 0$$

Rearrange:

$$KE_1 + PE_1 = KE_2 + PE_2$$

Substitute definitions:

$$((1/2)mv^2)_1 + (mgh)_1 = ((1/2)mv^2)_2 + (mgh)_2$$

Cancel m's:

$$((1/2)v^2)_1 + (gh)_1 = ((1/2)v^2)_2 + (gh)_2$$

This relationship can be used to solve unknown velocities and heights.

• **Example**: You and your friend decide to go up to the mountains for some sledding. If you begin your sled ride gliding 5 m/s on the top of a 50-m high hill, then slide down into a valley and up to the top of a 40-m high cliff, how fast will you be sliding when you fly off the cliff and head across the chasm toward the other slope (neglect friction)?

(Don't try this at home ☺)

50m

40m

You two and your sled have a combination of KE and PE throughout your ride. The sum of KE and PE at each position remains constant since we are neglecting friction.

$$KE + PE = constant$$

On each peak:

$$KE_{50} + PE_{50} = KE_{40} + PE_{40}$$

To determine the velocity at the 40 m peak, you need to isolate KE_{40}:

$$KE_{40} = KE_{50} + PE_{50} - PE_{40}$$

Substitute the definitions:

$$((1/2)mv^2)_{40} = ((1/2)mv^2)_{50} + (mgh)_{50} - (mgh)_{40}$$

Same mass throughout, so m's cancel:

$$((1/2)v^2)_{40} = ((1/2)v^2)_{50} + (gh)_{50} - (gh)_{40}$$

$$v^2_{40} = v^2_{50} + 2g(h_{50} - h_{40}) = (5 \text{ m/s})^2 + 2(9.8 \text{ m/s}^2)(10 \text{ m}) = 221 \text{ m}^2/\text{s}^2$$

$$v_{40} \approx 14.87 \text{ m/s}$$

The velocity of you, your friend, and the sled when you launch off the cliff is about 14.87 m/s (about 33 mph, yikes!). Notice that the mass canceled out and did not matter in this frictionless problem.

• **Example**: You survived your sledding expedition and decide to test the capabilities of your favorite rock launcher. Loading and readying the launcher involves compressing a spring with an average force of 100 N over 1 m. If you shoot a 1-kg rock straight up, what will be its velocity as it leaves the launcher? What will be its maximum height?

The compressed spring's potential energy is equal to the work done on it (assuming 100% efficiency).

$$PE = W = Fd = (100 \text{ N})(1 \text{ m}) = 100 \text{ J} = 100 \text{ kg·m}^2/\text{s}^2$$

The exit velocity of the rock occurs when all the potential energy in the spring is converted to kinetic energy: $PE = KE = (1/2)mv^2$.

Solve for v^2: $v^2 = PE / 2m = (100 \text{ kg·m}^2/\text{s}^2) / (2)(1 \text{ kg}) = 50 \text{ m}^2/\text{s}^2$

Take the square root: $v \approx 7.07 \text{ m/s}$ is the exit velocity.

If we have position 1 as just prior to launch when the spring compressed and position 2 as maximum height, by conservation of energy: $KE_1 + PE_1 = KE_2 + PE_2$, where the energy is all PE at position 1 and also at the top when the rock stops for a moment before descending under gravity at position 2.

This leaves: $PE_1 = PE_2$ or $100\ J = mgh_2$

Solve for h_2: $h_2 = (100\ kg{\cdot}m^2/s^2)\ /\ (1\ kg)(9.8\ m/s^2) \approx 10.2\ m$

The maximum height of a 1-kg rock is about 10.2 m (or about 33.46 ft). (Uh oh, the rock is coming back down—got to go!)

• When the forces acting on an object are **conservative**, energy is conserved. But what if non-conservative forces (such as friction) are also acting? When a **non-conservative force** such as friction does work on an object, energy is lost to the environment (as waste heat) and is not conserved.

• **Example**: After running from your rock launcher, you head back to the relative safety of your garage. There is an elevated section in your garage which can be accessed either by steps or a 30°-incline ramp. To take your mind off the near miss you just had with the rock, you decide to measure the coefficient of friction between the ramp and the 20-kg crate you had been moving in an earlier example. If the crate begins from rest and slides 3 m down the ramp in exactly 1 s (making it an average of 3 m/s), what is the coefficient of friction between the crate and ramp?

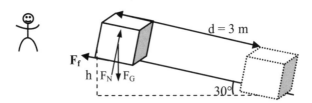

The energy lost through friction will be the difference between the crate's potential energy at the top and its kinetic energy at the bottom of the 3-m ramp. The initial potential energy is PE = mgh. (Remember to find h, we use $\sin \theta$ = opposite/hypotenuse or $\sin 30° = h/3$ m.)

$$PE = mgh = (20\ kg)(9.8\ m/s^2)(3\ m \times \sin 30°) = 294\ J$$

The kinetic energy 3 m down the ramp where the crate has a velocity of (3 m)/(1 s) or 3 m/s is:

$$KE = (1/2)mv^2 = (1/2)(20\ kg)(3\ m/s)^2 = 90\ kg{\cdot}m^2/s^2$$
$$PE - KE = 294\ J - 90\ J = 204\ J$$

The 204 J of energy lost to friction is the work done against the force of friction as the crate slid a distance d of 3 m, or:

$$W = 204 \text{ J} = 204 \text{ kg·m}^2/\text{s}^2$$

To determine the coefficient of friction, we write the work done against friction in terms of the force of friction: $W = F_f d$.

Remember, the force of friction F_f along the ramp and opposite to motion is the product of the kinetic friction coefficient μ_k and normal force F_N. F_N is the component of the weight of the crate perpendicular to the ramp at an angle $\cos 30°$ to F_G.

$$W = \mu_k F_N d = \mu_k mg (\cos \theta) d$$

Solve for μ_k:

$$\mu_k = W / (mg (\cos \theta) d)$$

$$\mu_k = (204 \text{ kg·m}^2/\text{s}^2) / (20 \text{ kg})(9.8 \text{ m/s}^2)(\cos 30°)(3 \text{ m}) \approx 0.4$$

The coefficient of friction between the crate and the ramp is 0.4.

• Where does lost energy go? Remember in the example several pages ago when you were doing work to push your 20-kg crate across your garage? The friction between the crate and ground caused a transfer of heat to the ground. What happened to the energy that was created as you moved the crate? Similarly, in the preceding example where the crate slid down the ramp, what happened to that energy? In the presence of non-conservative forces, energy is converted to other forms such as heat, internal thermal energy, or sound. The internal thermal energy reflects the kinetic and potential energy of atoms and molecules as they move. When you pushed the crate, both the garage floor and the crate gained internal thermal energy. There was also undoubtedly some sound wave energy created during the time the crate slid across the floor. Frictional forces are pervasive and produce heat, although sometimes friction's effect is small enough we can neglect it when solving a problem. Even electric energy produces heat. An example is an incandescent light bulb, which does not convert all the electrical energy to light and gets too hot to touch. Sound waves are also slowly converted into heat as air molecules vibrate and the sound fades.

5.10. Elastic and Inelastic Collisions

• When objects collide, momentum (p) can be transferred among the objects, but the **momentum** of the entire isolated system is **conserved** and remains constant. To understand what happens to energy during collisions, they are categorized as elastic or inelastic. In **elastic collisions** kinetic energy is conserved. In **inelastic collisions** some kinetic energy is converted to other forms of energy including internal energy, such as heat, or lost to the environment in the form of heat and sound waves.

• In an **elastic collision**, both conservation of kinetic energy and conservation of momentum hold. If there is no loss of kinetic energy during the collision, then no dissipative forces are present and the collision is completely or perfectly elastic. In this case, the initial kinetic energy before the collision is present after the collision. In an elastic collision, the objects bounce off one another without deforming each other permanently, the kinetic energy remains constant, and the internal energy does not change. For example, in an elastic collision a ball dropped onto a fixed surface will bounce up to its original height.

Elastic collision of ball bouncing off a fixed surface to its original height.

In reality, for collisions involving macroscopic objects you can see, such as a bouncing ball or sliding pucks on ice, there is some loss of kinetic energy into internal energy and other forms of energy such as sound, heat, or friction. If these losses are negligible, we can model the collision as elastic. At the atomic or subatomic level, elastic collisions can occur as long as the internal energy of the particles remains unchanged.

• Collisions between two objects that can be modeled as hard spheres may be nearly elastic, and by using **conservation of momentum** and **conservation of kinetic energy** you can calculate final velocities. The formulas for conservation of momentum and energy are used to predict the results of elastic collisions. Consider one sphere with mass m_1 and initial velocity v_{i1} which collides along one dimension, or straight on, with a second sphere of mass m_2 that is at rest or moving in the same dimension.

What will be their velocities v_{f1} and v_{f2} after the collision? Let's develop the equations for this *one-dimensional case* for v_{f1} and v_{f2}. Using:
Conservation of momentum: $m_1v_{i1} + m_2v_{i2} = m_1v_{f1} + m_2v_{f2}$
Conservation of KE: $(1/2)m_1v_{i1}^2 + (1/2)m_2v_{i2}^2 = (1/2)m_1v_{f1}^2 + (1/2)m_2v_{f2}^2$
Set the initial velocity of sphere 2, v_{i2}, to zero to simplify.
Rearrange conservation of momentum equation: $m_1(v_{i1} - v_{f1}) = m_2v_{f2}$
Rearrange conservation of KE equation: $m_1(v_{i1}^2 - v_{f1}^2) = m_2v_{f2}^2$
Use relation $(a^2 - b^2) = (a - b)(a + b)$: $m_1(v_{i1} - v_{f1})(v_{i1} + v_{f1}) = m_2v_{f2}^2$
Divide rearranged KE equation by rearranged momentum equation:
$$[m_1(v_{i1} - v_{f1})(v_{i1} + v_{f1}) = m_2v_{f2}^2] / [m_1(v_{i1} - v_{f1}) = m_2v_{f2}]$$
$$v_{i1} + v_{f1} = v_{f2} \quad \text{or rearrange to} \quad v_{f1} = v_{f2} - v_{i1}$$

Substitute $v_{f1} = v_{f2} - v_{i1}$ into conservation of momentum:
$$m_1 v_{i1} = m_1(v_{f2} - v_{i1}) + m_2 v_{f2}$$
$$m_1 v_{i1} + m_1 v_{i1} = m_1 v_{f2} + m_2 v_{f2}$$
$$2m_1 v_{i1} = v_{f2}(m_1 + m_2)$$
$$v_{f2} = v_{i1} 2m_1 / (m_1 + m_2) \textbf{ is sphere 2's final velocity}$$
Substitute $v_{i1} + v_{f1} = v_{f2}$ to find v_{f1}:
$$v_{i1} + v_{f1} = v_{i1} 2m_1 / (m_1 + m_2)$$
$$v_{f1} = [v_{i1} 2m_1 / (m_1 + m_2)] - [v_{i1}] = v_{i1}[2m_1 / (m_1 + m_2) - 1]$$
$$v_{f1} = v_{i1}[2m_1 / (m_1 + m_2) - (m_1 + m_2)/(m_1 + m_2)]$$
$$v_{f1} = v_{i1}[(m_1 - m_2)/(m_1 + m_2)] \textbf{ is sphere 1's final velocity}$$

If $v_1 = 1$ and if spheres are equal $m_1 = m_2$, then $v_{f2} = v_1 = 1$ and $v_{f1} = 0$. In this special case, all of the first sphere's energy and momentum is transferred to the second sphere. (Note in problems with $m_1 = m_2$, the conservation equations can be simplified quickly.)

What if $v_{i1} = 1$ and the masses differ? If the first sphere is 5 times the mass of the second, then:
$$v_{f2} = v_{i1} 2m_1 / (m_1 + m_2) = (1)(2)(5)/(5 + 1) = 5/3 \text{ and}$$
$$v_{f1} = v_{i1}[(m_1 - m_2)/(m_1 + m_2)] = (1)(5 - 1)/(5 + 1) = 2/3$$
The first sphere will continue moving in the same direction, but will lag the speed of the second by 1.

If $v_{i1} = 1$ and the second sphere is 5 times the mass of the first, then:
$$v_{f2} = v_{i1} 2m_1 / (m_1 + m_2) = (1)(2)(1)/(1 + 5) = 1/3 \text{ and}$$
$$v_{f1} = v_{i1}[(m_1 - m_2)/(m_1 + m_2)] = (1)(1 - 5)/(1 + 5) = -2/3$$
The first sphere bounces back in the direction from which it originated. Its speed is slower, but interestingly, the difference in the velocities of the two spheres after the collision again is 1.

• When some of the kinetic energy is converted into some other form of energy such as internal thermal energy, heat, or sound, the collision is called an **inelastic collision**. For example, in an inelastic collision a ball dropped onto a fixed surface will not bounce up to its original height.

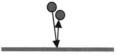

Inelastic collision of ball bouncing off a fixed surface and losing kinetic energy.

If two objects collide and stick to each other rather than rebounding, kinetic energy has been converted into internal energy and the velocity of the stuck-together objects after impact is determined by conservation of momentum. These types of collisions are sometimes referred to as **completely inelastic**. In inelastic collisions, momentum is conserved even though some of the kinetic energy is converted to other forms of energy. As mentioned, most collisions are inelastic to some degree because kinetic energy is transferred to other forms of energy or lost to the environment. If you are given a problem where you are asked to determine if a collision is elastic or inelastic, simply calculate the kinetic energy of the objects before and after the collision. If the kinetic energy remains the same, then the collision is elastic. Remember, momentum is conserved in both elastic and inelastic collisions.

• **Example**: Two round blobs of play putty, having masses $m_1 = 1$ and $m_2 = 3$, collide straight on and stick together. At impact, m_1 has velocity v_{i1} and m_2 has velocity $v_{i2} = 0$. What is their combined final velocity?

The combined mass after impact is: $m_1 + m_2 = 1 + 3 = 4$
Use conservation of momentum $m_1v_{i1} + m_2v_{i2} = m_1v_{f1} + m_2v_{f2}$, or:
$$1v_{i1} = 1v_f + 3v_f \quad \text{where } v_f \text{ is for the combined play putty blob}$$
$$v_{i1} = 4v_f \quad \text{or} \quad v_f = v_{i1}/4$$
After collision the combined play putty blob travels in the same direction as m_1's initial velocity v_{i1}, but with a new velocity v_f that is $1/4$ of v_{i1}.

• In a straight-line collision, the relative velocity after impact divided by the relative velocity before impact is called the **coefficient of restitution** e. This formula is developed using *conservation of momentum* and is:
$$e = (v_{f2} - v_{f1})/(v_{i1} - v_{i2})$$
This relative ratio of velocities is e = 1 for elastic, e < 1 for inelastic, and e = 0 when objects stick together (completely inelastic).

• If **collisions** are not straight on and occur in **two dimensions**, we model them in a similar manner to one-dimensional collisions. Momentum is still conserved, and for elastic collisions, kinetic energy is also conserved. The difference between one and two dimensional collisions is the need to account for the x- and y-components of the momentum so that momentum is conserved in both the x direction and the y direction.

For example, if a sphere m_1 traveling v_{i1} m/s hits another sphere of equal mass and is deflected at angle $\theta_1°$ (while the second sphere is sent at an angle of $\theta_2°$), you can find the direction and velocity of the hit sphere using the conservation of energy and momentum equations. If the initial velocity of m_2 is zero, the initial components of momentum p would be $p_x = m_1v_{i1}$ and $p_y = 0$, and the post-collision components would be:
p_x component: $m_1v_{i1} + m_2v_{i2} = m_1v_{f1} \cos \theta_1 + m_2v_{f2} \cos \theta_2$ or

$$m_1 v_{i1} = m_1 v_{f1} \cos \theta_1 + m_2 v_{f2} \cos \theta_2$$

p_y component: $m_1 v_{i1} + m_2 v_{i2} = m_1 v_{f1} \sin \theta_1 + m_2 v_{f2} \sin \theta_2$ or

$$0 = m_1 v_{f1} \sin \theta_1 + m_2 v_{f2} \sin \theta_2$$

For an elastic collision, conservation of energy is:

$$(1/2)m_1 v_{i1}^2 + (1/2)m_2 v_{i2}^2 = (1/2)m_1 v_{f1}^2 + (1/2)m_2 v_{f2}^2$$

When $m_1 = m_2$ and $v = 0$ the energy equation becomes:

$$v_{i1}^2 = v_{f1}^2 + v_{f2}^2$$

To solve final velocities and the unknown angle, we would use the simplified equations for the x and y components of momentum and the simplified kinetic energy equation.

• **Example**: If two spheres have an off-center collision, what is the final velocity of the second sphere v_{f2}, if the final velocity of the first sphere is $v_{f1} = 10$ m/s. Given $m_1 = 0.6$ kg, $m_2 = 0.4$ kg, $v_{i1} = 20$ m/s, $v_{i2} = -40$ m/s, and θ_1 deflects downward 30°.

Remember, from Section 1.7, the component vectors of **V** are $V_x = V \cos \theta$ and $V_y = V \sin \theta$. V has magnitude $|V| = [V_x^2 + V_y^2]^{\frac{1}{2}}$ and its direction is the angle V makes with x-axis, or: $\theta = \tan^{-1}(V_y/V_x)$. In this example, we first find the x and y component vectors of final velocity using conservation of momentum.

p_x component: $m_1 v_{i1} + m_2 v_{i2} = m_1 v_{f1} \cos \theta_1 + m_2 v_{f2} \cos \theta_2$ or

$$m_1 v_{i1} + m_2 v_{i2} = m_1 v_{f1} \cos \theta_1 + m_2 v_{xf2}$$

$$(0.6\,\text{kg})(20\,\text{m/s}) + (0.4\,\text{kg})(-40\,\text{m/s}) = (0.6\,\text{kg})(10\,\text{m/s}) \cos(-30°) + (0.4\,\text{kg})v_{xf2}$$

$$v_{xf2} = -23$$

p_y component: $0 = m_1 v_{f1} \sin \theta_1 + m_2 v_{f2} \sin \theta_2$ or

$$0 = m_1 v_{f1} \sin \theta_1 + m_2 v_{yf2}$$

$$0 = (0.6\,\text{kg})((10\,\text{m/s}) \sin(-30°)) + (0.4\,\text{kg})v_{yf2}$$

$$v_{yf2} = 7.5$$

The v_{f2} velocity is: $|v| = [v_{xf2}^2 + v_{yf2}^2]^{\frac{1}{2}} \approx 24$ m/s

Because θ_2 is measured from the minus x-axis, we use +23 for v_{xf2}:

$\theta_2 = \tan^{-1}(v_{yf2}/v_{xf2}) = \tan^{-1}(7.5/23) = 18°$ up from the negative x-axis

Note: When measured from the positive x-axis θ_2 is 162, which is also the arctan of (7.5/−23). You can test this by taking the tangent of 162.

• **Example**: Is the collision in the previous example elastic?

You can determine whether a collision is elastic or inelastic if kinetic energy was lost. Calculate the before and after KE.

$$(1/2)m_1v_{i1}^2 + (1/2)m_2v_{i2}^2 = (1/2)(0.6\,kg)(20\,m/s)^2 + (1/2)(0.4\,kg)(40\,m/s)^2 = 440\,J$$
$$(1/2)m_1v_{f1}^2 + (1/2)m_2v_{f2}^2 = (1/2)(0.6\,kg)(10\,m/s)^2 + (1/2)(0.4\,kg)(24\,m/s)^2 \approx 145\,J$$

No, the collision is inelastic because of the loss of kinetic energy.

5.11. Power

• **Power** is the rate work is done or energy is expended or changes its form or location. It is the work per time. Power is a **scalar** quantity.

$$P = work/time = W/t$$

Average power reflects the change in work or energy per time:

$$\text{Average power} = \Delta W/\Delta t = \Delta E/\Delta t$$
$$P_{Ave} = W/\Delta t = F(\Delta d/\Delta t) = F \cdot v_{Ave}$$

where Δt is the time interval in which the work is done.

A machine or engine may not be limited by how much work it can do, but rather by *how fast* it can do that work. If one machine can push a box up a ramp in 6 s and a second machine can push the box up the ramp in 2 s, the second machine works at 3 times the power of the first machine. Similarly, the power of an automobile depends on how quickly it can transform chemical potential energy to mechanical energy.

• **Units for power** are work/time or energy/time. In SI: Joules/second = $kg \cdot m^2/s^3$, which is a Watt (W), where 10^3 W is a kilowatt (kW) and 10^6 W is a megawatt (mW). In CGS: ergs/s. In English: ft·lbs/s or horsepower (hp) where 1 hp = 550 ft·lbs/s = 746 Watts. Be careful not to confuse the use of W for Watts with W for work.

• **Example**: Suppose your friend asks you to help him stack one-hundred 15-kg boxes of books from his warehouse floor onto a 2-m high shelf. If you each lift 50 boxes in 4 min what is your combined power working together and your individual power?

$$P = W/t = (\text{number of boxes})mgh/t$$
$$P_{together} = W/t = (100)(15\,kg)(9.8\,m/s^2)(2\,m)/(4\,min)(60\,s/min) = 122.5\,\text{Watts}$$
$$P_{alone} = W/t = (50)(15\,kg)(9.8\,m/s^2)(2\,m)/(4\,min)(60\,s/min) = 61.25\,\text{Watts}$$

Each of you works at half the power of both combined.

• **Example**: In an example in Section 5.1 you pushed a 30-kg sled up a 20° slope to the lodge. When friction was ignored and you put on your crampons and pushed the sled, the work was only against gravity:

$$W = F_Gd = (mg \sin \theta)d = (30 \text{ kg})(9.8 \text{ m/s}^2)(\sin 20°)(200 \text{ m}) \approx 20,111 \text{ J}$$

We also said this can be calculated $W = mgh$, where h is the total height.

$$W = mgh = (30 \text{ kg})(9.8 \text{ m/s}^2)(\sin 20°)(200 \text{ m}) \approx 20,111 \text{ J}$$

If it took you 20 min to push the sled up the slope, calculate the velocity using power.

The power is: $P = W/t = (20,111 \text{ J})/(20 \text{ min})(60 \text{ s/min}) = 16.76$ Watts

Power in terms of velocity is: $P_{Ave} = W/\Delta t = F(\Delta d/\Delta t) = F \cdot v_{Ave}$

Force is against gravity, so:

$$F_G = (mg \sin \theta) = (30 \text{ kg})(9.8 \text{ m/s}^2)(\sin 20°) = 100.6 \text{ N}$$

$$P_{Ave} = F \cdot v_{Ave} \quad \text{or} \quad v = P/F = (16.76 \text{ W})/(100.6 \text{ N}) = 0.17 \text{ m/s}$$

Check this result calculating v using 200-m slope in 20 min:

$$(200 \text{ m})/(20 \text{ min})(60 \text{ s/min}) = 0.17 \text{ m/s}$$

When a steady force is applied to an object, the change in the amount of work done on the object is the force times the change in the object's displacement. Therefore, we see that **power** *can be calculated using force and velocity*: $P = \Delta W/\Delta t = F(\Delta d/\Delta t) = F \cdot v$.

• **Example**: What power does it take to steadily push a 20-kg box 100 m over a surface with a friction coefficient of 0.2 in 10 min?

The applied force and displacement are parallel, angle θ is zero, so $W = Fd$. Remember from Chapter 3, the force required to keep the box moving at constant velocity must equal the force of kinetic friction, $F_{fk} = \mu_k F_N = \mu_k mg$. So, $W = Fd = \mu_k F_N d = \mu_k mgd$.

The power is given by $P = W/\Delta t = \mu_k mgd/\Delta t$. So:

$$P = (0.2)(20 \text{ kg})(9.8 \text{ m/s}^2)(100 \text{ m})/(10 \text{ min})(60 \text{ s/min}) = 6.53 \text{ Watts}$$

Alternatively, using velocity $v = (100 \text{ m})/(600 \text{ s}) = 0.1667$ m/s:

$$P = F \cdot v_{Ave} = \mu_k mgv_{Ave} = (0.2)(20 \text{ kg})(9.8 \text{ m/s}^2)(0.1667 \text{ m/s}) = 6.53 \text{ Watts}$$

5.12. Principles of Simple Machines

• A **simple machine** is an apparatus that is constructed to help perform **work**. Remember, $W = \mathbf{F \cdot d}$. *A simple machine is designed to reduce the force needed by increasing the displacement.* Simple machines include levers, pulleys, and incline planes. In each case, some required amount of work W is performed by increasing the distance d and thereby decreasing the force F needed. (The simple machine does not lower the total work required, it rather allows for less force by increasing distance.)

• A **lever** can lift a heavy object. When a force is applied (downward) to one end of the lever, it rotates about a pivot, and the other end of the lever exerts the force (upward) to do the work of lifting. When the lifting end of the lever is close to the pivot, the force that needs to be applied to the longer end is less than the force the lever is exerting on the object. This occurs because the applied force is acting over a longer distance.

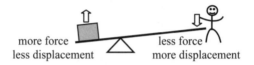

more force   less force
less displacement   more displacement

• A **pulley** or **system of pulleys** also reduces the amount of force required by increasing the distance over which the force is applied. If a free pulley wheel divides the weight of an object over two ropes, then, when lifting an object, the pulley exerts on each rope a force that is equal to half the weight of the object. For a simple one-pulley system, in order to lift an object a certain distance and therefore do a certain amount of work, twice as much displacement of the rope is required. In a one-pulley system, one half the force is applied over twice the displacement to lift the weight.

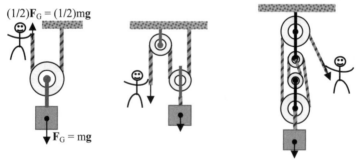

$(1/2)F_G = (1/2)mg$

$F_G = mg$

In Section 2.3 using the center figure above, we showed that 250 lb of force is required to lift 500 lb, and that the free end of the rope must be pulled 20 ft to raise the weight 10 ft. Adding more pulleys can further reduce the force required while proportionally increasing the required displacement of rope or cable. In question 2.3(d) at the end of Chapter 2 we found that in a two-free-pulley system which suspends a weight on 4 ropes, a 500 lb object can be lifted by pulling down with a force of 125 lb (figure on right above).

• An **incline plane** is also considered a simple machine that lifts an object with less force than its weight.

- The effectiveness of machines can be measured by calculating the **mechanical advantage**, which is the ratio:

mechanical advantage =
(force required to do work without machine)/(force required with machine)

For a lever, if the object is 1/4 the distance from the pivot as the point where the force is applied, the force needed would be 1/4 the object's weight, and the mechanical advantage would be 4. Note you have to move your end of the lever 4 times as far as the object moves. For a single free pulley, the load is spread over two ropes, so the mechanical advantage is 2. A frictionless incline plane with an angle of 10° from horizontal would provide mechanical advantage of about 5.8 $(1/\sin 10°)$.

5.13. Key Concepts and Practice Problems

- Work: $W = \mathbf{F} \cdot \mathbf{d} = Fd \cos \theta$.
- Energy is the ability to do work or produce an effect.
- Kinetic energy is the energy of motion. $KE = (1/2)mv^2$. $\Delta KE = W$.
- Potential energy PE is the energy of position.
- Gravitational PE: $PE_G = mgh$. For large h: $PE_G = (GMm)(1/r_1 - 1/r_2)$.
- $KE = (1/2)mv^2 = W = Fd = PE_G = mgh$.
- When net force is conservative: $W = \Delta KE + \Delta PE$.
- Electrostatic: $PE_E = Kq_1q_2/r$; $W_{12} = \Delta PE_E = (Kq_1q_2)(1/r_2 - 1/r_1)$.
- Electronvolt: $1 \text{ eV} = 1.602 \times 10^{-19}$ J.
- Potential difference: $V = W/q$.
- Rotational kinetic energy: $KE_{rot} = (1/2)I\omega^2$.
- Energy is conserved: $\Delta KE + \Delta PE = 0$; $KE_1 + PE_1 = KE_2 + PE_2$.
- Elastic collisions, KE and p conserved.
  Inelastic collisions, p conserved.
- Power: $P = \text{work/time} = W/t = \Delta W/\Delta t = \Delta E/\Delta t = F(\Delta d/\Delta t) = \mathbf{F} \cdot \mathbf{v}_{ave}$.

Practice Problems

5.1 Your boss tells you to push a box up a 10-m incline. You apply a force of 100 N parallel to the incline to overcome friction and gravity as you perform the task. As you finish, the boss returns and says he changed his mind, so you push the box back down the incline to where you started (displacement = 0). Pushing down was much easier, requiring a parallel force of only 20 N. How much work, if any, did you perform on the box?

5.2 (a) Standing on a sheer cliff, you throw a 1-kg rock straight up at 30 m/s. What amount of work did you perform on the rock and what is its kinetic energy as it leaves your hand? **(b)** The rock narrowly misses you on the way down and continues, falling past the cliff's edge and then hitting the ground below you at 150 m/s. Ignoring air resistance, how much work did gravity perform on the rock from the time you released the rock until it hit bottom?

5.3 (a) To wind a grandfather clock, you turn the 8-inch (in) winding crank 10 times with a force perpendicular to the radius of 1 lb. As a result you raise the 10-lb weight 4 ft. How much work did you perform turning the crank? **(b)** How much of that work was transformed into gravitational potential energy? **(c)** What happened to the rest of the work you expended?

5.4 (a) An icy bobsled run is 2,000 m long and has an average slope of 10°. As the team jumps on board, the bobsled is moving at 5 m/s, and when it crosses the finish line it is moving at 50 m/s. The combined mass of the bobsled and crew is 40 kg. What is the loss of potential energy? **(b)** What is the gain in kinetic energy? **(c)** What non-conservative forces were at work?

5.5 (a) You build a super rocket and shoot it straight up toward the moon. By the time it reaches the Moon's altitude it has run out of fuel and has zero velocity. Unfortunately, the Moon has moved to the other side of the Earth. Your rocket has little angular momentum and therefore falls almost straight back toward Earth. Assume it reached an altitude above the center of the Earth of 3.84×10^8 m and the Earth's upper atmosphere begins 6.8×10^6 m above Earth's center. If you consider only the effects of Earth's gravity, how much kinetic energy will your 10-kg rocket have when it enters the upper atmosphere? **(b)** At what speed will it be falling?

5.6 In the example in Subsection 5.6, if the distance between the electron and the proton is further increased from 53×10^{-11} m to an infinite distance, what is the ΔPE?

5.7 (a) What work is required to move a 1-C charge from the positive to the negative terminal of a 9-V battery? **(b)** What is this expressed in eV?

5.8 A metal hoop is released and rolls down a 50-m hill. How fast is it moving at the bottom (disregard air resistance and friction)?

5.9 (a) A baseball with a mass of 0.0145 kg is popped up above home plate. It leaves the bat at a vertical speed of 40 m/s and reaches a height of 70 m. How much of its initial kinetic energy was lost to air resistance?

(b) You shoot a hockey puck with a mass of 0.02 kg across the ice at 15 m/s. When it reaches the goal 20 m away, it is moving at 14 m/s. Assuming the only resistive force is friction with the ice, what is the coefficient of friction μ between the puck and the ice?

5.10 (a) Two spheres collide head on. Sphere A has a mass of 1 kg and an initial velocity of 10 m/s. Sphere B has a mass of 10 kg and an initial velocity of –1 m/s. If the collision is perfectly elastic, what are the spheres' final velocities? **(b)** If their coefficient of restitution is 0.5, what are their final velocities? **(c)** If the collision is totally inelastic, what are their final velocities?

5.11 If the same power required to light a 100-Watt bulb is used to power an electric winch that is 90% efficient (90% of power is converted into useful work), how long will it take the winch to raise a 10-kg mass 10 m?

5.12 (a) A wheel with a diameter of 2 ft turns an axle with a diameter of 1 in. What is the mechanical advantage? **(b)** If you attempt to turn the wheel by grasping the axle, what is the mechanical advantage?

Answers to Chapter 5 Problems

5.1 The vertical movement up and down (against and with the force of gravity) resulted in no net work, but pushing against friction in both directions did positive work in both directions. Solution 1: Compute work pushing both up and down. $W_{tot} = F_{up}d + F_{dn}d = (100\ N)(10\ m) + (20\ N)(10\ m) = 1{,}200\ J$. Solution 2: Find the frictional force F_f and gravitational force F_g. We know $F_f + F_g = 100$ and $F_f - F_g = 20$. Therefore, $F_f = 60$ and $F_g = 40$. No work is done vertically, so the only work done is pushing against F_f. $W = Fd = (60)(20) = 1{,}200\ J$.

5.2 (a) $W = (1/2)mv^2 = (1/2)(1)(30)^2 = 450\ J$. All the work was transformed to kinetic energy, which is also 450 J as you released the rock. **(b)** $W = (1/2)m(v_f^2 - v_i^2) = (1/2)(1)((-150)^2 - 30^2) = 10{,}800\ J$. (Alternatively: up 30 m/s to zero m/s; $(1/2)mv^2 = (1/2)(1)(30)^2 = 450\ J$; down zero m/s to 150 m/s: $(1/2)mv^2 = (1/2)(1)(150)^2 = 11{,}250\ J$; the difference is 10,800 J.)

5.3 (a) $W = Fd = F(10\ turns)(2\pi r) = (1)(10)(2\pi(0.667)) = 41.9\ ft\cdot lbs$. **(b)** $PE = F_g h = (10)(4) = 40\ ft\cdot lbs$. **(c)** Converted to heat (friction), clicking noise of ratchet, and similar losses.

5.4 (a) $PE = mgh = (40)(-9.8)(-2{,}000(\sin 10°)) \approx 136{,}140\ J$. **(b)** $KE_i = (1/2)mv_i^2 = (1/2)(40)(5^2) = 500\ J$. $KE_f = (1/2)(40)(50^2) = 50{,}000\ J$. $KE_f - KE_i = 50{,}000 - 500 = 49{,}500\ J$. **(c)** Air resistance, blade friction, brakes, hitting sides of channel, etc. Note how much PE was not converted to KE but lost.

5.5 (a) The rocket's loss of gravitational potential energy will equal its kinetic energy. $PE_g = G(m_e m_r)(1/r_1 - 1/r_2) =$
$(6.67 \times 10^{-11})(5.97 \times 10^{24})(10)[1/(3.84 \times 10^8) - 1/(6.8 \times 10^6)] =$
-5.7×10^8. Therefore $KE = 5.7 \times 10^8$ J. **(b)** $KE = (1/2)mv^2$.
$v = [2KE/m]^{1/2} = [2(5.7 \times 10^8)/10]^{1/2} \approx 10,700$ m/s $\approx 24,000$ mph.

5.6 $\Delta PE_e = (Kq_1 q_2)(1/r_2 - 1/r_1) =$
$(9.0 \times 10^9)(1.6 \times 10^{-19})(-1.6 \times 10^{-19})[1/\infty - 1/(53 \times 10^{-11})] = 4.35 \times 10^{-19}$ J.

5.7 (a) $W = qV = (1)(9) = 9$ J **(b)** $1 eV = 1.602 \times 10^{-19}$ J, so
$1 J = 1/(1.602 \times 10^{-19})$ eV $= 6.24 \times 10^{18}$ eV, and 9 J $= 5.62 \times 10^{19}$ eV.

5.8 All the potential energy is converted into kinetic energy so
$PE = mgh = KE$. $KE = (1/2)mv^2 + (1/2)I\omega^2$. $I = mr^2$ and $v = r\omega$.
$mgh = (1/2)mv^2 + (1/2)(mr^2)(v/r)^2$. $gh = (1/2)v^2 + (1/2)v^2 = v^2 =$
$(9.8)(50) = 490$. $v \approx 22.14$ m/s or about 50 mph.

5.9 (a) With no air resistance, all the initial KE would be transformed into PE. $(1/2)mv^2 = mgh$. $h = v^2/2g = (40)^2/(2)(9.8) = 81.63$ m. The difference in PE between 81.63 m and 70 m is $mg(81.63 - 70) =$
$0.0145(9.8)(11.63) = 1.65$ J. **(b)** $KE_i - KE_f = (1/2)mv_i^2 - (1/2)mv_f^2 =$
$(1/2)(0.02)(225) - (1/2)(0.02)(196) = 0.29$ J lost to friction.
$W = Fd = \mu F_n d = \mu mgd$. $\mu = W/mgd = 0.29/(0.02)(9.8)(20) = 0.074$.

5.10 (a) $m_A v_{iA} + m_B v_{iB} = m_A v_{fA} + m_B v_{fB}$, so $v_{fA} =$
$(m_A v_{iA} + m_B v_{iB} - m_B v_{fB})/m_A = (1(10) + 10(-1) - 10v_{fB})/1 = -10v_{fB}$.
So $v_{fA} = -10v_{fB}$. Also, $(1/2)m_A v_{iA}^2 + (1/2)m_B v_{iB}^2 = (1/2)m_A v_{fA}^2 +$
$(1/2)m_B v_{fB}^2$, so $(1/2)(1)(10)^2 + (1/2)(10)(-1)^2 = (1/2)(1)(v_{fA})^2 +$
$(1/2)(10)(v_{fB})^2$. $55 = (1/2)v_{fA}^2 + 5v_{fB}^2$. So $v_{fA}^2 = 110 - 10v_{fB}^2$. Using
substitution: $(-10v_{fB})^2 = 110 - 10v_{fB}^2$. $100v_{fB}^2 + 10v_{fB}^2 = 110$. $v_{fB} = 1$
and $v_{fA} = -10$. So, each sphere bounces back in exactly the opposite
direction and at its initial speed. **(b)** Using $v_{fA} = -10v_{fB}$;
$e = (v_{fB} - v_{fA})/(v_{iA} - v_{iB})$. $0.5 = (v_{fB} - v_{fA})/(10 - (-1))$. $v_{fB} - v_{fA} = 5.5$.
Substituting, $v_{fB} - (-10v_{fB}) = 5.5$, $11v_{fB} = 5.5$. $v_{fB} = 0.5$ and $v_{fA} = -5$.
So when $e = 0.5$, the spheres bounce back at one half the initial speed.
(c) The initial total momentum was zero, and it will be zero after the
totally inelastic collision. Therefore, the two spheres will each have zero
velocity and will appear to stick together.

5.11 $(100$ Watt$)(90\%) = 90$ Watts. Work $= mgh = (10)(9.8)(10) = 980$ J.
Power $=$ Work$/t$. $t =$ Work/Power $= 980/90 = 10.9$ s.

5.12 (a) One rotation of the wheel's circumference is 24 times the
distance of one rotation of the axle's circumference, so the required force
is 1/24 as large. Therefore the mechanical advantage is 24:1. **(b)** 1:24.

Chapter 6

ELASTICITY AND HARMONIC MOTION

6.1. Elasticity and Hooke's Law
6.2. Stress and Strain
6.3. Simple Harmonic Motion (SHM)
6.4. Energy and Simple Harmonic Motion
6.5. Simple Pendulum
6.6. Damped Oscillations and Resonance
6.7. Key Concepts and Practice Problems

> *"One cannot escape the feeling that these mathematical formulae*
> *have an independent existence and an intelligence of their own,*
> *that they are wiser than we are, wiser even than their discoverers,*
> *that we get more out of them than was originally put into them."*
> Attributed to Heinrich Hertz

6.1. Elasticity and Hooke's Law

• Some solid materials and objects exhibit the property of **elasticity** and can be stretched or compressed. These materials will resist changes to their size and shape and, when the deforming force is removed, will tend to return to their original size and shape.

The elastic property of solid materials is due to electric forces present in the atoms and molecules making up the material. When a force is applied that changes the atomic arrangement of the solid, work is done that raises potential energy. When the force is removed, the material returns to its original state of lower potential energy. If you stretch a piece of rubber, some electrons are pulled further from positive nuclei than in its natural state, causing a stronger attractive force until you stop stretching and allow the rubber to return to its original length.

• **Hooke's Law** states that *the force needed to stretch (or compress) a given spring is proportional to the distance the spring is stretched (or compressed)*. Said another way, if the degree to which a spring stretches is directly proportional to the force that deforms or stretches it, the spring obeys Hooke's Law. Other materials can also obey Hooke's Law, so *that the material's or object's deformation is directly proportional to the force that deforms it.* Mathematically **Hooke's Law** is:

$$F = kx$$

where k is the **spring constant** or **force constant** of the spring which depends on the material, size, shape, and type of spring, and x is the distance the spring is displaced from its *original natural or equilibrium position*. **Units** for F are Newtons N, and units for k are N/m.

spring at equilibrium
stretched spring

• **Example**: If a force of 1 N is needed to stretch a spring 0.1 m, how much force is needed to stretch that spring 0.2 m?

Use: $F = kx = 1\,N = k(0.1\,m)$
Determine k: $k = (1\,N) / (0.1\,m) = 10\,N/m$
When $x = 0.2\,m$, F is: $F = kx = (10\,N/m)(0.2\,m) = 2\,N$

• **Example**: What **work** W is required to stretch the spring in the previous example 0.2 m?

Use **average force** from initial to final positions: $F_{Ave} = (F_f - F_i)/2$
At initial position *zero* force is applied, so: $F_{Ave} = (1/2)F_f = (1/2)kx$
Work is: $W = F_{Ave}x = (1/2)F_f x = (1/2)kx^2 = (1/2)(10\,N/m)(0.2\,m)^2 = 0.2\,J$

In this example we showed **work** done by an external force **F** is:

$$W = F_{Ave}x = (1/2)kx^2$$

The force to stretch or compress a spring varies from F at initial position where $F_i = 0$ to F at final position, $F_f = kx$. We showed that since force increases linearly with x, we can use the **average force** between F_f and F_i, which is $F_{Ave} = (F_f - F_i)/2 = (1/2)F_f = (1/2)kx$, or:

$$F_{Ave} = (1/2)kx$$

The **potential energy** stored in a stretched or compressed spring is equal to the **work** required to stretch or compress it, $W = F_{Ave}x = (1/2)kx^2$. The **work** is stored as **potential energy**:

$$W = PE_s = (1/2)kx^2$$

When x = 0 and the spring is at equilibrium, $PE_s = 0$. When the spring is not in equilibrium, $PE_s > 0$. Whether the spring is stretched or compressed a given distance, it has the same potential energy.

• Materials and objects often possess an **elastic limit**. Beyond this elastic limit the internal structure of the solid is permanently distorted. When deformed past its elastic limit, a bent, stretched, or compressed material will not return to its original shape and may even reach a breaking point and fracture. The elastic limit depends on the material and may also depend on external factors such as humidity and temperature.

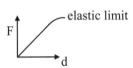

If deformation exceeds the **elastic limit**, the restoring force is no longer proportional to the displacement. The **restoring force** is the force that returns an object or material to its original position or shape. When a force F = kΔx is applied to compress a spring distance Δx, then according to **Newton's Third Law**, if the spring is held in a compressed position, the spring is also exerting an equal and opposite force $F_s = -k\Delta x$. This is referred to as a **linear restoring force** and is in the opposite direction of displacement. The sign of F_s indicates the spring is resisting either compression or stretching.

• **Example**: You decide to see how fast you can get a 1-kg block sitting on a table to slide by holding it securely against a spring that is compressed 0.1 m. If you let go, what will be the velocity of the block as it leaves the spring? (The spring constant is k = 75 N/m and the coefficient of friction between the block and table is 0.1. Assume mass-less spring.)

The energy and work are interchangeable so: W = ΔKE + ΔPE. Work is against friction. Initially all energy is PE:

$$-F_{kfriction}\, x = ((1/2)mv_f^2 - 0) + (0 - (1/2)kx^2)$$
$$-\mu_k mg\, x = (1/2)mv_f^2 - (1/2)kx^2$$
$$v_f^2 = 2((1/2)kx^2 - \mu_k mg\, x)\, /\, m$$
$$v_f^2 = 2[(1/2)(75\,\text{N/m})(0.1\,\text{m})^2 - (0.1)(1\,\text{kg})(9.8\,\text{m/s}^2)(0.1\,\text{m})]\,/\,(1\,\text{kg}) = 0.554$$
$$v_f \approx 0.74\,\text{m/s}$$

• **Example**: What is the equation for force if a mass is suspended vertically and hanging from a spring?

If the displacement is small and the spring's elastic limit is not exceeded, Hooke's Law is still valid. When the mass is in equilibrium, the net force on the system is zero:

$$F = kX - mg = 0$$

If the weight is pulled down by distance x, the net force is:

$$F = ma = k(X + x) - mg = kX + kx - mg$$

Since $kX - mg = 0$: $F = kx$.

Even in the presence of gravity, Hooke's Law can still be applied.

6.2. Stress and Strain

• Solids can be deformed in different ways, including compressing or stretching the length, twisting or shearing to change internal angles, and squeezing to change bulk volume. Liquids and gases are also subject to bulk deformations and volume changes. A **stress** is a force that causes deformation and a **strain** is the resulting change in a material.

• The **stress** applied to an elastic material is equal to the *applied force* **F** *divided by the affected cross-sectional area* A of the material. **Units** of **stress** are N/m^2.

$$\boxed{\text{Stress} = F/A}$$

A force that acts to stretch a material is a **tensile stress**. A force that tends to compress a material is a **compressive stress**. Tensile and compressive stresses act along straight lines.

tensile stress   compressive stress

• A stress produces an effect in a material called **strain**. The strain represents the fractional deformation of a material or substance. The strain caused by a tensile or compressive stress is equal to the *fractional change in the material's length L*, and is dimensionless having cancelled units. For **length deformations**:

$$\boxed{\text{Strain} = \Delta L/L}$$

where L is the original natural length of the material and ΔL is the change in length caused by the applied stress. Stretching may cause the cross-sectional area to slightly decrease while volume is constant (as depicted below).

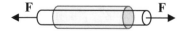

• When strain $\Delta L/L$ in a material is directly proportional to its stress F/A, Hooke's Law is obeyed. When an applied stress becomes too great, Hooke's Law will break down and the linear relationship will not continue. Stress and strain can be graphed for a given material.

Strain vs. Stress

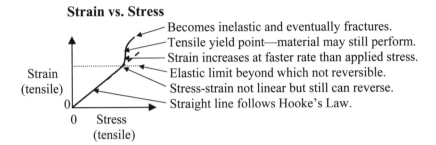

In an "ideal" material, *Hooke's Law* would hold for all stress values and the graph would be linear (dashed line in figure). In real materials, however, increasing stress eventually causes a material to no longer follow Hooke's Law's linear relationship, though it may still be able to return to its original shape until continued stretching exceeds its elastic limit. Beyond the elastic limit the material may still perform for a while, but cannot return to its original shape. Eventually increased stress will cause fracturing.

The stress-strain graph for a compressive force is similar for metal springs and metal objects, and the slope of the Hooke's Law line is also the same. Other materials, however, do not graph the same for tensile and compressive forces.

Shear Stress and Torsion Stress

• **Shear stress** occurs when an object is held fixed at one end and a lateral stress is applied to the opposite end. **Torsion stress** occurs when an object is held fixed at one end and a twisting stress is applied to the opposite end.

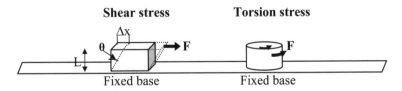

Shear stress is measured as force per area F/A and shear strain as $\Delta x/L = \tan \theta$, where Δx is the displacement parallel to the force and L is the length or in the above figure, the height. Shear stress and shear strain are related through the shear modulus G, where:

$$G = \text{(stress)} / \text{(strain)} = (F/A) / (\Delta x/L)$$

Torsion stress is more complicated and involves the polar moment of inertia, which is $\pi r^4/2$ for a solid cylinder.

Young's Modulus and Bulk Modulus

• If a material obeys Hooke's Law, the relationship of stress F/A to strain $\Delta L/L$ is linear, and the ratio of stress to strain is constant. The ratio of stress to strain is called the **elastic modulus** or **Young's modulus** Y:

> Young's modulus Y = stress/strain = $(F/A)/(\Delta L/L)$ = constant
>
> (stress) = (Y)(strain)

Units for **Young's modulus** are N/m^2. Young's modulus ranges include approximately 0.05×10^{10} N/m^2 for rubber to 20×10^{10} N/m^2 for steel.

• **Example**: When visiting your friend's warehouse, you notice a 10 m long aluminum rod with about a 2 cm diameter. It is securely suspended from the ceiling. You can't help but wonder if it would actually stretch if you held it and took your feet off the ground. If the Young's modulus for this aluminum is 6.9×10^{10} N/m^2 and your mass is 60 kg, how much will the rod stretch?

$$Y = (F/A)/(\Delta L/L) \quad \text{or} \quad \Delta L = (FL)/(AY) = (mgL)/(\pi r^2 Y)$$

$$\Delta L = (60\,\text{kg})(9.8\,\text{m/s}^2)(10\,\text{m})/\pi(0.01\,\text{m})^2(6.9 \times 10^{10}\,\text{N/m}^2) \approx 0.271\,\text{mm}$$

• While solids can undergo shear and length deformations, solids and liquids can undergo bulk deformations. When stress is applied uniformly to all surfaces of an object, such as when it is submersed in fluid or at a high atmospheric pressure, its volume can change. The **bulk modulus** involves this change in volume V and is the ratio of volume stress to volume strain. The **volume stress** on a substance or material is the applied force per area ($\Delta F/A$) or pressure change ΔP. The **volume strain** is the fractional change in volume $\Delta V/V$. The ratio of the applied pressure to the fractional change in volume is the **bulk modulus B**:

$$B = \text{pressure stress/volume strain} = (\Delta F/A)/(\Delta V/V) = -(\Delta P)/(\Delta V/V)$$
$$(\text{stress}) = -(B)(\text{strain})$$

The negative sign reflects the fact that volume decreases as pressure is applied. Bulk modulus ranges include approximately 0.22×10^{10} N/m^2 for water to 16×10^{10} N/m^2 for steel, and about 10^5 N/m^2 for air. Liquids in general are difficult to compress, which makes them useful for transferring forces in hydraulic systems.

• **Example**: After hanging from the aluminum rod, your friend asks you to help him with his experimental apparatus. He puts a block of aluminum in a small tank of liquid and applies a pressure of 500 atm as measured by a gauge he has hooked up. He tells you that the density ρ of aluminum is 2.7×10^3 kg/m^3, and its bulk modulus is 7.7×10^{10} N/m^2. He wants to find the fractional change in density of the aluminum.

You write the bulk modulus, $B = -\Delta P/(\Delta V/V)$, and solve it for $\Delta V/V$:

$$-\Delta V/V = \Delta P/B = -(500 \text{ atm})(101{,}325 \text{ N/m}^2/\text{atm}) / (7.7 \times 10^{10} \text{ N/m}^2)$$
$$\approx -0.00066 \text{ or } -0.066\%$$

You remember density = mass/volume or $\rho = m/V$, so $V = m/\rho$, or $m = V\rho$. Since mass m is constant $V_1\rho_1 = V_2\rho_2$:

$$\rho_2 = \rho_1(V_1/V_2) = \rho_1(V_1/(V_1 + \Delta V))$$

If we set V_1 as 1 unit: $V_1 + \Delta V = (1 + (-0.00066)) = 0.99934$
Substituting in:

$$\rho_2 = \rho_1(1/0.99934) = 2.7 \times 10^3(1/0.99934) = 2.70178 \times 10^3$$

Finally, the fractional change in density is:

$$\Delta\rho/\rho = (\rho_2 - \rho_1)/\rho_2 = (1.78)/(2.70178 \times 10^3) = 0.0006588 \approx 0.066\%$$

which is the same as the (negative of) fractional change in volume.
Note: A formula for very small fractional changes you may encounter is:
$\Delta\rho/\rho = -\Delta V/V$.

6.3. Simple Harmonic Motion (SHM)

• Materials are to some degree elastic and many elastic materials can vibrate or oscillate. If a guitar string is deformed or perturbed, it will vibrate or oscillate. A material can also oscillate independently of its elastic properties such as a pendulum. *Oscillations or vibrations occur because a **restoring force** is returning the material or system to its natural equilibrium.*

In an elastic material if a force is applied that causes deformation, and then the force is released, internal forces (restoring force) cause the material to return to its equilibrium shape. If momentum carries the oscillation through the equilibrium shape and beyond, the material becomes deformed again, but with the displacement in the opposite direction. Then, the restoring force again acts to return the material to equilibrium, when momentum carries it through to the original displacement, and the cycle continues. If energy did not dissipate, the oscillation would continue indefinitely.

We have just described **simple harmonic motion** (SHM), which involves oscillations or vibratory motion. The vibratory or oscillating motion of a mass attached to a spring is a good illustration of SHM and its equations. Suppose a block is at rest in its equilibrium position on a *frictionless* surface. If an external force displaces the block to the right, there will be an opposite **restoring force** F exerted on the block by the spring directed to the left.

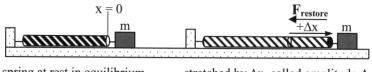

spring at rest in equilibrium   stretched by Δx, called amplitude A

If the displacement of the block by the amount Δx has not stretched the spring beyond the range where Hooke's Law is valid, Hooke's Law can be used to model the system. Often it is the restoring force rather than the applied force that is of interest in SHM, so a *negative sign* is used when writing **Hooke's Law** to highlight that the restoring force points opposite the displacement of the mass from its equilibrium position:

$$F = -kx$$

We can use Newton's Law, $F = ma$, to write an **equation** for the motion of an object experiencing **simple harmonic motion** as:

> Force $F = ma = -kx$
>
> Acceleration $a = -kx/m$

• If a block is released from its initial stretched position $+\Delta x$, it will be accelerated back to the left by restoring force $-F$. Maximum acceleration occurs at $x = +\Delta x$ as the block begins to move back toward $x = 0$. As the block approaches $x = 0$, its acceleration decreases to zero and its velocity reaches maximum. Passing $x = 0$, the momentum of the block will carry it to the left side of $x = 0$, or negative x, and the spring is compressed until it reaches $x = -\Delta x$. As it travels to the left of $x = 0$, the restoring force and acceleration are directed back to the right and the block slows. At $x = -\Delta x$, motion will stop and the acceleration which is still toward the right will cause the block to reverse its motion and move toward $x = +\Delta x$ again. The block oscillating or vibrating forward and back between $x = +\Delta x$ and $x = -\Delta x$ forms what is referred to as a cyclic, oscillatory, periodic, or SHM with an amplitude A of Δx.

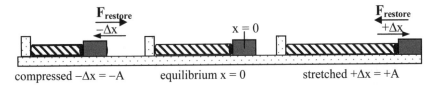

compressed $-\Delta x = -A$ equilibrium $x = 0$ stretched $+\Delta x = +A$

If a pen is attached to the block and a roll of paper is vertically scrolled as the block is oscillating, the motion can be recorded as a function of time. A **sinusoidal** pattern will be drawn showing the SHM. The figure below is a rotated view of the spring-mass above to show how it graphs as a sine wave.

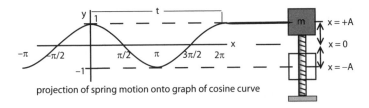

projection of spring motion onto graph of cosine curve

The graphs of the sine and cosine functions are described by:
$$y = \sin x \quad \text{and} \quad y = \cos x$$
The graph of $y = \cos x$ is the graph of $y = \sin x$ shifted by $\pi/2$ to the left, since $\cos x = \sin(x + \pi/2)$. (See *Master Math: Trigonometry.*) The **amplitude** A is the maximum deviation from the centerline (x-axis) along the y-axis in the cosine graph above. Changing the *amplitude* changes the y-component of a sine or cosine graph. The amplitude of a sine or cosine function is changed by multiplying the equations by **amplitude** A:
$$y = A \sin x \quad \text{and} \quad y = A \cos x$$

The **period** τ represents one complete cycle of the sine or cosine function along the horizontal x-axis. The periods of $y = \sin x$ and $y = \cos x$, including $y = A \sin x$ and $y = A \cos x$, are 2π or $360°$. The periods are changed when the functions of sine and cosine have a multiplier b of x and are written:
$$y = A \sin bx \quad \text{and} \quad y = A \cos bx$$

where the **period** τ is $2\pi/b$ and $b > 0$, or b is $2\pi/\tau$. When sine or cosine is graphed for displacement x of the block as a function of *time* t, or x(t), we can write the equation for the cosine graph of the block and spring above as x(t) = A cos bt, or:

$$x(t) = A \cos 2\pi t/\tau$$

This equation describes the **harmonic motion** as *displacement as a function of time*, and we will further develop it in the following pages. In this equation, τ is the period of the repeating motion, and $A = \Delta x$ is the maximum deviation from centerline, or amplitude. This equation describes the ideal case where friction is ignored and the oscillation occurs indefinitely with no decrease in amplitude. We have learned that when a material or system obeys Hooke's Law and the displacement of a block attached to a spring varies sinusoidally with time, the system is said to display simple harmonic motion. In SHM, the displacement, velocity, and acceleration all vary sinusoidally with time.

Looking at Simple Harmonic Motion Through Circular Motion

• In *simple harmonic motion*, or *oscillatory motion*, a particle or object moves back and forth between two fixed positions along a straight line. The connection between SHM and uniform **circular motion** can be visualized by projecting the image of a particle moving steadily in a circular path onto a screen (perpendicular to the plane of the circle). By projecting the circular path from its side, the projected image looks like a particle moving back and forth (or up and down) in a straight line. The *shadow of the particle translates to simple harmonic motion.* We see that the back-and-forth motion of a simple harmonic oscillator is the same as the projection of the motion of a particle traveling uniformly in a circular path.

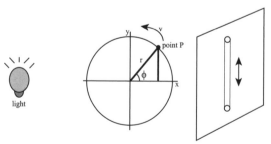

Another way to visualize the correlation between a particle moving around a circle in circular motion and its corresponding harmonic motion is to imagine the image of the moving particle projected onto the circle's y-axis or x-axis. The position of the particle can be represented by vector **r**, and the projection of **r** onto the x-axis (where the vertical line is drawn down from point P) is a certain length x(t) along the x-axis. The length x(t) varies with time as the particle moves around the circle. The *maximum displacement* occurs when the full length of vector **r** is projected on to the axis. This *maximum displacement* corresponds to the **amplitude** A. The vertical line drawn to the x-axis forms a right triangle with sides vector r and the x-axis itself, and corresponds to the particle's location. Remember, **trigonometric functions** can be defined using ratios of sides of a right triangle, angles in standard position in a coordinate system, and arc lengths on a circle of radius one (unit circle). (See *Master Math: Trigonometry*.) Also, remember for a right triangle that sin ϕ = opposite/hypotenuse and cos ϕ = adjacent/hypotenuse. Therefore, for the right triangle drawn in the circle, sin ϕ = y/r and cos ϕ = x/r, or rearranging, y = r sin ϕ and x = r cos ϕ. Since **r** is the maximum length that x or y can reach, it is the amplitude A. Therefore:

$$y = A \sin \phi \quad \text{and} \quad x = A \cos \phi$$

Let's look at x = A cos ϕ: Note that x is positive when the displacement is to the right and negative when the displacement is to the left where cos ϕ < 0. In the circle drawn on the previous page, the position of the particle is given by the angle ϕ, which is measured in radians. As we learned in Section 1.11, the **rate of change of ϕ is the angular velocity of the particle**. The **angular velocity** ω is the change in angle ϕ divided by the change in time t, or:

$$\omega = \Delta\phi / \Delta t$$

If the motion is uniform, angular velocity is constant, and the angular displacement is related to the angular velocity by:

$$\phi = \omega t$$

Substitute into x = A cos ϕ:

$$\boxed{x(t) = A \cos \omega t}$$

where A is the **amplitude** or maximum displacement of the motion from the equilibrium position. In Section 1.11 we learned the *time required for one revolution or period* is: $\tau = 2\pi/\omega$ and rearranging gives $\omega = 2\pi/\tau$. The ω in the SHM equation represents the **angular frequency**. It is related to the frequency f of the motion, and inversely related to the period τ, so that $\omega = 2\pi/\tau = 2\pi f$ and $f = 1/\tau$. The frequency is the number of oscillations per second and has units of **Hertz** (Hz). Substitute for ω:

$$\boxed{x(t) = A \cos 2\pi t/\tau}$$

This equation for projected circular motion is the same equation we derived for (one-dimensional) SHM of a spring! The **projection of circular motion is simple harmonic motion**.

Velocity in Simple Harmonic Motion

• In SHM, the **velocity** changes as a particle or object oscillates. When the displacement is at maximum amplitude, the velocity is zero, and when the displacement returns to zero, the velocity is maximum. In circular motion as a particle moves around a circle, the direction of motion and the velocity vector is *tangent* to the circle. The **velocity vector** tangent to the circle at point P has a cosine component cos t, a sine component sin t, and points in the direction the particle is moving. The acceleration a particle in circular motion experiences is **centripetal acceleration**, which points inward along the radius r line.

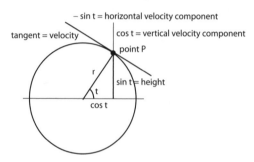

By comparing the motion of the particle moving around the circle with its projection (as shown in the previous figure), it is evident that even when the velocity of the particle is constant, the projected velocity of the particle slows to a stop at each end (top and bottom). By rotating a right triangle around a circle the relationships between sine, cosine, distance, and velocity can be visualized. The equation for displacement is $x(t) = A \cos \omega t$, where A is the *amplitude* or maximum displacement, and the **velocity** equation is:

$$v(t) = -A\omega \sin \omega t \quad \text{where } A\omega \text{ is the maximum speed}$$

Since $\omega = 2\pi/\tau$:

$$v(t) = -v_{max} \sin 2\pi t/\tau$$

We can also see the connection between displacement and velocity through the cosine and sine wave curves. *A particle moving around a circle* can be translated into a *particle moving along a cosine or sine curve.* Just as oscillatory displacement can be graphed as a cosine or sine wave, the **velocity of the oscillatory motion** can also be related to a *sine or cosine wave pattern.* By definition, the *velocity is the rate of change of distance, and the slope of a cosine or sine curve at a given point represents the velocity at that point.* It turns out that the slope or velocity at each point along a *cosine curve graph* is equal to the corresponding value of the displacement of that point on the *sine curve graph* and vice versa. In calculus, the derivative (or rate of change) of distance is velocity. Also, the derivative of sine is cosine and the derivative of cosine is negative sine.

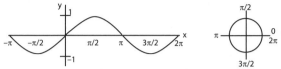

Sine graph--slope per point of cosine and projection of circular motion onto y-axis

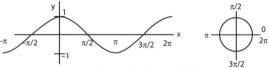

Cosine graph--slope per point of sine and projection of circular motion onto x-axis

Acceleration in Simple Harmonic Motion

• The **acceleration** also oscillates in SHM and therefore can be described by a sine wave or sinusoid. In the SHM of a mass on a spring, when the displacement is zero, the spring is applying no force and the acceleration is also zero. When the displacement is maximum, the spring is applying maximum force and the acceleration is maximum. The force applied by the spring is in the opposite direction as the displacement. The **acceleration for simple harmonic motion** is:

$$a(t) = -A\omega^2 \cos \omega t \quad \text{where the maximum acceleration is } A\omega^2$$

Since $\omega = 2\pi/\tau$:

$$a(t) = -a_{max} \cos 2\pi t/\tau$$

We can also develop this equation using the particle revolving around the circle, where the acceleration acting on the particle is centripetal acceleration a_c. If we project the revolving particle as we did above, we can write:

$$a(t) = -a_c \cos \omega t$$

where the negative sign reflects the opposite directions of the acceleration and displacement. The *centripetal acceleration* of the revolving particle is $a_c = v^2/r$, and a_c is related to velocity and radius (maximum amplitude) by:

$$a_c = v^2/r = v_{max}^2/A$$

Since the *speed* of the object moving in a circular path of radius r is $v = r\omega$, and the radius r corresponds to the amplitude A:

$$a_c = (A\omega)^2/A = A\omega^2$$

Acceleration for simple harmonic motion is therefore given by:

$$a(t) = -A\omega^2 \cos \omega t \quad \text{where the maximum acceleration is } A\omega^2$$

• We have also described acceleration of a mass undergoing SHM using the restoring force described in the *spring-mass* system as: $a = -kx/m$, where k is the spring constant and m is mass. If we substitute x with the displacement $x(t) = A \cos \omega t$ for SHM, **acceleration for simple harmonic motion** can also be written:

$$a = -kx/m = -(kA/m) \cos \omega t = -(kA/m) \cos 2\pi t/\tau$$

where kA/m is the maximum acceleration, and the negative sign shows acceleration in the opposite direction from displacement.

Notes on Amplitude, Period, and Frequency

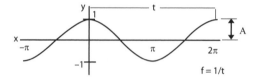

• The **amplitude A** is the maximum deviation or displacement from the centerline of a horizontally drawn repeating or oscillating graph in the vertical direction. In an oscillating system, the amplitude is the magnitude of a wave. In a sine or cosine function, the amplitude is changed by multiplying the equations by A: $y = A \sin x$ and $y = A \cos x$. Period and frequency in SHM are independent of amplitude.

• The **period** τ represents one complete cycle of a repeating or oscillating function. In SHM the period τ is the time it takes to go through one complete cycle of the oscillating motion. In the figure of the cosine graph of the spring-mass system above, τ is time interval between any two successive corresponding points on the sinusoidal curve. The periods of $y = \sin x$ and $y = \cos x$, including $y = A \sin x$ and $y = A \cos x$, are 2π or 360°. In *circular motion* the period τ is the time required for one revolution or period. (Also, see following paragraphs.)

$$\tau = 2\pi/\omega$$

• In SHM **frequency** f is the number of oscillations or cycles per time (seconds). If there are 10 complete cycles of an oscillatory motion each second, the frequency is $f = 10$ cycles per second. Since a cycle is not a physical quantity or unit, units of frequency are in s^{-1}, called a Hertz (Hz). Therefore, 1 cycle per second = 1 Hz. If the frequency of an oscillation is

10 Hz or 10 s^{-1}, each complete cycle takes 1/10 s. Therefore, **frequency** f and the **period** τ are related. (Also, see following paragraphs.)

$$f = 1/\tau$$

The **angular velocity** ω is also the **angular frequency**, and is distinguished from frequency f in that f describes the number of complete oscillations or cycles per second, while ω describes the number of radians the angle ϕ in circular motion completes per second. In circular motion each complete cycle corresponds to a change in ϕ of 360° or 2π radians, so that $\omega = 2\pi$ radians. Angular frequency, or angular velocity, ω is measured in radians per second, while frequency f is measured in Hz.

• In our discussion of acceleration above, we showed that acceleration of a mass undergoing SHM can be written: $a = -kx/m = -(k/m)\,A\cos\omega t$. We also described acceleration in SHM as: $a(t) = -A\omega^2\cos\omega t$. By comparing these two equations,

$$a(t) = -A\omega^2\cos\omega \quad \text{and} \quad a = -(k/m)\,A\cos\omega t$$

we see that ω^2 must equal k/m:

$$\omega^2 = k/m \quad \text{or} \quad \omega = [k/m]^{1/2}$$

This shows the relationship between the frequency ω and mechanical properties described by the spring (or force) constant k and mass m. Since period $\tau = 2\pi/\omega$ and $\omega^2 = k/m$, the **period for simple harmonic motion** can be written:

$$\tau = 2\pi/\omega = 2\pi[m/k]^{1/2}$$

Also, since frequency $f = 1/\tau$, **frequency for single harmonic motion** can be written:

$$f = 1/\tau = \omega/2\pi = (1/2\pi)[k/m]^{1/2}$$

Equation Summary

$$x(t) = A\cos\omega t = A\cos 2\pi t/\tau$$

$$v(t) = -A\omega\sin\omega t = -v_{max}\sin 2\pi t/\tau$$

$$a(t) = -A\omega^2\cos\omega t = -(kA/m)\cos 2\pi t/\tau$$

$$\omega = 2\pi/\tau = [k/m]^{1/2}$$

$$\tau = 2\pi/\omega = 2\pi[m/k]^{1/2}$$

$$f = \omega/2\pi = (1/2\pi)[k/m]^{1/2}$$

6.4. Energy and Simple Harmonic Motion

• When an oscillating mass is displaced from its equilibrium position, the work needed to displace it a distance x is: $W = F_{Ave} x$, where F_{Ave} is the average force applied over the displacement, which is $F_{Ave} = (F_f - F_i)/2$. At the initial position, *zero* force is applied, so: $F_{Ave} = (1/2)F_f$. If we substitute **Hooke's Law** $F = kx$ (see Section 6.1), F_{Ave} becomes:

$$F_{Ave} = (1/2)F_f = (1/2)kx$$

The equation for the **work** done by an external force becomes:

$$W = F_{Ave} x = (1/2)kx^2$$

Since the **work** required to stretch (or compress) a spring or system is stored as **potential energy**, $W = PE$:

$$\boxed{W = PE = (1/2)kx^2}$$

The **energy** of a simple harmonic oscillator translates between potential and kinetic energy with: $PE + KE =$ constant. The **kinetic energy** is: $KE = (1/2)mv^2$. As the oscillator moves, x and v change, and energy shifts to the extremes of kinetic and potential. At $x = A$, or maximum displacement (amplitude), the velocity is zero and energy is all PE. When x returns to zero, $x = 0$, velocity is maximum and energy is all KE. Therefore, the **energy of the simple harmonic oscillator** is:

$$\boxed{\begin{array}{c} (1/2)kx^2 = (1/2)mv^2 \\ (1/2)kA^2 = (1/2)mv_{max}^2 \end{array}}$$

Solving for **maximum velocity**:

$$\boxed{v_{max}^2 = kA^2/m \quad \text{or} \quad v_{max} = A[k/m]^{1/2}}$$

• **Example**: In a frictionless horizontal spring-mass system where the spring constant is 2,000 kg/s^2, if the 2.5-kg mass is displaced 2 cm so that the spring is compressed, what is the maximum velocity and maximum force on the mass?

Maximum velocity occurs at $x = 0$, or equilibrium position, and is:
$$v_{max}^2 = kA^2/m = (2,000 \text{ kg/s}^2)(0.02 \text{ m})^2/(2.5 \text{ kg}) = 0.32 \text{ m}^2/s^2$$
$$v_{max} \approx 0.57 \text{ m/s}$$

Maximum force occurs at $x = A$ and is given by Hooke's Law:
$$F = kx = kA = (2,000 \text{ kg/s}^2)(0.02 \text{ m}) = 40 \text{ N}$$

• **Example**: You decide it's time to test out all this spring stuff for yourself. Suppose your mass is 50 kg and at equilibrium with you holding on, the spring stretches 3.5 cm. If your friend pulls down on you, stretching the spring an additional 3.0 cm, and then lets go, what is your displacement, velocity, and acceleration 7 s later?

The equations we can use are:

$$x(t) = A \cos 2\pi t/\tau \quad v(t) = -v_{max} \sin 2\pi t/\tau \quad a = -(kA/m) \cos 2\pi t/\tau$$

We need to find k and τ. Note that the cosine angles are in *radians, so set your calculator.* We can use values at equilibrium displacement:

$$k = F/x = mg/x = (50 \text{ kg})(9.8 \text{ m/s}^2)/(0.035) = 14{,}000 \text{ kg/s}^2$$

$$\tau = 2\pi/\omega = 2\pi[m/k]^{\frac{1}{2}} = 2\pi [50 \text{ kg} / 14{,}000 \text{ kg/s}^2]^{\frac{1}{2}} \approx 0.375 \text{ s}$$

Maximum displacement A was at t = 0, so displacement at t = 7 s is:

$$x(t) = A \cos 2\pi t/\tau = (0.030 \text{ m}) \cos(2\pi(7\text{s})/(0.375\text{s})) \approx -0.015$$

Velocity at t = 7 s, where $v_{max} = A[k/m]^{\frac{1}{2}}$ is:

$$v(t) = -v_{max} \sin 2\pi t/\tau = -A[k/m]^{\frac{1}{2}} \sin 2\pi t/\tau$$

$$v(t) = -(0.030\text{m})[14{,}000\text{kg/s}^2/50\text{kg}]^{\frac{1}{2}} \sin(2\pi(7\text{s})/(0.375\text{s})) \approx -0.435 \text{ m/s}$$

The negative sign shows velocity opposite displacement.

Acceleration at t = 7 s is a(t) = −(kA/m) cos 2πt/τ, or:

$$a(t) = -[(14{,}000\text{kg/s}^2)(0.030\text{m})/(50\text{kg})] \cos(2\pi(7\text{s})/(0.375\text{s})) \approx 4.2 \text{ m/s}^2$$

6.5. Simple Pendulum

- In a **simple pendulum** all of the mass is considered to be the same distance from the support point, and the rope, string, chain, or cable holding the mass is considered mass-less. *A simple **pendulum** can be approximated as a simple harmonic oscillator when the angular amplitude is small.* A mass on the end of a string will oscillate back and forth if disturbed, and if the amplitude of the swing motion is small it can approximate SHM.

θ is in radians
L is length of pendulum
x is displacement

When angle θ and therefore the amplitude is small, this pendulum (depicted above) approximates a right triangle so, sin θ = x/L, or x = L sin θ. When x is sufficiently small:

$$x \approx L\theta$$

Angle θ can be up to 0.25 radians (14.3°) before the difference between θ in radians and $\sin \theta$ reaches 1%. Gravity is the source of the *restoring force* that brings the pendulum to its equilibrium position after it is disturbed. The **restoring force** is $F = -mg \sin \theta$, or **for small θ:**

$$F \approx -mg\theta$$
$$F \approx -(mg/L)x$$

Because the *restoring force is proportional to the angular displacement* θ and also displacement x, when the amplitude is small the motion of a pendulum is consistent with **Hooke's Law** and approximates SHM.

$$F = -kx$$

We can develop the equation for the pendulum force constant, which is (restoring force)/(displacement) by combining the previous two equations to give the **force constant for a pendulum**:

$$k = mg/L$$

• The **period** for a simple pendulum approximating SHM can be derived using the period for SHM:

$$\tau = 2\pi/\omega = 2\pi[m/k]^{1/2} = 2\pi \, [m/mg/L]^{1/2} = 2\pi[L/g]^{1/2}$$

Therefore, the period for a simple pendulum is:

$$\tau = 2\pi[L/g]^{1/2}$$

where $\omega = [k/m]^{1/2} = [mg/Lm]^{1/2} = [g/L]^{1/2}$.

• **Example**: If a pendulum having a 1-kg mass takes 1.5 s for each sideways motion, what is the pendulum's length?

If each sideways motion takes 1.5 s then the period is 3 s.
We can solve $\tau = 2\pi[L/g]^{1/2}$ for L:

$$L = (\tau/2\pi)^2 g = (3 \, s/2\pi)^2(9.8 \, m/s^2) \approx 2.2 \, m$$

6.6. Damped Oscillations and Resonance

• *Simple and damped harmonic motion and resonance* are all described by **sinusoids**. If the amplitude of the motion remains constant (no frictional forces), then the motion is *simple harmonic motion*. If the amplitude decreases over time (frictional loss), the motion is called *damped harmonic motion*. If the amplitude increases over time, the motion is called *resonance*.

Damped Oscillations

• So far in this chapter we have been modeling systems that are considered ideal in that no energy is dissipated from friction or other forces. In the real world, friction is present in an oscillating system, and the motion can become damped. In a system undergoing **damped oscillation**, energy is dissipated (mostly as heat), and the motion eventually comes to a stop.

In a system with *weak damping* the oscillations persist for many cycles before stopping. A weakly damped system can somewhat approximate SHM. Even though the amplitude gradually decreases, the period remains constant. In a weakly damped system the SHM formulas for period may be useful, although the equations for displacement, velocity, and acceleration may not be valid. A system undergoing *strong damping* is not a good approximation of SHM. Nevertheless some SHM modeling may be useful. Strong damping is often deliberately introduced into mechanical systems in order to restrain and control oscillations.

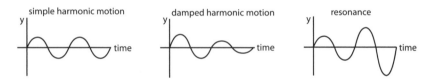

Resonance

• If the amplitude of an oscillating system increases over time, the motion is called **resonance**. Vibrating or oscillating systems possess one or more natural frequencies at which they will naturally oscillate. The period τ, in fact, depends on the intrinsic characteristics of an oscillator.

If an external force is applied to an oscillator (such as a pendulum or spring-mass system) with a frequency other than its natural frequency, the oscillator will vibrate with the frequency of the applied force providing that force is much greater than the natural internal restoring force of the oscillator. If an external force pushes on an oscillator forcing it to oscillate at some frequency, and the applied force is comparable to the natural restoring force of the oscillator, the oscillator will have a vibration in which the amplitude varies according to whether the two oscillations are in or out of phase. If the natural frequency and the applied external force have the same period and are in phase, the applied force and restoring force will be acting together and the amplitude will increase—especially if the applied force occurs at the same point in every cycle and is in phase. When the applied and natural forces have the same frequency or period and are in phase, the system is in resonance. In other words, when an oscillator with a natural frequency is acted on by a sinusoidal force whose frequency is close to the natural frequency, the amplitude of the oscillations will become larger. When the applied force equals the natural frequency, the oscillator will be in resonance with the acting force.

6.7. Key Concepts and Practice Problems

- Hooke's Law $F = kx$: deformation is proportional to deforming force.
- Stress is a force that causes deformation; strain is the resulting change.
- The ratio of stress to strain is the elastic modulus or Young's modulus.
- SHM occurs because a restoring force returns a system to natural equilibrium.
- In SHM: $x(t) = A \cos \omega t$; $v(t) = -A\omega \sin \omega t = -v_{max} \sin \omega t$; $a(t) = -A\omega^2 \cos \omega t = -(kA/m) \cos \omega t$; and $\omega = 2\pi/\tau = [k/m]^{\frac{1}{2}}$, $\tau = 2\pi/\omega = 2\pi[m/k]^{\frac{1}{2}}$, $f = \omega/2\pi = (1/2\pi)[k/m]^{\frac{1}{2}}$.
- Energy of a simple harmonic oscillator translates between PE and KE.
- $W = PE = (1/2)kx^2$; $(1/2)kx^2 = (1/2)mv^2$; $(1/2)kA^2 = (1/2)mv_{max}^2$; $v_{max} = A[k/m]^{\frac{1}{2}}$.
- A simple pendulum approximates SHM for small angular amplitude, and $F \approx -mg\theta \approx -mgx/L$; $k = mg/L$; $\tau = 2\pi[L/g]^{\frac{1}{2}}$; $\omega = [k/m]^{\frac{1}{2}} = [g/L]^{\frac{1}{2}}$.
- In simple harmonic motion—amplitude remains constant.
- In damped harmonic motion—amplitude decreases (frictional loss).
- In resonance—amplitude increases.

Practice Problems

6.1 If your truck has rear springs with a combined spring constant $k = 50,000$ N/m, what mass can you load over the axle without lowering the rear more than 10 cm?

6.2 You submerge a volleyball with a volume of 5,270 cm^3 into a column of water to a depth of 10 m. It displaces a volume of water equal to only 4,640 cm^3. You look up water pressure at a depth of 10 m and find it is 2 atm. What is the bulk modulus B of the volleyball?

6.3 (a) A puck with a mass of 1 kg is attached to the free end of a perfectly elastic horizontal spring affixed to a frictionless surface. You push the puck 1 m, compressing the spring, and then release the puck. You observe that it takes 1 s to shoot out 2 m and return to the release point. What is the spring's constant k? **(b)** What is the puck's maximum velocity and acceleration? **(c)** What will the period τ be if the experiment is repeated with a 2-kg puck? **(d)** With a 0.5-kg puck?

6.4 (a) In the example in Subsection 6.4, you notice that after 50 oscillations, your amplitude has decreased from 3 cm to 2 cm due to energy losses to the environment. How much energy has been lost? **(b)** What has happened to your frequency of oscillation? **(c)** What is your new maximum velocity?

6.5 You see a big branch overhanging a pond. You shimmy up the tall tree and out the branch and attach a 100-ft rope that doesn't quite reach down to the water. You climb back down the tree, retrieve the free end of the rope, bring it up the bank, and stretch it taught. Then like Tarzan you grab onto the rope so your center of gravity is 102 ft from the end tied to the tree branch and launch yourself over the pond. How long does it take to swing the maximum distance out over the water? (Assume small θ.)

6.6 How is a high-pitched screech when a microphone is placed in front of a speaker in a PA system an example of resonance?

Answers to Chapter 6 Problems

6.1 $F = mg = kx$, so $m = kx/g = (50{,}000)(0.10)/9.8 = 510$ kg.

6.2 $B = -\Delta P/(\Delta V/V) = -2/((4{,}640 - 5{,}270)/5{,}270) = 16.73$ N/m^2.

6.3 (a) $\tau = 2\pi[m/k]^{\frac{1}{2}}$, so $k = m(2\pi)^2/\tau^2 = 1(2\pi)^2/1^2 \approx 39.48$ N/m.
(b) $v_{max} = A\omega = (1)(2\pi) \approx 6.28$ m/s; $a_{max} = kA/m = (39.48)(1)/1 = 39.48$ m/s^2, (also $a_{max} = A\omega^2 = (1)(2\pi)^2 = 39.48$).
(c) $\tau = 2\pi(m/k)^{\frac{1}{2}} = 2\pi[2/39.48]^{\frac{1}{2}} \approx 1.414$ s, or greater by $2^{\frac{1}{2}}$.
(d) $\tau = 2\pi[0.5/39.48]^{\frac{1}{2}} \approx 0.707$ or less by the square root of 1/2.

6.4 (a) $E_{lost} = (1/2)kx_i^2 - (1/2)kx_f^2 = (1/2)(14{,}000)(0.03^2 - 0.02^2) = 3.5$ J.
(b) Frequency is the same; it does not vary with amplitude.
(c) $v_{max} = A\omega = A[k/m]^{\frac{1}{2}} = 0.02[14{,}000/50]^{\frac{1}{2}} \approx 0.33$ m/s.

6.5 $\tau = 2\pi[L/g]^{\frac{1}{2}} = 2\pi[102/32]^{\frac{1}{2}} \approx 11.2$ s. Since reaching max distance out takes one-half of your pendulum's period, you reach max distance after 5.6 s.

6.6 If the amplified sound wave coming from the speaker has a greater amplitude (louder) than the sound initially entering the microphone, a positive feedback loop is created where a louder sound begets an ever louder sound until the volume (amplification) is turned down or the microphone is moved away from the speaker.

Chapter 7

FLUIDS

"It is not certain that everything is uncertain."

"Nature is an infinite sphere of which the center is everywhere and the circumference nowhere."

Both attributed to Blaise Pascal

• Except when a solid is deformed, its atoms and molecules remain mostly fixed in place by strong intermolecular forces. The intermolecular forces in a fluid, however, are weaker and its molecules do not remain fixed, but are able to move more freely. A **fluid** has no rigid structure and no fixed shape or form, and has the ability to flow. Both *liquids* and *gases* are considered to be fluids, but have some important differences. In **liquids** the binding forces between molecules are strong enough that the molecules tend to sustain fixed distances from each other so that the liquid possesses a specific volume while allowing molecules to still move past one another. We see this when placing an amount of liquid into a container and it fills to a certain level. In a **gas** the forces between molecules are so weak that molecules can move much more independently. Because of this gases do not have a definite volume, so that when an amount of gas is placed into a container it will expand to uniformly fill the container. Also, because gas molecules are farther apart and do not have a fixed spacing, they can be compressed more easily than liquids.

7.1. Pressure and Static Fluids

• **Pressure** is force per unit area.

$$P = F/A$$

Force per area measures pressures in solids and fluids. For example, a 1-mm diameter nail tip driven into a solid surface with 10 N of force exerts a pressure: $P = F/A = F/\pi r^2 = (10\,\text{N})/\pi(0.0005\,\text{m})^2 \approx 1.3 \times 10^7\,\text{N/m}^2$.

• There are many **units** for measuring **pressure** including N/m^2 and Pascal (Pa), where $1\,\text{N/m}^2 = 1\,\text{Pa}$. Other units include atmospheres (atm), where: $1\,\text{atm} = 1.013 \times 10^5\,\text{N/m}^2 = 1.013\,\text{bar} = 1.013 \times 10^6\,\text{dyn/cm}^2 = 14.7\,\text{lb/in}^2 = 760\,\text{torr} = 760\,\text{mm Hg} = 1.013 \times 10^5\,\text{Pa}$.

• **Fluids**, including liquids and gases, exert **forces**. A fluid that fills a container exerts forces that are distributed over the container's surface where the fluid contacts that container. (As we will see, the pressure increases with depth.) A fluid has no rigidity so that when a fluid is static or at rest it cannot exert a force *parallel* to a surface. The *force a static fluid exerts* on the surface of a container is always *perpendicular to the surface*. In the figure of two completely filled containers the arrows show force perpendicular to the surface and increasing with depth.

Atmospheric Pressure

• **Atmospheric pressure** is pressure in the surrounding air near the Earth's surface. The atmospheric pressure varies with temperature and altitude above sea level. The **atmosphere** is a layer of air around Earth which is held down by gravity. Because gas is easily compressed, approximately 90% of the air in our atmosphere is concentrated within about 16 km (or 9.9 mi or just over 52,000 ft) of the surface. The air around Earth has weight and exerts a force on the surface. At sea level this force is 1.013×10^5 N per square meter and is defined as 1 atmosphere or just 1 atm. This means that under normal conditions, the *pressure the atmosphere exerts on the surface* is:

$$1\,\text{atm} = 1.013 \times 10^5\,\text{N/m}^2 = 1.013 \times 10^5\,\text{Pa}$$

• **Example**: Your friend asks you to estimate the mass of the air above Earth's surface.

If you assume the force of gravity remains at 9.8 m/s^2 and weight = F = mg, then *per square meter*, the mass of the atmosphere is: m = F/g = (10^5 N)/(9.8 m/s^2) ≈ 10^4 kg for each square meter of surface. "Wow! That's really heavy," your friend exclaims. "How come we don't get crushed?" You think for a moment. "I guess because our tissues must have the same pressure pushing back."

Pascal's Principle

• In the seventeenth century Blaise Pascal noted that the pressure at any depth of a container or body of water is the pressure of the weight of the water above plus the atmospheric pressure acting on the water's surface. In addition, the pressure at any point at any depth acts equally in all directions. **Pascal's principle** goes further to state that: *a pressure applied to the surface of a confined fluid is transmitted undiminished by the fluid in all directions and to all points within the fluid.* This principle is used in the development of hydraulic systems which use fluids to transmit pressure and force.

• The container below has two openings and is filled with liquid so that the levels are equal. If two negligible mass pistons are placed on the fluid in the two openings and a force F_1 is applied to the small piston having area A_1, the pressure exerted on the fluid in this opening and therefore the pressure exerted by the fluid on the piston will be $P = F_1/A_1$. Providing no other external forces are applied, the application of the force F_1 will cause the other piston to rise. To hold both pistons at their original level, a force F_2 must also be applied to the second piston so that the pressure is the same: $P = F_1/A_1 = F_2/A_2$. In this equation, since area A_2 is larger than A_1, F_2 must also be correspondingly larger than F_1. Therefore, when the fluid in this system is static, the pressure exerted by the fluid on each piston is the same and $P = F_1/A_1 = F_2/A_2$.

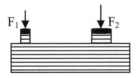

If we apply a pressure to the smaller piston $P = F_1/A_1$, it will be transmitted to all the inner surfaces of the container including the larger piston. Each point in the container will experience a *change* in pressure by $P = F_1/A_1$. This is described by **Pascal's principle** which we stated above: a pressure applied to the surface of a confined fluid is transmitted undiminished by the fluid in all directions and to all points within the fluid. Because of Pascal's principle, a downward force F_1 on the small piston is converted into a larger *upward* force F_2 on the larger piston. The force is multiplied by the ratio of the areas of the pistons but acts over a displacement that is decreased by the same ratio.

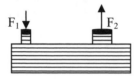

If the small piston is pushed down, the pressure exerted on all sides of the container is equal to the applied force divided by the area of the small piston. The resulting force on the large piston equals the pressure times the area of the large piston (since $F = PA$), so that the force exerted on the small piston has been multiplied by the ratio of the piston areas:

$$F_1/A_1 = F_2/A_2 \quad \text{or} \quad F_1 = F_2 A_1/A_2$$

These types of hydraulic systems allow a relatively small force to create a large force.

• **Example**: Suppose you want to build a platform that will lift you 2 m. If your mass is 50 kg, the piston's radius is 20 cm, and the diameter of the round lift is 100 cm, what is the force on the piston needed, and how far must the piston be displaced?

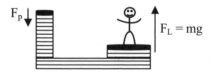

The pressure is the same throughout: $P = F_p/A_p = F_L/A_L$
The force on the piston needed is F_p:

$$F_p = F_L A_p/A_L = (50 \text{ kg})(9.8 \text{ m/s}^2)(\pi)(0.2 \text{ m})^2 / (\pi)(0.5 \text{ m})^2 = 78.4 \text{ N}$$

The ratio of the areas of the lift to piston is:

$$A_L/A_p = (\pi)(0.5 \text{ m})^2 / (\pi)(0.2 \text{ m})^2 = 6.25$$

Therefore the ratio of displacements is also 6.25, so the displacement of the piston required is $6.25 \times 2 \text{ m} = 12.5 \text{ m}$.

Density and Specific Gravity

• Average **density** is mass divided by volume.

$$\boxed{\rho = m/V}$$

The density of solids and liquids remains mostly constant except for some slight variations if the material is compressed or deformed or if temperature is changed. In fact, the density of the water at the bottom of the ocean is only slightly greater than the density of the water at the surface. In contrast, the density of gases is susceptible to changes in pressure and temperature, and gases compress easily. The air contained in a vertical tube held at the surface of the Earth will be more compressed (by the weight of the air above it) at the bottom of the tube than at the top. Similarly, the atmosphere of the Earth has a much higher density near the surface than at high altitudes. Because of their sensitivity to temperature and pressure, the density of gases is often stated with reference to particular temperatures and pressures.

• **Units** for density are in mass per volume and include kg/m^3 and g/cm^3, where $1,000 \text{ kg/m}^3 = 1 \text{ g/cm}^3$. The **density of water** is approximately $1 \text{ g/cm}^3 = 1,000 \text{ kg/m}^3$.

• Density can be used to identify an unknown substance. For example, if an unknown liquid has a measured mass of 18.5 g (or 18.5×10^{-3} kg) while taking up a volume of 23.4 milliliters (or $23.4 \times 10^{-6} \text{ m}^3$), its density is:

$$\rho = (18.5 \times 10^{-3} \text{ kg}) / (23.4 \times 10^{-6} \text{ m}^3) = 791 \text{ kg/m}^3$$

After determining this value you can search for substances with this same density. You can perform other measurements and tests to confidently identify the substance in question.

• The **specific gravity** is *the ratio of the density of a substance to the density of water* (at a specified temperature).

$$\text{Specific gravity } SG = \rho_{\text{substance}} / \rho_{\text{H2O}}$$

Since the *density of water* in CGS (centimeter-gram-seconds) units is 1 g/cm^3, the *specific gravity* of a substance in g/cm^3 units works out to be the same as the value of the density. In MKS (meter-kilogram-seconds) units the *density of water* is $1,000 \text{ kg/m}^3$. A reference temperature for the density of water that is often used is 4 °C. An object with a specific gravity less than 1 g/cm^3 will float in water. If its specific gravity is greater than 1 g/cm^3, it will sink in water.

• **Example**: If the density of silver is $10,490 \text{ kg/m}^3$, or 10.49 g/cm^3, what is the specific gravity of silver?

$$SG = \rho_{\text{substance}} / \rho_{\text{H2O}} = (10,490 \text{ kg/m}^3) / (1,000 \text{ kg/m}^3) = 10.49$$

Variation of Pressure With Depth

• The **pressure in a fluid** at any depth in a container or body of water is the pressure due to the weight of the water above plus the atmospheric pressure acting on the water's surface. In certain cases the atmospheric pressure may be negligible compared to the fluid's weight.

If you have a container of fluid filled to a height h, the fluid will exert a force equal to its weight mg on the bottom of the container. The pressure exerted by a liquid because of its weight (ignoring overhead air pressure) can be determined if you know its density and volume. For example, the pressure at the bottom of a tank with vertical sides having *length* l, *width* w, and *height* h is the weight of the water in (above the bottom of) the tank.

⊢ h $F = mg$
$P = F/A = mg/A = (\rho lwh)g/lw = \rho gh$

The volume of water is area times height, or: $V = Ah = lwh$.
Since density ρ is mass/volume, mass is: $m = \rho V = \rho Ah = \rho lwh$.
The weight or force of the water is: $F = mg = \rho Vg = \rho Ahg = \rho lwhg$.
The **pressure** is: force/area $= P = F/A = \rho Ahg/A = \rho glwh/lw$.
Since area cancels, **pressure exerted by a liquid** is:

$$P = \rho gh$$

Therefore the **fluid pressure** *at any depth* **of h with density ρ is ρgh.** This equation can be thought of as including a column of liquid h high with a cross-sectional area of $l \times w = $ one $\times$ one.

• **Example**: The Mariana Trench is the deepest area in Earth's oceans. The deepest point within the trench based on a number of measurements is about 11,000 m. What is an estimate of the pressure?

Let's make our estimate assuming the water is not compressed and has a density of 1,000 kg/m³. Also assume the atmospheric pressure of the overhead air is negligible. We can calculate the pressure using:
$P = \rho gh = (1{,}000 \text{ kg/m}^3)(9.8 \text{ m/s}^2)(11{,}000 \text{ m}) = 1.08 \times 10^8 \text{ Pa}$
The pressure is therefore about 1.1×10^8 Pa (or 16,000 psi).

Measuring Air Pressure

• **Air pressure** is measured using different methods and under differing circumstances. Two illustrative methods include a mercury barometer and an aneroid barometer.

• In a **mercury barometer** a tube containing mercury (Hg) is inserted upside-down in an open reservoir of mercury so that the pressure from the weight of the mercury in the tube is pressing on the mercury trapped in the tube, and the weight of air is pushing on the mercury in the open reservoir. Note that the top end of the inserted tube develops a vacuum, so only the weight of mercury in the tube is pressing down. In a mercury barometer the height of mercury in the evacuated tube is determined by the atmospheric pressure acting on the mercury in the reservoir. At normal sea-level atmospheric pressure of 1 atmosphere, the height of a tube of mercury is: 760 mm Hg = 1 atm = 1.013×10^5 Pa.

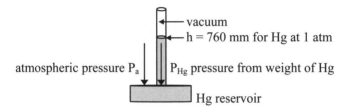

atmospheric pressure P_a

vacuum

h = 760 mm for Hg at 1 atm

P_{Hg} pressure from weight of Hg

Hg reservoir

Since the pressure in a fluid is $P = \rho gh$ and the **density of mercury** is 1.36×10^4 kg/m^3, we see that the height in the tube at atmospheric pressure is h = $P/\rho g$, or:

$$h = (1.013 \times 10^5 \text{ Pa}) / (1.36 \times 10^4 \text{ kg/m}^3)(9.8 \text{ m/s}^2) \approx 0.76 \text{ m or } 760 \text{ mm}$$

• Another type of barometer is called an **aneroid barometer**, which has an evacuated chamber with flexible walls that deflect as ambient pressure changes. The deflection is linked with a pointer which moves across an indicating scale. This type of barometer is portable and is used in portable instruments and aircraft altimeters.

Gauge Pressure

• Mechanical **pressure gauges** generally measure the difference in pressure between some volume the gauge is connected to and the ambient atmosphere. The Standard Atmospheric Pressure is defined at sea level and is 1 atm = 1.013×10^5 Pa. In English units the Standard Atmospheric Pressure is 14.696 psi or about 14.7 lb/in^2. The **gauge pressure** is the pressure measured by a gauge and reflects the pressure difference between some system or volume and the surrounding atmosphere. The **gauge pressure** is:

$$p_g = p_s - p_a$$

with p_g, p_s, and p_a being *gauge*, *system*, and *atmospheric* pressures.

7.2. Archimedes' Principle and the Buoyant Force in Static Fluids

• **Archimedes' principle** states that *the buoyant force on a submersed object is equal to the weight of the fluid it displaces.* To understand this, imagine a piece of driftwood floating in the water. There is a buoyant force acting on the wood that holds, or buoys, it up. But how does the buoyant force work? First, remember that due to the weight of the liquid above, the pressure in a body of liquid increases with depth. Also, remember that the pressure at any depth acts equally in all directions. Because of the increasing pressure "below" and the fact that pressure "pushes" equally in all directions, we can imagine there would be an upward-acting pressure on any submerged object which would counteract the weight of the object.

Going further, the buoyant force F_B on any submersed object reflects the difference between the force due to pressure acting downward on the upper surface (which is at lower pressure) and the upward force due to pressure acting on the lower surface (which is at higher pressure). This net upward force caused by the pressure difference between the upper and lower ends of a submerged object is equal to the *weight of the volume of liquid* that the object takes up or displaces. **Archimedes** (in the third century B.C.) discovered this **buoyant force** on submersed objects, finding it to be equal to the *weight of the fluid displaced.*

• We can develop an equation for the **buoyant force** on a submersed object. First, remember that pressure is: $P = F/A$.
In a fluid, pressure is: $P = \rho g h$.
It follows that force is pressure times area: $F = PA = \rho g h A$.
Also, in a static fluid, the horizontal forces acting on the object cancel each other so we only need to consider the vertical forces. To develop an equation, let's use a cylinder submerged in a fluid having density ρ_f.

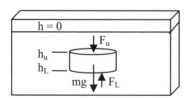

The downward pressure due to the fluid acting on the upper surface at depth h_u is:

$$P_u = \rho_f g h_u$$

The downward fluid force acting on the upper surface at depth h_u is:

$$F_u = (\rho_f g h_u)A$$

The upward pressure due to the fluid acting on the bottom surface at depth h_L is:

$$P_L = \rho_f g h_L$$

The upward fluid force acting on the lower surface at the depth h_L is:

$$F_L = (\rho_f g h_L)A$$

The buoyant force F_B is the difference between the forces due to pressure acting on the upper and lower surfaces. With A representing the area of the upper and lower surfaces, buoyant force is:

$$F_B = P_L A - P_u A = \rho_f g(h_L - h_u)A$$

Since the volume of the cylinder is $V_c = (h_L - h_u)A$, we can write:

$$\boxed{F_B = \rho_f V g}$$

which describes the **buoyant force** of a cylinder or similar object.

• It is also useful to understand the *net force* F_{net} acting on the cylinder: (The downward direction is negative.)

$$F_{net} = -mg - F_u + F_L = -mg + (F_L - F_u) = -mg + \rho_f g(h_L - h_u)A$$

Since the volume of the cylinder is $V_c = (h_L - h_u)A$ and the mass of the fluid displaced by the cylinder is $m_f = \rho_f V$:

$$F_{net} = -mg + \rho_f g(h_L - h_u)A = -mg + \rho_f g V_c = -mg + m_f g = -mg + w_f$$

where w_f is the weight of the fluid displaced by the cylinder. This gives the **net force** acting on the cylinder as:

$$\boxed{F_{net} = -mg + w_f}$$

where w_f is the **buoyant force**, which is the amount that the downward force due to gravity mg is reduced. Since the *buoyant force is the weight of fluid displaced by the object*, net force can be written:

$$\boxed{F_{net} = -mg + F_B}$$

where

$$F_B = w_f = \rho_f V g$$

• The net force equation,

$$F_{net} = -mg + w_f$$

reveals that if the buoyant force w_f exceeds the weight of the object mg, then F_{net} is positive or *upward* and the object will move upward in the fluid. This occurs when the density ρ of the submersed object is less than the fluid density ρ_f. If the top of the object begins to rise above the surface, the volume of liquid displaced by the object will decrease until the buoyant force decreases to a value that equals mg. When mg = w_f, the net force is zero, $F_{net} = 0$, and the object will remain partially submerged and floating.

An equilibrium is reached when the buoyant force is equal to the weight of the object—this is the level where the object floats. We see this with floating ice. Since the density of ice is about 0.9 g/cm^3, which is slightly less than the density of liquid water, 1 g/cm^3, only about ten percent of the volume of an iceberg protrudes above the surface as it floats. When an object is partially submerged and floating, the buoyant force is equal to the weight of the volume of liquid displaced by the *submerged portion* of the object. As a floating object sinks down in a fluid, the buoyant force becomes greater than its weight, and will push the object upward. If a floating object rises above the equilibrium position, the buoyant force becomes less than the weight and the object will again sink to a lower level. Eventually an equilibrium level will be reached.

• **Example**: Humans have a density that is about the same as water—with muscle tissue being slightly more dense and fat tissue being slightly less dense. Suppose your mass is 50 kg and you are very muscular without much excess fat on your body. Since a muscular person has a density of just slightly more than that of water, you can estimate your density to be about 1.03 g/cm^3, providing you have not taken a deep breath (which increases your volume more than your weight). What would be your effective weight in water?

Since your density ρ_y is about 1,030 kg/m^3, your volume is about:
$$V = m / \rho_r = (50 \text{ kg}) / (1{,}030 \text{ kg/m}^3) \approx 4.85 \times 10^{-2} \text{ m}^3$$

Since the density ρ_w of water is about 1,000 kg/m^3, the buoyant force is:
$$F_B = w_f = \rho_f V g = (1{,}000 \text{ kg/m}^3)(4.85 \times 10^{-2} \text{ m}^3)(9.8 \text{ m/s}^2) \approx 475 \text{ N}$$

Your effective weight w_y in water is your weight in air minus the buoyant force:
$$w_w = w_a - F_B = (50 \text{ kg})(9.8 \text{ m/s}^2) - (475 \text{ N}) = 490 \text{ N} - 475 \text{ N} = 15 \text{ N}$$

which is about 3.37 lb.

Finding the Density of an Irregularly Shaped Object

• We developed **Archimedes' principle** using a cylindrical object since we could easily calculate the pressure difference between its upper and lower surfaces. While we can develop equations using other geometric shapes, there is a straightforward method that shows Archimedes' principle works for objects other than cylinders. Consider an **irregularly shaped object** with volume V that is submerged in a static fluid with no forces acting on it other than its own weight and the buoyant force. Imagine an equal size and shape blob or volume of fluid next to it at rest at the same depth. Since the object and the equivalent blob of liquid are at the same depth and are in equilibrium, the weight and the buoyant force must be equal. This is stated by Archimedes' principle—that the buoyant force on the submerged object must be equal to the weight of an equal volume of fluid. This principle helps us determine the density of an object.

We learned earlier in this chapter that the average **density** ρ of any object is mass divided by volume: $\rho = m/V$. If you know the mass and volume, you can quickly calculate density. *But what if you need to determine the density of an object that has an irregular shape?* It can be a challenge to determine the volume of an irregularly shaped object. It turns out that we can use Archimedes' principle to find the density of an irregular object without needing to determine its volume. To do this we measure the weight of the object in air w_a (ignoring negligible buoyancy effects of air). We also measure the object's immersed weight w_i while it is submerged in a fluid such as water having known density ρ_f.

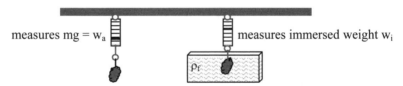

measures $mg = w_a$ measures immersed weight w_i

ρ_f

We have learned that according to Archimedes' principle the **buoyant force**, $F_B = w_f = \rho_f V g$, equals the amount that the downward force due to gravity (object's weight mg) is reduced when it is submerged. The buoyant force is also the weight of fluid displaced by the object. It follows that an object's immersed weight w_i must be equal to its weight in air, $w_a = mg$, minus the buoyant force, $F_B = w_f$, pushing upward.

$$w_i = w_a - w_f$$

For any object having a volume V, the weight w_f of the fluid the object displaces is:

$$w_f = \rho_f g V$$

Therefore, $w_i = w_a - w_f$ becomes $w_i = w_a - \rho_f gV$, or:

$$\rho_f gV = w_a - w_i$$

$$\boxed{V = (w_a - w_i) / (\rho_f g)}$$

which is the **volume of the object** in terms of its weight in air, weight in fluid, gravity, and fluid density. Since the density ρ of the object is:

$$\rho = m/V = (w_a/g)/V$$

Substituting for V gives:

$$\rho = (w_a/g) / [(w_a - w_i) / (\rho_f g)]$$

Rearranging gives:

$$\boxed{\rho = (w_a)(\rho_f) / [(w_a - w_i)]}$$

which is the **density of the object in terms** of its weight in air, its weight in fluid, and the density of that fluid. We do not need to directly know the mass or volume of the object!

• **Example**: Imagine a friend comes to you with a bracelet her boyfriend has just given her. She said the boyfriend claimed it was "pure gold," but she is suspicious and said that "it does not seem to be pure gold and undoubtedly contains some silver." She had heard the story of how Archimedes' had tested a crown to see if it was pure gold and asked if you could help her determine if the bracelet was really pure gold.

You weigh the bracelet at 1.06 oz, which converts to about 0.03 kg. You look up the density of gold as 19,300 kg/m³. If the bracelet was pure gold it would have a volume of:

$$V_g = m_g/\rho_g = 0.03 \text{ kg} / 19,300 \text{ kg/m}^3 = 1.6 \times 10^{-6} \text{ m}^3$$

To determine if the bracelet is pure gold, you weigh it in air:

$$w_a = mg = 0.03 \text{ kg} \times 9.8 \text{ m/s}^2 = 0.294 \text{ N}$$

Then you weigh it while it is immersed in water: $w_i = 0.274$ N.
The bracelet's density is $\rho_b = (\rho_{water})(w_a) / [(w_a - w_i)]$:

$$\rho_b = (1,000 \text{ kg/m}^3)(0.294 \text{ N}) / (0.294 \text{ N} - 0.274 \text{ N}) = 14,700 \text{ kg/m}^3$$

The bracelet's measured volume is $V_b = (w_a - w_i)/(\rho_{water}g)$:

$$V_b = (0.294 \text{ N} - 0.274 \text{ N}) / [(1,000 \text{ kg/m}^3)(9.8 \text{ m/s}^2)] = 2.0 \times 10^{-6} \text{ m}^3$$

"Wow!" your friend says. "The bracelet's measured volume is definitely greater than it would be if it were pure gold, which would have a volume of $V_g = 1.6 \times 10^{-6}$ m³. It must have silver or some other alloy in it." Her face became serious. "Do you think I should tell him the truth about the bracelet? What would I say?" You thought for a moment. "Just tell him gold's density is 19,300 kg/m³ and silver's is 10,490 kg/m³, so gold has a greater density that silver, and therefore for the same weight, its takes up less volume."

She then asked, "So how did Archimedes figure out the crown's purity?" "Ok, here's the story as I've heard it," you respond. "Archimedes made two masses—one gold and one silver—which were the same weight as the crown. He then filled a container to the very top with water and put the mass of silver into it, noting that the volume of water that overflowed was equal to that of the mass of silver. In the same way Archimedes filled the container again and put in the mass of gold. He noted that less water overflowed with the gold. This is because gold has a greater density than silver, and therefore gold takes up less volume for the same weight. He again refilled the container and put in the crown." "I bet more water overflowed for the crown than for the equal-weight mass of gold," your friend surmised. "You've guessed right," you said. "Archimedes figured the gold had been blended with silver."

• **Example**: You and your friend decide to go on another mountain climbing adventure. You really want to go but are feeling a bit tired. You notice the tank of helium in the corner of your lab and have a brainstorm. What if you attach a large helium balloon to yourself to reduce your effective weight? Even a "lift" that would reduce your weight by 20% would feel good! If your mass is 60 kg, how big of a helium balloon would you need?

Just like with liquid, the buoyant force is the weight of air displaced:

$$F_B = w_f = \rho_f V g$$

You estimate the mountain temperature and look up the density of air as about 1.22 kg/m^3. You want a buoyant force of 20% of your weight:

$$(0.2)(60 \text{ kg})(9.8 \text{ m/s}^2) = 117.6 \text{ N}$$

You then need to determine the volume of air that weighs 117.6 N. Since the density of air is 1.22 kg/m^3, volume is:

$$V = w_f / \rho_f g = (117.6 \text{ N}) / (1.22 \text{ kg/m}^3)(9.8 \text{ m/s}^2) \approx 9.84 \text{ m}^3$$

The volume of a sphere is $V = (4/3)\pi r^3$, so the radius would be:

$$r^3 = 3V/4\pi = (3)(9.84 \text{ m}^3)/(4\pi) \approx 2.35 \quad \text{or} \quad r = [2.35]^{\frac{1}{3}} \approx 1.33 \text{ m}$$

"There you have it!" you exclaim. "My helium balloon needs to have a radius of about 1.33 m."

"Wow, that's a diameter of about 8.7 ft! What about the weight of the helium balloon?" your friend says.

"Good point," you respond. You estimate the density of helium in a balloon at Standard Temperature and Pressure (20 °C, 760 mm Hg) as: $\rho_h = 0.18$ g/L = 0.18 kg/m^3. The balloon's weight is:

$$\rho_f V g = (0.18 \text{ kg/m}^3)(9.84 \text{ m}^3)(9.8 \text{ m/s}^2) \approx 17 \text{ N}$$

"Let's add a bit for the Mylar balloon itself and estimate an extra 20 N," you suggest. "This adds to my 588 N weight. So even accounting for a few percent extra weight for the balloon itself, I will still get a decent lift." "Yes, and that huge balloon will also scare wild animals away!" your friend teases.

7.3. Surface Tension and Capillary Action

Cohesive Forces and Surface Tension

• There are intermolecular **cohesive forces** that hold together molecules of the same type in a solid or liquid. These cohesive forces are attractive forces that exist between nearby molecules of the same type. It is these attractive cohesive forces between molecules in a liquid that keep it in a condensed state and create the phenomenon called **surface tension**.

While the cohesive forces between molecules within a liquid are shared with molecules all around them, the molecules *on* the surface have no liquid molecules above the surface and consequently form stronger cohesive forces with neighboring molecules *along the surface*. This forms a "skin" of like molecules along the surface that is a bit stronger and more impenetrable than other sections within of the fluid below the surface. This property of stronger intermolecular forces along the surface of a fluid is called **surface tension**.

  Cohesive forces between molecules shown by the small arrows.

Additionally, because there are no liquid molecules above the surface to exert attractive forces that counteract the attractive forces between molecules below the surface, molecules on the surface experience a slight net downward force. This downward force creates a slight compression of the surface that minimizes the surface area of a liquid.

Cohesive forces and surface tension are on display when you observe a droplet of water. A *water droplet* is held together generally by cohesive forces and has an especially strong layer of cohesive forces along its surface creating surface tension. Another demonstration of surface tension occurs when a liquid surface supports small objects that are denser than the fluid. We observe this when certain insects walk across a pond or when we place a needle on a water surface and it "floats".

• Because *energy is required to stretch the surface area*, the surface has some amount of potential energy. This **potential energy** is referred to as **surface energy**. Surface energy increases as the surface area is increased. The *potential energy per unit of area* of the surface is defined as the **surface tension** or the **coefficient of surface tension**.

$$\text{coefficient of surface tension} = \sigma = PE/A$$

Consistent with this equation, the coefficient of **surface tension** has **units** in the MKS system as J/m^2. The coefficient of surface tension is not only a measure of potential energy per unit area, but because work can represent the change in energy, it is also the amount of work needed to increase the surface area by some amount. Suppose you have a *rectangle of thin wire*, and you create a thin layer of liquid across it by dipping it in soapy water. If the size of your wire rectangle can be increased by sliding out one side, its area can then be increased. This will enable you to measure not only the increase in surface area but also the **work** performed, (which equals the increase in potential energy of the liquid surface). The **coefficient of surface tension** is then given by:

$$\text{coefficient of surface tension} = \sigma = PE/A = W/\Delta A$$

where W is the **work** performed, and ΔA is the increase in **area**.

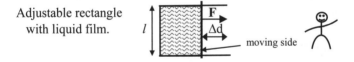

Adjustable rectangle with liquid film. l F Δd moving side

In this model the work required to increase the surface area of the water film is equal to the force times the displacement of the adjustable side. The **surface tension** σ is described in terms of force per unit length:

$$\text{coefficient of surface tension } \sigma = W/\Delta A = F\Delta d/l\Delta d = F/l$$

where force F is required to overcome the surface tension, and the increase in area is equal to the width l of the rectangle times Δd displacement.

• **Example**: Your friend points out that there are actually two surfaces of the soapy water film in the figure above (front and back) and suggests the force F should be multiplied by 2. Is the previous equation for σ correct?

Yes there are actually two surfaces of the soapy water film in the figure above. You can consider the force needed to pull the adjustable side of the rectangle as **2F**, one **F** on each surface. Therefore, as you move the wire by Δd, the work done is W = 2FΔd. The increase in surface area of the film would then be $2l\Delta d$ and the work per unit area and the surface tension come out the same: $\sigma = W/\Delta A = 2F\Delta d/2l\Delta d = F/l$ is still correct.

• We can visualize the fact that increasing the surface area of a liquid requires work by imagining dipping a glass tube into soapy water and allowing a film of the liquid to spread across one end of the tube. If you were to gently blow on the other end of the tube, you would *increase the surface area* of the liquid film by creating a protruding soapy bubble. By blowing you have exerted a force on the film and displaced it a certain distance. You have done **work** against the cohesive forces in the film in order to increase the surface area. The bubble you created now possesses some amount of **potential energy**. When you stop blowing, the film will return to its original state of minimum potential energy.

• *Surface tension* is measured in **units** of dynes/cm, as the force in dynes needed to break a surface skin of length 1 cm. For example, if the surface tension of water is 72.8 dynes/cm at 20 °C, it would take a force of 72.8 dynes to break a surface film of water 1 cm long. Surface tension is also described as a surface energy in units of ergs per square centimeter. Examples of surface tension values include: 72.8 dynes/cm for water at 20 °C; 22.3 dynes/cm for ethyl alcohol; and 465 dynes/cm for mercury. In MKS units surface tension is in $J/m^2 = N/m$. The value for the surface tension of a liquid depends on the type of gas the liquid contacts.

• **Example**: Why are water droplets nearly spherical? Why are falling raindrops not perfectly spherical?

The surface of a water droplet is in a state of minimum potential energy, which is reflected in it being in a state of minimum surface area. The surface area will be smallest when it takes the shape of a sphere. Tiny raindrops mostly retain their spherical shape, but as they grow in size, air pressure, air resistance, and other external forces cause them to flatten and change shape. (They are not teardrop shaped.)

• **Example**: Why does hot water clean better than cold water? Why does adding detergents and soaps clean even better than just hot water?

The *surface tension of water*, which is due to the polarity of water molecules, decreases with increasing temperature. (Polar molecules possess an uneven distribution in the density of their electrons allowing electrostatic attraction between the more negative region of one polar molecule and the more positive region of another.) Because the surface tension decreases with rising temperature, hot water can permeate smaller cracks and holes in a surface and acts as a better cleaning agent than cold water. Adding detergents and soaps further reduces water's surface tension for even better penetration.

Capillary Action or Capillarity

• Have you ever seen a liquid in a small tube climb up the walls so that its surface forms a **meniscus**? This is called **capillary action** or **capillarity** and is a result of surface tension and the adhesive forces between the liquid and the tube. We learned there are **cohesive forces** between *like* molecules in a liquid that hold the molecules in the liquid together and create the surface tension. There are also attractive forces between *unlike* molecules called **adhesive forces.**

Adhesive forces can exist between molecules in a liquid and the molecules in a container holding the liquid that attract the liquid molecules to the container walls. The adhesive forces between water and the walls of a tube cause an upward force on the water at the water-tube interface, which result in an upward bending meniscus. The surface tension holds the water surface together as the water at the water-tube interface is reaching upward onto the walls of the tube.

In the presence of these adhesive forces, the surface of a liquid in a tube can either curve upward or downward at the wall depending on whether the *adhesive* forces between the liquid molecules and the wall are greater or less than the *cohesive* forces between the liquid molecules.

Adhesive forces dominate.   Cohesive forces dominate.

When the adhesive forces between the liquid and the wall are greater than the cohesive forces between liquid molecules, the liquid will curve upward forming a **meniscus**. This is what we see with water in a glass tube. An upward meniscus will rise until the upward force due to adhesion equals the weight of the upward sloping liquid.

• If you insert one end of a tiny open-ended glass tube into a cup of water, the water will rise to a certain height in the tube. The upper surface of the water in the tube will have a concave shape that is lower in the center and higher along the edges where it touches the tube's wall. The water in the tiny tube will climb up the wall until the upward adhesive force due to the surface tension is balanced by the weight of the raised water. The rise of a liquid in a tube is called capillary action or capillarity. The up and down movement of the water in the tube depends on which is stronger: the cohesive forces between water molecules or the adhesive forces between the water molecules and the molecules in the glass tube. The upper surface of the water meets the glass wall at a certain *angle* θ shown in the figure.

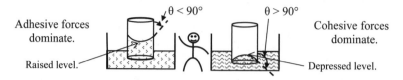

When the adhesive forces dominate and a liquid adheres to a tube and rises, the angle θ that the fluid makes with the tube wall will be less than 90°. The greater its adherence to the tube, the smaller angle θ becomes and the fluid *rises*. Conversely, if the cohesive forces between molecules in a liquid are large compared to the adhesive forces between the liquid molecules and wall, the angle θ will be greater than 90°. In this case, the level of the liquid in the tube will *fall below the level of fluid in the container*. This can be seen with mercury in a glass tube.

When adhesive forces are large and a liquid can easily rise in a tube, the liquid is said to **wet** the tube wall. The better a fluid can wet, the smaller the angle θ between the liquid and wall. For example, water and ethyl alcohol, which will completely wet a clean glass tube, have an angle θ of near 0°. Mercury does not wet glass well and has an angle θ near 140°.

• We can develop an *equation for how high* h a liquid will rise (or be depressed) in a tiny tube, or capillary tube, if one end is submerged in a liquid. The force pulling the liquid upward is due to surface tension and depends on the angle θ.

To develop the equation we begin by setting the *upward force* equal to the *downward force*, since height h is where the liquid stops moving and **equilibrium** is reached. The upward force due to the surface tension balances the downward force, which is due to the gravitational force acting on the water column. (For the upward force imagine strong adhesive forces between the tube and the liquid pulling on the surface of the liquid and dragging it upward.) The *downward force* is the gravitational force on the cylindrical water column, which is its weight:

$$\text{downward force} = mg = \rho Vg = \rho(\pi r^2 h)g$$

The *upward force* due to the surface tension at the edge of the meniscus acts along the surface of the liquid. At the liquid-tube interface, the surface tension makes an angle with the tube that is equal to the angle θ between the liquid and the tube wall. Since angle θ is between the direction of the surface tension (parallel to the surface) and the vertical side of the tube, the *upward force* is equal to the upward component of the force per unit length, which is the vertical component of the surface tension (σ cos θ) times the length of the line of contact which is the circumference of the meniscus (2πr). In other words, the *upward force* is equal to the cosine of θ times the surface tension (σ = W/ΔA = FΔd/lΔd = F/l), where *l* is the length of the line of contact which in this case is the circumference of the tube (2πr):

$$\text{upward force} = (2\pi r)\,\sigma\cos\theta$$

Then we set the upward and downward forces equal:

$$2\pi r\sigma\cos\theta = \rho\pi r^2 hg$$

Solve for the **height h a liquid will rise (or be depressed)** in a tiny tube:

$$\boxed{h = 2\sigma\cos\theta\,/\,\rho gr}$$

When the surface tension is strong and lifts the edges of the meniscus up so they are nearly vertical, the angle θ will be very small and cos θ will be nearly 1. In this case, when *angle θ is small*, h becomes:

$$h = 2\sigma\,/\,\rho gr$$

• **Example**: You have two identical glass 0.15-mm-radius tubes in your air-conditioned laboratory. What would be the heights if you stick the lower end of one into a container of water and the other into a container of mercury? Assume angle θ is about 1° for water and 140° for mercury.

Because angle θ for your water tube is small, height h is:

$$h = 2\sigma\cos\theta\,/\,\rho gr \approx 2\sigma\,/\,\rho gr$$

You look up the surface tension for water σ as about 0.073 N/m, and the density ρ as 1,000 kg/m³, then calculate height using $h = 2\sigma\,/\,\rho gr$:

$$h = (2)(0.073\,\text{N/m})/(1{,}000\,\text{kg/m}^3)(9.8\,\text{m/s}^2)(1.5\times10^{-4}\,\text{m}) \approx 0.099\,\text{m} = 9.9\,\text{cm}$$

This is a large distance due to the tiny diameter of the tube.

For mercury angle θ is 140°, so height h is: $h = 2\sigma\cos\theta\,/\,\rho gr$.
You look up the surface tension for mercury σ as about 0.465 N/m, and the density ρ as 1.36×10^4 kg/m³ and calculate height using:

$$h = 2\sigma\cos\theta\,/\,\rho gr$$
$$= (2)(0.465\,\text{N/m})(\cos140°)\,/\,(1.36\times10^4\,\text{kg/m}^3)(9.8\,\text{m/s}^2)(1.5\times10^{-4}\,\text{m})$$
$$\approx -0.0356\,\text{m} = -3.56\,\text{cm}$$

The negative sign indicates that the "height" is actually the amount that the mercury in the tube is below the level in the container.

• Capillary action causes a number of phenomena we observe in our environment such as the wicking action of liquids onto various fabrics and materials and even the movement of groundwater in soil. While water spreads through a cotton fabric via capillary action, some materials such as wool are not easily wet by water and have the property of being more "water resistant."

7.4. Fluid Flow, Continuity Equation, and Bernoulli's Equation

• The study of fluids in motion is a broad and complex subject with many practical uses, including understanding blood flow through veins and arteries, the delivery of water and other fluids through pipes and nozzles, and flows around objects such as aircraft and ships. This section provides an introduction to the smooth (laminar) flow of fluids that does not involve turbulence or high viscosity and require complicated mathematics. Note that **laminar flow** of a fluid is smooth, non-turbulent, streamlined, and is often described as flowing smoothly in parallel layers, without rotating eddies or mixing.

The Equation of Continuity

• When a fluid is in motion, it moves in a way that conserves mass. A basic principle of fluid flow is that the amount of fluid that enters one end of a pipe equals the amount of fluid that exits the other end. Going further, when fluid flows smoothly and in a steady state, continuous manner, the rate of flow (flow rate) into one end of a pipe equals the rate of flow out the other end. This **flow rate** can be defined as the *mass of fluid per second* that passes some point in the pipe.

$$\boxed{\text{flow rate} = m/t = \rho A d/t = \rho A v}$$

where m is the mass of fluid, ρ is the density of the fluid, d is the distance a fluid particle travels during time t, A is the cross-sectional area of the pipe, and $\rho A d$ is the mass of fluid that flows past some point at time t. Note that if we set the last two terms equal we have **flow velocity** v as:

$$\boxed{v = d/t}$$

For pipes with a variable cross-sectional area, *a liquid must flow with a faster speed in a narrow section* than in a wider section (in order to have the same mass-per-time flow rate).

If a pipe has a cross sectional area A_1 at point 1 and area A_2 at point 2, the **flow rate** must be the same at each point:

$$\rho_1 A_1 v_1 = \rho_2 A_2 v_2$$

which is the **continuity equation** for steady, laminar, one-dimensional flow. Since most liquids are not very compressible, the density does not change, so that $\rho_1 = \rho_2$:

$$A_1 v_1 = A_2 v_2$$

This is the **equation of continuity for an incompressible liquid** and describes the relationship between flow velocity and cross-sectional area. We see this principle in a flowing river or stream, which has higher flow velocities at narrow sections and slower flow velocities at wide sections.

• **Example**: You decide to test this "flow rate in equals flow rate out" concept by setting up an apparatus with one large tube connected to 10 small tubes. The radius of the large tube is 2 cm and each small tube is 0.5 cm. If you deliver a flow velocity of water in the large tube at $v_1 = 0.2$ m/s, what is the velocity v_2 in one of the small tubes?

We can use $A_1 v_1 = A_2 v_2$, where A_1 is the cross-sectional area of the large tube and A_2 is the combined cross-sectional areas of the 10 small tubes. The flow in each small tube is: $A_{st} = (0.1)A_2$, or $(10)A_{st} = A_2$, so the equation becomes:

$$A_1 v_1 = (10)A_{st} v_2$$

Solve for v_2:

$$v_2 = v_1(A_1/(10)A_{st}) = v_1(0.1)(A_1/A_{st})$$

Since a tube's cross-sectional area is πr^2: $(A_1/A_2) = (\pi r_1^2/\pi r_2^2) = (r_1^2/r_2^2)$
Velocity v_2 in one of the small tubes becomes:

$$v_2 = v_1(0.1)(r_1/r_{st})^2 = (0.2 \text{ m/s})(0.1)(2 \text{ cm}/0.5 \text{ cm})^2 = 0.32 \text{ m/s}$$

The velocity in the large tube was delivered as 0.2 m/s and we found each of the 10 small tubes should have a velocity 0.32 m/s.

Bernoulli's Equation

• The behavior of many fluids in motion can be described by Bernoulli's equation, developed by **Daniel Bernoulli** in the 1700s. It is primarily applied to liquids, but can be applied to gases when pressure differences are very small and compression is minimal. Bernoulli's equation relates the pressure, density, and velocity of a flowing fluid and can be derived by assuming that the fluid is *incompressible* and is flowing smoothly (*laminar flow*) without turbulence or viscous effects.

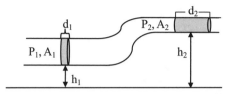

We can derive Bernoulli's equation by considering fluid flow in a pipe. Since an external force is required to drive the flow, this force is doing **work** on the fluid. In turn, work is also being done by the fluid on other fluid elements in the pipe. In the diagram we can imagine that work is done on the fluid at cross-section point 1, and work is done by the fluid at cross-section point 2. The net work W done on the fluid causes a change in the potential energy and kinetic energy of the fluid:

$$W = \Delta PE + \Delta KE$$

At point 1 the work done on the fluid involves moving a volume V_1 of fluid with force F_1 for a net displacement of d_1 at a constant velocity v_1. Therefore, the **work** done on the fluid is $W = F_1 d_1$ and volume $V_1 = A_1 d_1$ has been displaced distance d_1. At point 2, because the fluid is incompressible, the volume displaced through point 1 must equal the volume displaced at position 2, so that a volume of fluid with cross-sectional area A_2 and a length d_2 is displaced d_2 at a constant velocity v_2, where $A_1 d_1 = A_2 d_2$. Since the volume of fluid moves by pushing on the next element of fluid in front of it, the **work** done on the fluid at point 2 is $F_2 d_2$. The net work done on the fluid is:

$$W = F_1 d_1 - F_2 d_2$$

Since $F_1 = P_1 A_1$ and $F_2 = P_2 A_2$:

$$W = P_1 A_1 d_1 - P_2 A_2 d_2$$

Since $V = V_1 = A_1 d_1 = V_2 = A_2 d_2$:

$$W = (P_1 - P_2)V$$

As the fluid is moved from point 1 to point 2, a mass of fluid ($m = \rho V$) at a pressure P_1, traveling with velocity v_1 at height h_1, becomes an equal mass of fluid ($m = \rho V$) at pressure P_2, moving with velocity v_2 at height h_2. Since work done between points 1 and 2 is the change in kinetic and potential energy:

$$\Delta PE = \Delta(mgh) = mg(\Delta h) = \rho V g(h_2 - h_1)$$

and

$$\Delta KE = \Delta((1/2)mv^2) = (1/2)m\Delta v^2 = (1/2)\rho V(v_2^2 - v_1^2)$$

If we substitute $\Delta PE = \rho Vg(h_2 - h_1)$, $\Delta KE = (1/2)\rho V(v_2{}^2 - v_1{}^2)$, and $W = (P_1 - P_2)V$, into $W = \Delta PE + \Delta KE$ we get:

$$(P_1 - P_2)V = \rho Vg(h_2 - h_1) + (1/2)\rho V(v_2{}^2 - v_1{}^2)$$

Divide by V and multiply:

$$P_1 - P_2 = \rho gh_2 - \rho gh_1 + (1/2)\rho v_2{}^2 - (1/2)\rho v_1{}^2$$

Rearranging to set point 1 conditions equal to point 2 conditions gives the **Bernoulli equation**:

$$\boxed{P_1 + \rho gh_1 + (1/2)\rho v_1{}^2 = P_2 + \rho gh_2 + (1/2)\rho v_2{}^2}$$

Since points 1 and 2 in the figure could have been chosen elsewhere, Bernoulli's equation shows that the value of $P + \rho gh + (1/2)\rho v^2$ will be the same for any location in a pipe and connects the fundamental characteristics P, ρ, and v at different points in the flowing fluid. The Bernoulli equation has numerous practical uses and applications.

• If velocity is zero and the fluid is static, the velocity terms in Bernoulli's equation are zero, and the equation reduces to:

$$P_1 + \rho gh_1 = P_2 + \rho gh_2$$
$$P_1 - P_2 = \rho g(h_2 - h_1)$$

This mimics the familiar equation: $P = F/A = mg/A = \rho Vg/A = \rho gh$.

• You may be wondering if you can model a body of water such as a pond or a fish tank that is leaking fluid through a pipe or holes using Bernoulli's equation. To test this, imagine you decide to measure the flow velocity of fluid leaving the drain in the bottom of your rather large 2 m long, 1 m wide, 1.5 m high fish tank. If the drain tube that leaves the fish tank has a 1-cm diameter, what would the flow velocity in the tube be once you open the valve?

You can model this as a pipe with a variable cross-sectional area where the open top of the tank is the inlet of a pipe (cross-section 1) and the exit of the drain tube as the exit of a pipe (cross-section 2). This allows you to use **Bernoulli's equation**:

$$P_1 + \rho gh_1 + (1/2)\rho v_1{}^2 = P_2 + \rho gh_2 + (1/2)\rho v_2{}^2$$

Since the top of the tank and the exit of the drain tube are both open to atmospheric pressure, $P_1 = P_2$:

$$\rho gh_1 + (1/2)\rho v_1{}^2 = \rho gh_2 + (1/2)\rho v_2{}^2$$

Because flow rate in will equal flow rate out:

$$A_1v_1 = A_2v_2$$

Rearrange the modified Bernoulli equation and substitute for $v_1 = v_2(A_2/A_1)$:

$$(1/2)\rho v_2{}^2 - (1/2)\rho[v_2(A_2/A_1)]^2 = \rho gh_1 - \rho gh_2$$

$$(1/2)\rho v_2{}^2[1 - (A_2/A_1)^2] = \rho g(h_1 - h_2)$$

If the surface area of the body of water, in this case the upper surface of water in the fish tank (A_1), is much greater than the surface area of the exit pipe (A_2), the ratio of areas (A_2/A_1) will be minuscule. Since $A_2 = \pi(0.005 \text{ m})^2 = 7.85 \times 10^{-5} \text{ m}^2$ and $A_1 = 2 \text{ m} \times 1 \text{ m} = 2 \text{ m}^2$, $(A_2/A_1)^2 = 1.5 \times 10^{-9}$, which is near zero. The equation becomes:

$$(1/2)\rho v_2{}^2[1] = \rho g(h_1 - h_2)$$

or

$$\boxed{v_2{}^2 = 2g(h_1 - h_2)}$$

This simplification of Bernoulli's equation is called **Torricelli's equation**, because it was discovered by E. Torricelli before Bernoulli's work. Interestingly, the flow velocity depends on height just as an object falling through height $h = h_1 - h_2$, in the familiar equation $v^2 = 2gh$.

We now can calculate the flow velocity from our 1.5 m high fish tank:

$$v_2{}^2 = 2(9.8 \text{ m/s}^2)(1.5 \text{ m}) = 29.4$$

Taking the square root: $v_2 \approx 5.4$ m/s.

• Another variation of Bernoulli's equation involves understanding the dynamics of liquid flowing through a **horizontal pipe** that has segments with *different cross-sectional areas*. When the *height is constant*, so that $h_1 = h_2$, those terms cancel and the Bernoulli equation becomes:

$$P_1 + (1/2)\rho v_1{}^2 = P_2 + (1/2)\rho v_2{}^2$$

We can see from this equation that where the flow velocity is high (in segments with smaller cross sections), the pressure will be lower. We can rearrange the equation to isolate the pressure difference between two points in a horizontal pipe corresponding to v_1 and v_2:

$$\boxed{P_1 - P_2 = (1/2)\rho(v_2{}^2 - v_1{}^2)}$$

The interesting property is that *narrower regions of a pipe have higher flow velocities and lower pressures exerted on the pipe walls*, as shown by the height differences in the liquid in the side tubes off the pipe:

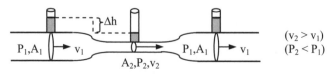

• **Example**: If you want to see this for yourself and set up an apparatus like the above figure, how would you determine the difference in the heights of liquid from the vertical tubes off the wide section 1 and the narrow section 2?

First, the difference in the heights Δh the liquid rises in the vertical tubes is proportional to the pressure difference, and obeys the equation for fluid pressure:

$$P_1 - P_2 = \rho g \Delta h$$

You can substitute into $P_1 - P_2 = (1/2)\rho(v_2^2 - v_1^2)$ for a horizontal pipe:

$$\rho g \Delta h = (1/2)\rho(v_2^2 - v_1^2)$$
$$\Delta h = (1/2g)(v_2^2 - v_1^2)$$

which is the difference in the heights of the of liquid in the vertical tubes. Note that if the fluid is an incompressible liquid like water, and you can calculate v_2 using: $A_1 v_1 = A_2 v_2$.

• Just like a fluid flowing through a pipe, when a fluid flows across a surface, it exerts less pressure when the flow velocity is high than when it is low. This principle is often presented as a simplified version of how lift is generated for airplane flight. The explanation of lift is more complicated and involves the turning of gas (air) by a solid object (airplane wing) such that when the flow of air is turned in one direction lift is generated in the opposite direction. See NASA's website or a similar reliable source if you are interested in an explanation of lift.

7.5. Viscosity, Reynolds Number, and Terminal Velocity

Viscosity

• All real fluids exhibit some amount of viscosity, though in some cases a fluid can be considered to have negligible viscosity. *Ideal fluids* by definition have no viscosity. The **viscosity** of a fluid is the result of internal interactions, or internal friction, among the molecules of a liquid or gas. Viscosity can be thought of as a measure of a fluid's ability to flow. It causes a fluid to be resistant to flow, so that the higher the viscosity the more resistant it is to flow. A viscous fluid will also resist the motion of an immersed object through it. A fluid like honey is more viscous than water.

Viscosity in liquids results from cohesive forces between liquid molecules and adhesive forces between the liquid and walls of containers or pipes through which it flows. Because the interaction between **gas** molecules is weaker than that between liquid molecules, gases are much less viscous. Gases do, however, have frictional forces acting within them and, like liquids, vary in their viscosity. Viscosity varies with **temperature**. This can be observed by placing cooking oil in the refrigerator—it will become more thick and viscous. If you subsequently heat it, it will become thinner and less viscous. For liquids, an increase in temperature leads to lower viscosity. This is not the case for gases.

• The degree of viscosity in a fluid can be evaluated by the amount of lateral force required to make it move sideways. This is called shearing and can be done by horizontally sliding two **plates** separated by a layer of fluid. In this setup, the coefficient of viscosity η can be found by measuring the force required to slide the upper plate at velocity v across the stationary lower plate. The viscous force F_v resisting motion is proportional to the velocity v and surface area A of the plates and inversely proportional to plate separation d:

$$F_v = \eta\, vA/d$$

where η is the **coefficient of viscosity**. (Note that μ is also often used to represent the coefficient of viscosity.) Values for η are relatively small for less viscous substances such as air, and greater for more viscous substances such as oil or honey. The coefficient of viscosity decreases with increasing temperature in liquids, but it increases with rising temperature for gases.

Units of η in SI are $N \cdot s/m^2 = Pa \cdot s$. In CGS units are $dyne \cdot s/cm^2 = Poise$ (P) after the French physicist J. Poiseuille. $1\ Pa \cdot s = 10\ P$.

Turbulence and Reynolds Number

• Fluid flow through a pipe is generally smooth and laminar at low velocities. If flow velocity increases, a critical speed is reached when the laminar flow is no longer consistently smooth, and transient circular eddies begin to appear. This transition velocity where flow is no longer laminar depends on fluid properties and pipe diameter. As velocity increases further, the flow becomes turbulent. **Turbulent flow** is generally associated with the presence of eddies and chaotic motions. Turbulence commonly occurs with low viscosity fluids at higher flow velocities. The onset of turbulence in flow has been found experimentally to obey the **Reynolds number** (R):

$$R = 2rv\rho_f/\eta = Dv\rho_f/\eta$$

where v is the average flow velocity, ρ is the fluid density, r is the pipe radius, D is diameter, and η is the coefficient of viscosity. The Reynolds number has no units and is called a dimensionless number. It reflects the ratio of *inertial* forces to *viscous* forces. The Reynolds number is used to determine whether a flow will be laminar or turbulent. Approximate Reynolds number values for flow in a pipe range from R < 2,300 for laminar flow, R > 4,000 for turbulent flow, and transient effects occur between about 2,300 and 4,000. In turbulent flow, inertial forces dominate viscous forces. In laminar flow, viscous forces dominate. The Reynolds number also applies to an object moving through a fluid, where D represents the size in one dimension, and ρ, and η refer to the fluid. Just as with fluid flowing through a pipe, when an object moves through a fluid, the flow around it can be either laminar or turbulent.

• **Example**: If the water leaving your 3/4-in garden hose is flowing at 10 gallons/min, is it laminar or turbulent?

We can use the Reynolds number $R = 2rv\rho_f/\eta = Dv\rho_f/\eta$.
Convert: 10 gal/min = 6.3×10^{-4} m^3/s and a 3/4-in hose = 0.019 m.
We need flow velocity which relates to volume flow rate by v = V/A where A is cross-sectional area:
$$v = (6.3 \times 10^{-4} \text{ m}^3/\text{s}) / \pi(0.0095 \text{ m})^2 \approx 2.22 \text{ m/s}$$
The density and viscosity of water are 1,000 kg/m^3 and 0.001 Pa·s.
$$R = Dv\rho_f/\eta = (0.019 \text{ m})(2.22 \text{ m/s})(1,000 \text{ kg/m}^3)/(0.001 \text{ Pa·s}) = 42,180$$
Definitely turbulent!

Terminal Velocity

• One of the effects of *viscosity* is that it retards the motion of an object falling through a fluid until **terminal velocity** is reached. When an object falls through a fluid such as water, oil, honey, or even air, the rate that it falls is slowed by viscous effects. If a rock falls through water or oil under the acceleration of gravity it experiences the downward force of gravity and the retarding upward forces due to buoyancy and viscous friction. As the rock accelerates downward, the viscous force increases with increasing velocity until equilibrium is reached between the up and down forces when the net force becomes zero. At this point *terminal velocity* is reached, there is no further acceleration, and the rock continues falling at its constant, *terminal velocity*. (See Section 1.6.)

The terminal velocity rates for the rock falling through water, oil, or honey will differ. For the same distance, the rock falls through water much faster than through the more viscous oil due to higher frictional, viscous forces in the oil. As the rock falls, it initially gathers speed, accelerating under gravity while the viscous force increases until the frictional force begins to balance its weight. This is the point where maximum (terminal) velocity is reached. As the rock continues down, it will fall at this terminal velocity with no further acceleration. If the rock falls through air, which has relatively low viscosity, it will accelerate to a greater speed until it reaches its terminal velocity in air, which is greater than what it reaches in water, oil, or honey. When terminal velocity is reached it will stop accelerating.

The frictional or viscous force that retards the motion of an object that is either falling or moving through a fluid, such as the motion of an aircraft or submarine, is called **drag**. When the flow of fluid around a moving object is **laminar**, the drag or viscous force is proportional to the velocity.

7.6. Key Concepts and Practice Problems

- Fluid (liquid or gas) has no rigid, fixed structure or shape and flows.
- Pressure is: $P = F/A$. Fluid pressure at depth h is: $P = \rho gh$.
- Pascal's principle: pressure applied to the surface of a fluid is transmitted undiminished in all directions and to all points within the fluid.
- Average density $\rho = m/V$ and specific gravity $SG = \rho_{substance} / \rho_{H2O}$.
- Archimedes' principle: buoyant force equals weight of fluid displaced.
- Buoyant force of cylinder or similar object: $F_B = \rho_f Vg$.
- Volume of irregular object: $V = (w_{air} - w_{immersed}) / (\rho_{fluid}g)$.
- Surface tension is due to cohesive forces between molecules in a liquid.
- Capillary action results from surface tension and adhesive forces.
- Height h liquid rises (or lowers) in a tiny tube: $h = 2\sigma \cos \theta / \rho gr$.
- When a fluid is in motion, it moves in a way that conserves mass.
- Laminar pipe flow: flow rate in = flow rate out. Flow rate = $m/t = \rho Av$.
- Continuity equation, laminar 1-dimensional: $\rho_1 A_1 v_1 = \rho_2 A_2 v_2$.
  If $\rho_1 = \rho_2$, $A_1 v_1 = A_2 v_2$.
- Bernoulli equation: $P_1 + \rho gh_1 + (1/2)\rho v_1^2 = P_2 + \rho gh_2 + (1/2)\rho v_2^2$.
- Viscosity results from internal friction among fluid molecules.
- Reynolds number: $R = 2rv\rho_f/\eta = Dv\rho_f/\eta$ reveals laminar or turbulent.

Practice Problems

7.1 (a) A basketball is filled to a pressure of 12 lb/in^2 as measured by a mechanical gauge. If transported to the Moon, what would the gauge's pressure reading be for the basketball? **(b)** You blow up a balloon to a pressure of 3 atm. If you make a pinprick 1 mm in diameter in the balloon, what will be the pressure of the escaping air in Pa? **(c)** You estimate that 70% of a duck on a pond floats above the water. Based on this observation, estimate the duck's specific gravity.

7.2 (a) If the 50-kg person discussed in the example in Section 7.2 is submerged in the Dead Sea (density of 1,240 kg/m³ due to dissolved minerals), what will be his effective weight? **(b)** When in equilibrium, what percentage of his body would float above the surface?

7.3 (a) In the example in Section 7.3 of 0.15-mm tubes in water and mercury, if you use a tube with a radius of 1.5 mm rather than 0.15 mm, how high will the water rise (assume cos θ remains close to 1)? **(b)** What height will the water reach in a tube with a radius of only 0.015 mm? **(c)** Why should the heights be different?

7.4 (a) If water is flowing through a 1-in garden hose at 2 ft/s, what will be its exit velocity through a nozzle with diameter 0.1 in? **(b)** If a large reservoir with a surface elevation of 1,500 m is drained through a 3-m diameter conduit with an exit point at elevation 700 m, what will be the velocity of the water exiting the conduit (ignoring friction and viscosity)?

7.5 In a wind tunnel of 0.1 m diameter, at what velocity will laminar flow begin to break down? Assume air density of 1.3 kg/m³ and a coefficient of viscosity of 1.73×10^{-5} Pa·s.

Answers to Chapter 7 Problems

7.1 (a) $p_{gauge} = p_{bb} - p_{atm}$ so $p_{bb} = p_{gauge} + p_{atm} = 12 + 14.7 = 26.7$. On the Moon, where atmospheric pressure is effectively zero, the gauge would read 26.7 psi. **(b)** The pressure is the same regardless of the size of the hole because pressure is measured as force per area, such as N/m². Since the outside pressure is 1 atm, the pressure difference at the small hole is $3 - 1 = 2$ atm, but as the air leaves the balloon its pressure immediately drops to 1 atm. **(c)** Since a volume of water that occupies 30% of the duck's volume weighs the same as the duck, the duck's density is about 30% the density of water, which is 1,000 kg/m³. Therefore $SG_{duck} = \rho_{duck}/\rho_{H2O} = 300/1,000 = 0.30$.

7.2 (a) $F_B = \rho_{DS}Vg = (1,240 \text{ kg/m}^3)(4.85 \times 10^{-2} \text{ m}^2)(9.8 \text{ m/s}^2) \approx 589$ N. $w_w = w_a - F_B = 490 - 589 = -99$ N or upward "weight" of 22 lbs. **(b)** Equilibrium is reached when the weight of the displaced Dead Sea water equals his body weight. (1,240)(percent submerged) = 1,030. Percent submerged = 83%, so 17% of his body would float above the water.

7.3 (a) Using $h = 2\rho/\rho gr$, if r is increased by a factor of 10, h is reduced by a factor of 10, so $h = 0.99$ cm. **(b)** If r is decreased by a factor of 10, h is increased by a factor of 10, so h = 99 cm or about 39 in. **(c)** This is an example of scaling. In (a) the diameter along which the lifting force works is 10 times longer, but the area of water to be lifted (πr^2) is 100 times greater, so the ratio of the lifting force to the amount of liquid being raised is 1/10 as great.

7.4 (a) Since the cross-sectional area of the nozzle is 1/100 that of the hose, the exit velocity must be 100 times as great to conserve mass, or 200 ft/s. **(b)** Using Torricelli's equation, $v^2 = 2g(h_1 - h_2) = 2(9.8)(1,500 - 700) = 15,680$. Therefore, $v \approx 125$ m/s.

7.5 The threshold Reynolds number for the breakdown of laminar flow is near 2,300. $R = Dv\rho/\eta$ so $v = R\eta/D\rho = (2,300)(1.73 \times 10^{-5})/(0.1)(1.3) \approx 0.31$ m/s.

Chapter 8

HEAT AND TEMPERATURE

"Do not imagine that mathematics is hard and crabbed, and repulsive to common sense. It is merely the etherealization of common sense."

"Overwhelmingly strong proofs of intelligent and benevolent design lie around us..."

Both attributed to Lord Kelvin

8.1. Temperature

• While we think of temperature as a measure of hotness or coldness, it is telling us the average kinetic energy of the molecules that make up a substance or material. Heat and temperature are often used in tandem as we think of a hotter substance as having a higher temperature. There is, however, a difference between heat and temperature. Heat is a form of energy (thermal energy), while temperature is a property of a material which depends on that material. More specifically, **heat** is the energy that flows due to temperature differences, and **temperature** is a measure of the average internal or kinetic energy of the atoms and molecules in a substance or system. We will discuss heat later in this chapter.

The Scales—Fahrenheit, Celsius, and Absolute Kelvin

• Scales and measurement devices have been developed to study and understand temperature. Temperature is measured on three scales: **Fahrenheit, centigrade** or **Celsius**, and **Kelvin.**

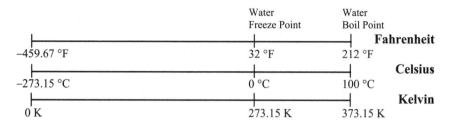

The **Celsius**, or **centigrade, scale** is clearly related to the boiling/vaporizing point and the freezing/melting point of water. The absolute, or **Kelvin scale**, is thought of as the more fundamental, having an absolute zero, 0 K, which is the lowest theoretical temperature a material or substance can have.

• The **Fahrenheit scale** (which is not metric) is the most commonly used scale in the U.S., though Celsius is the predominant scale used throughout the world and with many U.S. scientists, particularly physicists. D.G. Fahrenheit originally devised the Fahrenheit scale so that negative numbers would not often occur. The scale was developed with the value of 32° corresponding to the freezing point of water, and the value of 212° corresponding to the boiling point of water (both at sea-level atmospheric pressure). In the Fahrenheit scale, comfortable room temperature is about 70° and the human body temperature is just under 100° (98.6°).

• The **Celsius, or centigrade, scale**, which is metric, is set directly to the freezing and boiling points of water at sea-level pressure. It has the freezing point of water or melting point of ice corresponding to 0 °C and the boiling point of water corresponding to 100 °C.

• By subtracting the differences in boiling and freezing points, we see that 1 centigrade degree is (212 − 32)/(100 − 0) = 180/100 = 9/5 larger than 1 Fahrenheit degree. To convert between Fahrenheit and Celsius, we use the fact that the freezing point of water is 0 °C and 32 °F, and a change in temperature of 9 °F corresponds to a change of 5 °C. The equations relating the two (which are worth memorizing) are:

$$T_C = (5/9)(T_F - 32)$$
$$T_F = (9/5)T_C + 32$$

where T_C represents the Celsius temperature and T_F the Fahrenheit temperature. Using from $T_C = (5/9)(T_F - 32)$ we see that comfortable room temperature, 70 °F, corresponds to about 21.1 °C, and a normal human body temperature of 98.6 °F is 37.0 °C.

• **Example**: Find the temperature where the Fahrenheit and Celsius scales are equal.

Set $T_F = T_C = T$ in either $T_C = (5/9)(T_F - 32)$ or $T_F = (9/5)T_C + 32$ and solve for T.

$$T = (5/9)(T - 32) = (5/9)(T) - (5/9)(32)$$
$$(9/9)T - (5/9)(T) = -(5/9)(32) \quad \text{or} \quad T = -(5/9)(32) / (4/9) = -40$$

Or:

$$T = (9/5)T + 32 \quad \text{or} \quad T - (9/5)T = 32 \quad \text{or} \quad T(5/5 - 9/5) = 32$$
$$T = 32/(-4/5) = -40$$

• The **Kelvin, or Absolute, scale,** was developed by British scientist Lord Kelvin and is used by scientists. Absolute zero or zero Kelvin is the coldest theoretical temperature. At 0 K atoms and molecules would possess the least possible amount of thermal energy. The Kelvin scale therefore has zero set to what is believed to be the lowest possible temperature a material can theoretically have, which is 0 K = −273.15 °C = −459.67 °F. Note that zero on the Celsius scale is: 0 °C = 273.15 K. On the Kelvin scale, water freezes at 273.15 K and boils at 373.15 K. Temperature on the Kelvin scale is always a positive number.

The Kelvin scale is metric having the *same degree size as the centigrade scale* so that a change of 1 °C is the same as a change of 1 K. Because a degree size or change on the absolute scale equals a degree size or change on the Celsius scale, we can convert between Celsius temperature T_C to Kelvin temperature T_K:

$$\boxed{T_K = T_C + 273}$$

Note that 273.15 K is often rounded to 273 K and the degree symbol (°) is not shown. To convert between the Kelvin and Fahrenheit scales we can combine:

$$T_C = (5/9)(T_F - 32) \quad \text{or} \quad T_F = (9/5)T_C + 32 \quad \text{with} \quad T_K = T_C + 273$$

Thermometers

• **Thermometers** are devices used to measure temperature. There are various types of thermometers, though they are typically designed to take advantage of the expansion or contraction of a given material or substance in the presence of temperature changes. A thermometer can then be calibrated by the amount of thermal expansion and contraction that occurs within the substance it uses. Following are well-known types of thermometers:

The common **"bulb" thermometer** is usually a glass tube containing a liquid, such as mercury or alcohol. The liquid level in the tube rises or falls with changing temperature. These thermometers generally contain a reservoir or "bulb" of mercury or other liquid at the bottom end of the thermometer. As the mercury or other liquid expands or contracts with temperature change, that effect is amplified in the small-bore calibrated tube that extends from the bulb thereby showing the mercury moving up and down the tube. Temperature is read using calibrated marks along the tube.

The **bi-metal strip thermometer** converts a temperature change into mechanical displacement by moving a pointer back and forth along a calibrated scale. This thermometer consists of a strip of two layers of metals which have different coefficients of thermal expansion and bend (moving the pointer) as the temperature changes. While metals do not expand by a large amount, if a thin coiled strip of two bonded metals is heated, the end of the strip can bend enough to direct the pointer. These thermometers are used in thermostats and air temperature thermometers.

Based on the voltage difference generated between the ends of a wire that is in a temperature gradient, a **thermocouple** works by measuring the voltage, or temperature difference, between two joined dissimilar metals (usually wires) that are each experiencing the same temperature gradient. A **thermistor** is a temperature-sensing device that exhibits a change in resistance with changing temperature. The **silicon bandgap temperature sensor**, which uses the principle that the forward voltage of a silicon diode is temperature dependent, lends itself to being included in silicon integrated circuits.

The popular **Galileo's thermometer** has several spheres or weights having various densities which are suspended in a clear liquid contained in a sealed glass cylinder. As the temperature of the liquid changes, the density of the liquid also changes, causing the suspended weights to rise and fall, so that each weight remains at the position where its density is equal to the density of the surrounding liquid (according to Archimedes' principle). If the weight's density is less than the liquid it will float, and if it has greater density than the liquid it will sink to the level of equal density. Because the weights each have a specific density, they will sink at a particular temperature, thereby showing the temperature of its surroundings. The temperature is read from engraved numbers on discs hung from each weight.

8.2. Thermal Expansion

• You may have observed the phenomena of materials and substances expanding and contracting with temperature changes. While there are exceptions for certain substances at certain temperatures, when most substances are heated they expand and when cooled they contract. The thermal expansion of materials must be considered in the design of structures and precision instruments. For example, bridges are designed with expansion joints and sidewalks are laid with gaps to allow for expansion and contraction. The phenomenon of **thermal expansion** applies not only to the change in length of a solid object but also to the change in volume of solids, liquids, and gases.

Linear Expansion

• Substances and materials vary in their amount of **thermal expansion** when exposed to changes in temperature. While an object can expand or contract in all dimensions, it may be that just one length dimension is important to the integrity of a larger structure. When a change in temperature causes a solid object to have a change in length, it is referred to as **linear expansion**, even though other dimensions also expand. It has been shown that the linear expansion, or change in length, of most solid materials is a function of the initial length L_0, the temperature change ΔT, and the material's coefficient of linear expansion. We write this:

$$\Delta L = \alpha L_0 \Delta T \quad \text{or} \quad \Delta L/L_0 = \alpha \Delta T$$

where ΔL is the change in length, L_0 is the initial length, ΔT is the temperature change, and α is a proportionality constant called the **coefficient of linear expansion** which is characteristic of the particular material. This coefficient is equal to:

$$\alpha = \Delta L/L_0 \Delta T$$

reflecting *the fractional change in length per degree of temperature change*. The **coefficient of linear expansion** α is normally evaluated for the centigrade scale, with **units** in inverse Celsius degrees $°C^{-1}$ or in inverse degrees Kelvin K^{-1}. The value of α depends only on the material and varies from one substance to another. Some typical values include: $2.9 \times 10^{-5} \, °C^{-1}$ for lead, $1.2 \times 10^{-5} \, °C^{-1}$ for concrete, and $0.06 \times 10^{-5} \, °C^{-1}$ for quartz.

Values for the coefficient of linear expansion usually have some variation with temperature, though in many cases it is small enough to be considered constant. A few substances, however, have negative linear expansion coefficients for certain temperature ranges. These substances shrink as the temperature increases and expand when temperature decreases.

Because the linear expansion for most solids is directly proportional to the change in temperature ΔT, you can plot length vs. temperature for a given material, such as a rod made of iron or copper, and the graph will be a straight line. As the temperature of the rod is raised the *change in the length ΔL of the rod* will *increase linearly with the change in temperature* ΔT. The change in length is also proportional to the rod's original length, as reflected in: $\Delta L = L_0 \alpha \Delta T$. After a change in temperature, the rod's new length becomes:

$$L_{new} = L_0 + \Delta L = L_0 + \alpha L_0 \Delta T = L_0(1 + \alpha \Delta T)$$

• **Example**: Temperatures in Fairbanks, Alaska, can range from as low as about $-50\,°C$ in the winter to as high as about $35\,°C$ in summer, a ΔT of $85\,°C$. If you were building a highway, what seasonal variation in length would you need to allow for in each 10-m long section of concrete?

Using a coefficient of expansion of $1.2 \times 10^{-5}\,°C^{-1}$ for concrete, the change in length you would want to allow for is:

$$\Delta L = \alpha L_0 \Delta T = (1.2 \times 10^{-5}\,°C^{-1})(10\text{ m})(85\,°C) = 0.0102\text{ m} = 10.2\text{ mm}$$

You would need to consider that each 10 m long section could expand 10.2 mm in length.

• **Example**: If a metal rod is heated and stretches according to the figure, what is its coefficient of expansion?

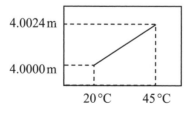

$\alpha = \Delta L / L_0 \Delta T = (4.0024\text{ m} - 4.0000\text{ m}) / (4.0000\text{ m})(25\,°C) = 2.4 \times 10^{-5}\,°C^{-1}$

which is close to the coefficient of expansion for aluminum.

• **Example**: Your friend holds up two 1 m long rods. One is copper with a coefficient of expansion of $1.7 \times 10^{-5}\,°C^{-1}$, and the other is brass with a coefficient of 1.9×10^{-5} °C^{-1}. He says they fit into an apparatus pointing toward each other, and at room temperature of about 21°C there is a 1-mm gap between them. He asks you if you can determine the temperature at which the rods would expand enough to make contact.

For brass $\Delta L_b = \alpha_b L_{0b} \Delta T$ and for copper $\Delta L_c = \alpha_c L_{0c} \Delta T$.

As the rods are heated, they will expand and touch when the sum of the two changes in length equals the initial width of the gap. Therefore:

$$\Delta L_b + \Delta L_c = \alpha_b L_{0b} \Delta T + \alpha_c L_{0c} \Delta T = 0.001 \text{ m}$$

Solve for temperature change:

$$\Delta T(\alpha_b L_{0b} + \alpha_c L_{0c}) = 0.001 \text{ m}$$

$$\Delta T = 0.001 \text{ m} / (\alpha_b L_{0b} + \alpha_c L_{0c})$$

$$\Delta T = 0.001 \text{ m} / [(1.9 \times 10^{-5} \text{ °C}^{-1})(1 \text{ m}) + (1.7 \times 10^{-5} \text{ °C}^{-1})(1 \text{ m})] \approx 27.78 \text{ °C}$$

Since the initial temperature was 21 °C and ΔT is 27.78 °C, the final temperature is about 48.78 °C.

Thermal Stress

• In Section 6.2 we discussed stress, strain, and the elastic, or Young's, modulus. Remember a **stress** is a force that causes deformation and a **strain** is the resulting change in a material. The ratio of stress to strain is called the **elastic modulus** or **Young's modulus** Y and is:

Young's modulus $Y = (F/A) / (\Delta L/L)$ or $F/A = Y\Delta L/L$

We discussed stress and strain in terms of a physical pulling, pushing, or twisting deformation of a material, but a stress can also be created when an object or material expands or contracts due to a *change in temperature*, especially when different sections of the object expand or contract at different rates. We can imagine a glass object cracking when it is suddenly heated or cooled and the thicker sections cannot expand or contract as quickly as the thinner sections. If an object has a cross-sectional area A, an elastic modulus Y, and a coefficient of linear expansion α, we can combine the **linear expansion** with **Young's modulus** Y by substituting $\Delta L/L = \alpha \Delta T$ into $F/A = Y\Delta L/L$ to get an equation for **thermal stress**:

$$F/A = \alpha \Delta T Y$$

where stress now depends on the cross-sectional area, the temperature change, coefficient of linear expansion α, and Young's modulus Y.

• **Example**: In the example in the previous subsection you calculated the expansion of the 10 m long sections of concrete to allow for temperature changes in the highway you are constructing in Fairbanks, Alaska. What will happen if you construct your highway out of 10 m long sections of concrete in Fairbanks, Alaska, without leaving a gap for expansion? Suppose you lay your highway when the temperature is about 5 °C and it rises to 35 °C. The compression strength for concrete is 2×10^7 N/m^2, Young's modulus for concrete is 2×10^{10} N/m^2, and the coefficient of expansion for concrete is 1.2×10^{-5} °C^{-1}.

The stress created during the summer when the concrete heats up is:

$$F/A = \alpha \Delta TY = (1.2 \times 10^{-5} \, °C^{-1})(30 \, °C)(2 \times 10^{10} \, N/m^2) = 7.2 \times 10^6 \, N/m^2$$

You find that the stress, 7.2×10^6 N/m^2, is less than the compression strength for concrete, 2×10^7 N/m^2, so your highway will not crumble. (Your friend tells you it is still a good idea to put in the extra space between sections in case there is a weak spot in the concrete.)

Volume Expansion

• When an object, material, or substance is heated or cooled it can expand or contract in all three dimensions. We studied the change in length above, but sometimes it is important to consider the change in volume, especially when the substance is a liquid or gas.

The equation for **volume expansion** is similar to the equation for linear expansion ($\Delta L = \alpha L_0 \Delta T$) and is:

$$\boxed{\Delta V = \beta V_0 \Delta T \quad \text{or} \quad \Delta V / V_0 = \beta \Delta T}$$

where ΔV is the change in volume, V_0 is the initial volume, ΔT is the change in temperature, and β is the **coefficient of volume expansion**. Rearranging, we see the coefficient of volume expansion:

$$\beta = \Delta V / V_0 \Delta T$$

The **units** for β, like that of the linear coefficient α, are normally given in inverse Celsius °C^{-1} or Kelvin K^{-1}. Typical values of β range from 8.7×10^{-5} °C^{-1} for lead, 3.6×10^{-5} °C^{-1} for concrete, 0.18×10^{-5} °C^{-1} for quartz, to 21×10^{-5} °C^{-1} for water (at about room temperature). Note that as water is cooled and reaches the range of 0 °C to 4 °C, its thermal expansion coefficient drops to zero and then becomes negative.

Just as with linear expansion, the equation for the **new volume** V_{new} after a change in temperature is similar to the equation for the new linear thermal expansion ($L_{new} = L_0 + \Delta L = L_0(1 + \alpha\Delta T)$), and is written:

$$V_{new} = V_0 + \Delta V = V_0 + \beta V_0 \Delta T = V_0(1 + \beta\Delta T)$$

For solids that expand or contract uniformly in all three dimensions, for small temperature changes the **volume expansion coefficient** β is approximately equal to 3 times the coefficient of linear expansion α.

$$\beta = 3\alpha$$

Over small temperature ranges, thermal expansion for length, area, and volume can be written in terms of the linear expansion coefficient α as: linear $\Delta L = \alpha L_0 \Delta T$, area $\Delta A = 2\alpha A_0 \Delta T$, and volume $\Delta V = 3\alpha V_0 \Delta T$.

• **Example**: If you completely fill your 16-gallon steel gas tank on a cool 10 °C morning and park the car without driving, what will happen if the tank heats up to 40 °C during the hot summer afternoon? The linear expansion coefficient for steel is about 1.2×10^{-5} °C^{-1}, and the volume expansion coefficient for gasoline is about 95×10^{-5} °C^{-1}.

The expansion of the steel tank is: $\Delta V = 3\alpha V_0 \Delta T$.

$$\Delta V_{tank} = (3)(1.2 \times 10^{-5}\,°C^{-1})(16\,gal)(30\,°C) \approx 0.0173\,gal$$

The expansion of the gasoline in the tank is: $\Delta V = \beta V_0 \Delta T$.

$$\Delta V_{gas} = (95 \times 10^{-5}\,°C^{-1})(16\,gal)(30\,°C) = 0.456\,gal$$

The gasoline in the tank will expand more than the tank by:

$$\Delta V_{gas} - \Delta V_{tank} = 0.456 - 0.0173 \approx 0.439\,gal$$

Therefore, about 0.44 gallons of gasoline will spill from the tank. Note as gas is spilled, it will stop contributing to the expansion.

Volume Changes Affect Density

• While we have been discussing the effect of temperature changes on volume, you may have been realizing that for most substances most of the time, as volume increases, density decreases, and as volume decreases, density increases. This occurs because, while changes in temperature can affect volume, the amount of mass remains constant. The relationship between mass m, volume V, and density ρ is:

$$\rho = m/V \quad \text{or} \quad V = m/\rho$$

We can see that if mass is constant, an increase or decrease in volume will cause an inverse change in density. Therefore, a fractional change in volume $\Delta V/V$ corresponds inversely to a fractional change in density

$\Delta\rho/\rho$. For example, if there is a 1 percent increase in volume, approximately a 1 percent decrease in density occurs. As we learned a change in temperature leads to a fractional volume change:

$$\Delta V/V_0 = \beta\Delta T$$

In Section 6.2 we determined the fractional change in density of an aluminum block using the bulk modulus and the formulas $m = \rho V$, $V_1\rho_1 = V_2\rho_2$, and $\Delta\rho/\rho = (\rho_2 - \rho_1)/\rho_2$. This led to a result that was consistent with an equation that applies for *very small **fractional changes***:

$$\Delta\rho/\rho = -\Delta V/V$$

In cases where this applies we can write an equation relating changes in **temperature, volume, and density**:

$$\Delta\rho/\rho_0 = -\Delta V/V_0 = -\beta\Delta T$$

where the negative sign reflects an increase in the change in density $\Delta\rho$ when the change in volume ΔV is decreasing (shown by negative sign).

Temperature Changes and Gases

• The behavior of **gases** in the presence of temperature increases is more complicated than solids and liquids since gases will expand as much as pressure or volume will allow. If the pressure of a gas is fixed and temperature is increased, the gas will expand uniformly with increasing temperature. While liquids and solids have a range of expansion coefficients, gases respond more uniformly to changes in temperature. The coefficients of expansion for gases are, therefore, nearly the same, though they vary with temperature. At constant pressure and 0 °C the *volume expansion coefficient* β for any gas is approximately 3.67×10^{-3} °C^{-1} and at 100 °C it is about 2.68×10^{-3} °C^{-1}.

• **Example**: If you have an unknown gas at 0 °C held in a container by a floating piston, which provides a constant pressure but allows temperature and volume to increase, what temperature increase would cause the gas to increase in volume by 10%?

We can use $\Delta V = \beta V_0 \Delta T$. Let $V_0 = 1$. Therefore:

$$\Delta T = \Delta V/\beta V_0 = (0.1)/(3.67 \times 10^{-3} \text{ °C}^{-1})(1) \approx 27.25 \text{ °C}$$

Because significant volume expansion can occur in gases, it is important to be vigilant when gases are in closed containers. This is why propane tanks should not be completely filled, as they can explode if temperature rises during a hot afternoon. Gases are affected by changes in temperature, pressure, and volume, which we will discuss in the next chapter.

8.3. Heat as Energy

Internal Energy, Temperature, and the First Law

• Molecular movement occurs continually within the atoms and molecules that make up all types of matter, including solids, liquids, and gases. The molecular movement reflects that atoms and molecules possess **internal thermal energy** in the form of the **kinetic energy**. Of the three primary types of matter, solids have the least kinetic energy as their molecules vibrate in place. In liquids, molecules have more kinetic energy and can move around one another allowing flow. Molecules in a gas have even more kinetic energy and are able to float freely and disperse enough to expand and fill a container. The thermal properties of a substance or material are based on the movements, or kinetic energy, of its individual atoms and molecules. At absolute zero temperature, molecules possess the least possible amount of kinetic energy.

Temperature is a measure of the average internal or kinetic energy of the atoms and molecules of a substance. In other words, the temperature is related to the degree to which the atoms and molecules in a substance move. Higher temperatures reflect more rapid movements. Higher temperatures also generally cause thermal expansion and lengthening distances between atoms and molecules in a material or substance.

In situations involving work or energy transfer, the internal energy of an object or substance only needs to be considered when it changes during the process. In these cases the internal energy of an object may be raised (reflecting increased movement of atoms and molecules) by a frictional force such as the internal energy in a train track when a train is screeching to a halt or when a metallic material is being hit or crushed by an outside force. The internal energy increase will be evident by an increase in temperature. In processes that are considered ideal, frictionless, or perfectly elastic, the internal energy does not need to be considered.

• *Internal energy is increased by adding **heat**.* If a material or substance is heated, its atoms and molecules move faster, whereas when it is cooled, its atoms and molecules move more slowly. If you place a metal rod over a flame, you will add energy to the rod in the form of heat and increase its internal energy. **Heat *is a form of energy*** that can be transferred to an object using a difference in temperature. In other words, **heat** is the thermal energy transferred from one object or system to another due to *differences in temperature*. Just as an object does not possess work but rather does work, an object does not possess heat but rather transfers heat due to a temperature difference. Both work and heat are measures of energy transfer. A hot object has an amount of thermal energy which it can transfer to a cooler object as heat. Note that when two objects or systems are at the *same temperature*, there is *no net heat transfer* between them, and they are in **thermal equilibrium** with each other.

• When the thermal properties of a material are examined and its internal energy changes during a process, we need to consider the **internal energy** U of the material and the amount of **heat or thermal energy** Q that is transferred to the material. If a material or substance absorbs *heat* Q, its *internal thermal energy* U increases by that amount, so that:

$$Q = \Delta U$$

If an object or substance *does work* W when *heat Q is added to it*, its *internal energy* U will decrease by the amount of work done by the object or substance. The *law of conservation* of energy applies to heat and work so that if **heat** Q is added to a substance or object and the substance or object does **work** W, the increase in **internal energy** U plus the **work** W done by the substance or object will equal the amount of heat added:

$$Q = \Delta U + W$$
$$\Delta U = Q - W$$

By conservation of energy: the change in an object's or substance's internal energy ΔU equals the heat received Q minus the work done by the object or substance W. This equation describes the principle of energy conservation when thermal energy is included and is called the **First Law of Thermodynamics** (discussed in the next chapter).

When we say the system "does work," we are referring to the definition of work: force times the distance moved in the direction of the force. An example of a thermodynamic system doing work is a piston in a cylinder that is driven upward by an expanding gas. When the gas is heated it expands, thereby doing work on the piston by lifting it. If the piston is pushed down, it is doing work on the gas (while the gas does negative work on the piston).

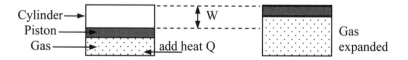

Measuring Heat

• Because heat is a form of energy, it is measured in the same **units** used to **measure energy**, such as Joules (J). The British thermal unit BTU is the amount of heat required to raise 1 lb of water 1 °F. 1 BTU ≈ 1055 J. Another popular unit for measuring heat is **calories** (cal) using lower case "c": 1 cal is the amount of heat required to raise 1 g (gram) of water by 1 °C. Also, 1 cal = 4.186 J. (Note there are different defined values ranging from 4.184 to 4.19.) Alternatively, heat is measured in **Calories** (Cal) with a capital "C": 1 Cal is the amount of heat required to raise the temperature of 1 kg of water by 1 °C. This unit, Calorie, is more properly called the **kilocalorie** (kcal), kilogram calorie, or large calorie, and is 1,000 calories. 1 Cal = 4,186 J = 4.186 kJ. Depending on your weight, if you run a 10 min mile (6 mph) for 1 hr, you will burn 500 to 800 Cal.

• **Example**: Your friend is doing some reading on James Prescott Joule and tells you that Joule developed sensitive thermometers to use in his experiments on heat and work. He told you about one of Joule's experiments in which he let a 5-kg mass, which was attached to a paddle wheel immersed in a container of water, fall a distance of 2 m. As the mass fell it turned the paddle in the water, which generated heat in the water from the friction of the rotating paddle. Your friend said the mass of the water in the container was 1 kg and asked you to figure out the increase in water temperature.

You quickly realize that the work done on the water by the falling mass can be described by the potential energy of the mass:

$$W = mgh = (5 \text{ kg})(9.8 \text{ m/s}^2)(2 \text{ m}) = 98 \text{ J}$$

Since 1 Cal = 4,186 J:

$$(98 \text{ J}) / (4,186 \text{ J} / 1 \text{ Cal}) \approx 0.0234 \text{ Cal}$$

It takes 1 Calorie or 1 kcal of heat to raise the temperature of 1 kg of water by 1 degree. There is 1 kg of water in the container so: 0.0234 Cal corresponds to 0.0234 °C increase in temperature of the water.

Specific Heat

• In general terms, the **specific heat,** or **specific heat capacity,** c is the amount of heat required to raise the temperature of a unit of mass of a substance by a unit change in temperature. In more narrow terms, the **specific heat** c of a substance is the amount of heat energy required to raise the temperature of 1 kg of the substance by 1 degree Celsius or Kelvin. The specific heat relates heat and temperature:

$$Q = mc\Delta T$$

where Q is the amount of heat that needs to be supplied to or removed from a material or substance having mass m and a specific heat of c in order to produce a temperature change of ΔT. The **specific heat** c can be written:

$$c = Q / m\Delta T$$

The specific heat capacity of a material or substance depends on the properties of the material or substance, such as molecular structure and phase (solid, liquid, or gas). This equation does not apply if a phase change occurs since, during the change, the heat added or removed is not changing the temperature.

• **Units** for specific heat c are: (heat energy)/(mass)(temperature change) = J/kg·°C = cal/g·°C = Cal/kg·°C. For example, the **specific heat of water** is 1 Calorie (kcal) per kilogram per degree, or c_{water} = 1 Cal/kg·°C = 1 cal/g·°C = 4,186 J/kg·°C = 4.186 J/g·°C. This means it takes 4,186 J of heat to raise 1 kg of water by 1 °C. Since the BTU is the amount of heat required to raise 1 lb of water 1 °F, the specific heat of water can be expressed as 1 BTU/lb·°F. The specific heat of water is higher than many other substances including metals, which makes water useful in processes that require temperature regulation.

While the **specific heat** of a substance can vary with temperature, near room temperature it remains mostly constant. Each material or substance has its own specific heat value. Substances that are easily heated have lower values, such as *copper* (0.0921 Cal/kg·°C) or *lead* (0.038 Cal/kg·°C), whereas substances that are difficult to heat have a higher values, such as *ice* (0.5 Cal/kg·°C) or *plastic* (0.4 kcal/kg·°C).

• **Example**: What is the amount of heat required to raise the temperature of 1 liter (L) of water 10 °C if it is at an initial temperature of 10 °C?

We can use $Q = mc\Delta T$, but first we need the mass m of water:
$$m = \rho V = [(1,000 \text{ kg/m}^3)(1 \text{ m}^3/1,000 \text{ L})](1 \text{ L}) = 1 \text{ kg}$$
Since the specific heat of water is 4,186 J/kg·°C:
$$Q = mc\Delta T = (1 \text{ kg})(4,186 \text{ J/kg·°C})(10 \text{ °C}) = 41,860 \text{ J}$$

• **Example**: If your friend places 10 oz of hot gold at 70 °C into the container holding 1 kg of water, which is now at 20 °C, what will be the equilibrium temperature, providing no heat is lost? (Specific heat of water is 4,186 J/kg·°C, gold is 130 J/kg·°C, and 10 oz ≈ 0.28 kg.)

You realize that if no heat is lost, energy will be conserved and the heat will be transferred from the gold to the water until a new equilibrium temperature T_f is reached—somewhere between 20 °C and 70 °C. You decide to balance the heat transfer from the gold to the water and solve for T_f:

$$m_{gold}c_{gold}\Delta T_{gold} = m_{water}c_{water}\Delta T_{water}$$

$$m_{gold}c_{gold}(T_{gold} - T_f) = m_{water}c_{water}(T_f - T_{water})$$

$$(0.28\ kg)(130\ J/kg{\cdot}°C)(70\ °C - T_f) = (1\ kg)(4,186\ J/kg{\cdot}°C)(T_f - 20\ °C)$$

$$2,548 - (36.4)(T_f) = (4,186)(T_f) - 83,720$$

$$86,268 = (4,222.4)(T_f)$$

$$T_f \approx 20.43\ °C$$

T_f is only slightly higher because the specific heat of water is much larger than the specific heat of gold and there is a greater mass of water.

• **Example**: After you go through your calculation in the previous example and are satisfied with your answer, your friend says it would be more accurate to include the 0.2-kg aluminum container in which the water and gold are held. If the specific heat of aluminum is about 900 J/kg·°C, recalculate the final equilibrium temperature.

$$m_{gold}c_{gold}\Delta T_{gold} = m_{water}c_{water}\Delta T_{water} + m_{Al}c_{Al}\Delta T_{Al}$$

$$(0.28\ kg)(130\ J/kg{\cdot}°C)(70\ °C - T_f) =$$
$$(1\ kg)(4,186\ J/kg{\cdot}°C)(T_f - 20\ °C) + (0.2\ kg)(900\ J/kg{\cdot}°C)(T_f - 20\ °C)$$

$$2,548 - (36.4)(T_f) = (4,186)(T_f) - 83,720 + (180)(T_f) - 3,600$$

$$89,868 = (4,402.4)(T_f)$$

$$T_f \approx 20.41\ °C$$

You point out to your friend that 20.41 °C is pretty close to 20.43 °C

• **Specific heats of gases** can be affected by changes in pressure or volume, which can occur with temperature changes. When examining specific heats of gases, we can look at what happens when we add heat while either the volume remains constant or the pressure remains constant. If heat Q is added to a certain mass of gas m and the temperature changes by ΔT while the volume is held constant, the **specific heat at constant volume** is:

$$c_v = Q/m\Delta T_v$$

When volume is held constant, the heat added to the gas increases the internal energy of the gas and therefore the temperature of the gas.

Alternatively, if heat Q is added to a certain mass of gas m and the temperature changes by ΔT while the volume is allowed to expand so that the pressure does not increase, the **specific heat at constant pressure** is:

$$C_p = Q/m\Delta T_p$$

When *pressure remains constant*, some of the heat added to the gas does **work** as the gas expands so there is less of an increase in internal energy and therefore temperature for the same Q. This means that:

$$T_p < T_v$$

For a given value of Q and m that means:

$$c_p > c_v$$

This shows that for **gases**, the specific heat at constant pressure is greater than the specific heat at constant volume. Typical values of c_p and c_v near room temperature are: $c_p = 1,010$ J/kg·°C and $c_v = 718$ J/kg·°C for air; $c_p = 1,040$ J/kg·°C and $c_v = 743$ J/kg·°C for N_2; $c_p = 919$ J/kg·°C and $c_v = 659$ J/kg·°C for O_2; $c_p = 5,190$ J/kg·°C and $c_v = 3,120$ J/kg·°C for He; and $c_p = 14,320$ J/kg·°C and $c_v = 10,160$ J/kg·°C for H_2.

Phase Change and Latent Heat

• When heat is added to or removed from a substance, the result may be a temperature change, a phase change, or both (sequentially). While the *specific heat* reflects the heat required to increase the temperature of a substance, the **latent heat of transformation**, Q_L, reflects the heat required to *change the phase* of a substance. A **phase change** occurs when a substance changes between existing as a solid, liquid, or gas. For example, as liquid water forms ice or steam a phase change is occurring. During the process of a phase change when the internal structure of the substance is changing, the temperature does not change until the substance has changed its phase. This suggests that a certain amount of heat is added or removed for a phase change to occur.

The heat that is necessary for the phase change to occur (without a change in temperature) is called the **latent heat** or **specific latent heat**. The word "latent" means hidden and signifies that there is no change in temperature during the phase transition. The **specific latent heat** is the quantity of heat energy required to change the state or phase of a unit of mass of a substance:

$$Q_L = mL$$

where Q_L is the heat transferred in Joules, m is the mass in kilograms, and L is the **latent heat** in Joules per kilogram. Values for latent heat L are tabulated for *latent heat of fusion* and *latent heat of vaporization* for various substances.

When the phase change concerns the transition between *a **liquid and a solid***, that is, melting or solidification (freezing), it involves the **(specific) latent heat of fusion**. The latent heat of fusion is the quantity of heat energy removed or released when 1 kg of a *liquid solidifies*, or *fuses*, without its temperature changing. The latent heat of fusion is also the quantity of heat energy required to transform 1 kg of a *solid into a liquid*. For example, during this phase change, the temperature of pure water remains constant at 0 °C.

When the phase change is from *a **liquid to a gas*** or *a **gas to a liquid***, it involves the **(specific) latent heat of vaporization**. The latent heat of vaporization is the quantity of heat energy required to vaporize 1 kg of a liquid into a gas without its temperature changing. The latent heat of vaporization is also the quantity of heat energy removed or released when 1 kg of a substance is transformed from a gas to a liquid. For example, during this phase change, the temperature of pure water remains constant at 100 °C.

Certain substances, such as solid carbon dioxide, transition directly from solid phase to gas phase at standard pressure in a process called **sublimation**. The heat required to sublimate is the sum of the latent heat of fusion and the latent heat of vaporization of that substance.

- **Units** for **latent heats** are in J/kg, kcal/kg, or cal/g. Values for latent heats depend on the properties of the substance. For water, the latent heat of fusion is about 3.33×10^5 J/kg or 80.0 kcal/kg, and the latent heat of vaporization is about 2.260×10^6 J/kg or 539 kcal/kg.

- **Example**: With thermometer in hand, your friend asks what the final temperature would be if you added half of a kilogram of ice (at 0 °C) to 2 liters of hot water at 70 °C? (The specific heat of water is 4,186 J/kg·°C, and the latent heat of fusion of water is 3.33×10^5 J/kg.)

You need to write an equation with two terms for the heat gained by the ice. One term represents the heat required to melt the ice having the latent heat of fusion for the ice L_i and mass of the ice m_i. The second ice-related term describes the heat required $Q = m_i c_w \Delta T$ to raise the temperature of the melted-ice water m_i from 0 °C to T_{final}. (If the ice were even colder than its freezing point there would be a third ice-related term for the heat gained by the ice as its temperature warmed up to the melting point.) The two ice-related terms must balance the heat lost by the initial container of 70 °C liquid water $Q = m_w c_w \Delta T$. We set these equal to find the final equilibrium temperature. First find the mass of water:

$$m_w = \rho_w V_w = [(1{,}000 \text{ kg/m}^3)(1 \text{ m}^3/1{,}000 \text{ liter})](2 \text{ liter}) = 2 \text{ kg}$$
$$L_i m_i + m_i c_w (T_{final} - 0\,^\circ\text{C}) = m_w c_w (70\,^\circ\text{C} - T_{final})$$
$$(3.33 \times 10^5 \text{ J/kg})(0.5 \text{ kg}) + (0.5 \text{ kg})(4{,}186 \text{ J/kg}\cdot^\circ\text{C})(T_{final} - 0\,^\circ\text{C})$$
$$= (2 \text{ kg})(4{,}186 \text{ J/kg}\cdot^\circ\text{C})(70\,^\circ\text{C} - T_{final})$$
$$166{,}500 \text{ J} + 2{,}093 \text{ J } T_{final} = 586{,}040 \text{ J} - 8{,}372 \text{ J } T_{final}$$
$$10{,}465 \text{ J } T_{final} = 419{,}540 \text{ J}$$
$$T_{final} \approx 40.09\,^\circ\text{C}$$

• The **melting point** temperature of a substance is where it changes (melts) from solid to liquid as heat is applied. The **boiling point** temperature is where it **vaporizes** from a liquid to a gas as heat is applied. Certain substances, such as solid carbon dioxide, can transition from solid to gas (without passing through a liquid phase) depending on pressure in a process called **sublimation**. When a gas is cooled (heat removed) through its boiling point, it will **condense** into a liquid. When a liquid is cooled (heat removed) through its melting point, it will solidify. Certain substances in a gas phase can **deposit** directly into a solid when cooled, depending on pressure. Transitions points vary with pressure.

		sublimation		
	melt ─────────────────→		vaporize	
SOLID	melting point	LIQUID	boiling point	GAS
	Solidifies ←─────────────		condense	
		deposition		

Endothermic and Exothermic

• Note that in addition to phase changes there are other processes, such as chemical reactions, that involve the transfer of heat energy. In fact, chemical reactions occurring between atoms and molecules either require or release heat energy. Reactions that require or absorb heat energy are called **endothermic reactions** and can be described by:
$$A + B + \text{heat} \rightarrow C$$
where A, B, and C are atoms or molecules. And reactions that release heat energy are called **exothermic reactions** and can be described by:
$$A + B \rightarrow C + \text{heat}$$
Note that when a chemical reaction occurs, the heat energy absorbed or released is a term in the energy conservation equation.

A familiar example of an exothermic reaction is combustion, or burning, which is the rapid oxidation of a material or substance. Combustion can occur under the right conditions involving temperature, oxygen availability, and the heat of combustion of the material or substance. The **heat of combustion** of a substance or material is the amount of heat energy (in Joules or calories) released per unit mass (kg) when combustion occurs. An example of a simple exothermic reaction is the combustion of hydrogen and oxygen (one type of rocket propellant):

$$2H_2 + O_2 \rightarrow 2H_2O(vapor) + heat$$

8.4. Heat Transfer

• Remember, temperature reflects the average internal energy of atoms and molecules in a substance or system, and heat is the energy that flows due to temperature differences. Heat energy flows naturally and spontaneously from a hotter object or substance toward a colder object or substance (according to the Second Law of Thermodynamics). The heat transfer from hotter to colder changes the internal energy of the objects or substances involved (according to the First Law of Thermodynamics). If two objects having different temperatures are placed in contact with each other, heat will flow from the hotter object to the colder until they both reach the same temperature and are therefore in **thermal equilibrium**. Because of energy conservation, the heat that flows out of the hotter object will equal the heat that flows into the colder object. Using energy conservation we can calculate equilibrium temperature.

• Heat can be transferred from one object or substance to another by three primary processes: **conduction, convection**, and **radiation**. Conduction involves contact, convection involves fluid motion, and radiation involves electromagnetic waves.

Conduction

• When two objects or substances having different temperatures come in contact with each other, heat energy is primarily transferred through conduction. If an object is heated at one point the heat will eventually spread throughout the object by conduction. Conduction is a common process of heat transfer and occurs through matter by direct contact. More specifically, **conduction** involves the transfer of heat energy by intermolecular collisions. While heat is transferred by molecular agitation within a material or substance, there is no net motion or displacement of the material as a whole that is responsible for the heat transfer.

For example, if one end of a metal rod is at a higher temperature, then energy will be transferred by conduction down the rod toward the colder end due to collisions of the higher speed molecules with the slower ones resulting in a net transfer of energy to the slower ones. If no further heating of one end of the rod occurs, an equilibrium will be reached, and the average kinetic energy per molecule will be the same throughout the rod. Similarly, to boil water on a stove, heat is transferred through the bottom of the pot by conduction where water molecules at the bottom transfer their kinetic energy to the molecules above them through collisions. This process continues along with convective processes (discussed below) until all of the water is at thermal equilibrium.

• Heat flows from hot to cold in response to temperature differences. In fact, in most substances the **rate of heat flow** at any point is proportional to the temperature gradient. For example, if heat energy is flowing down a thin rod that is heated at one end and wrapped in insulation so that the heat does not escape from the surface, the heat energy flow can be thought of as Joules per second passing some fixed point in the rod. This can be written:

$$\Delta Q/\Delta t \propto \Delta T/L$$

where Q is in Joules, ΔT in degrees Celsius or Kelvin, L is the length in meters along the rod, and $\propto$ means proportional. The quantity $\Delta T/L$ is the temperature gradient along the rod. The heat flow rate $\Delta Q/\Delta t$ is in Joules per second, or Watts. Note that if the cross-sectional area of the rod was doubled, the rate of heat flow would also double. Also, if heat is supplied at a steady rate to one end of the rod and is simultaneously released from the other end, the temperature distribution along the rod will become $\Delta T/L$ = constant, which reflects a linear change along the rod between the two ends. Because the amount of *heat conducted per time* (rate) $\Delta Q/\Delta t$ between the two ends of the rod is directly proportional to the *temperature gradient* $\Delta T/L$ for heat flow across a certain *cross-sectional area* A (in square meters), we can write the **rate of conduction heat transfer** as:

$$\boxed{\Delta Q/\Delta t = \kappa A \Delta T/L \quad \text{or} \quad Q = \kappa A t \Delta T/L}$$

where κ is the proportionality constant called the ***coefficient of thermal conductivity*** of the substance. From this equation we see that thermal conductivity κ will be large for materials that are good heat conductors and small for materials that are poor conductors or good insulators.

• **Units** for the *coefficient of thermal conductivity* κ are J/m·s·°C or Watts/m·°C or cal/cm·s·°C. Approximate κ values in J/m·s·°C at about room temperature for select substances include: 401 for copper, 429 for silver, 310 for gold, 0.6 for water, 1.0 for glass, 0.17 for oak wood, and 0.025 for air. Metals are generally good conductors of heat compared to wood or air. Not surprisingly, air in the atmosphere heated by conduction is heated by direct contact with the Earth's surface. Materials that conduct heat well are often also good conductors of electricity.

• **Example**: Your friend walks into your lab with a silver rod having a diameter of 10 mm and a length of 30 cm. He wants to know how long it will take to melt a 2-kg, well-insulated, cubic, 0 °C block of ice if he fixes one end of the (insulated) rod to a 100 °C heat source and the other end to the ice block.

You first realize you need to determine the heat energy Q to produce the phase change and melt the ice. You look up the *latent heat of fusion* L of water (given in the previous section), which is 3.33×10^5 J/kg. The *latent heat to melt the ice* is:

$$Q_L = m_iL_i = (2 \text{ kg})(3.33 \times 10^5 \text{ J/kg}) = 6.66 \times 10^5 \text{ J}$$

The *coefficient of thermal conductivity* κ for silver given above is 429 J/m·s·°C. You assume the temperature gradient remains uniform along the length of the rod. Therefore, we can find the *time it will take to melt the ice* using:

$$Q = \kappa At\Delta T/L \quad \text{or} \quad t = QL/\kappa A\Delta T$$

$$t = (6.66 \times 10^5 \text{ J})(0.30 \text{ m}) / (429 \text{ J/m·s·°C})(\pi \times 0.005^2 \text{ m}^2)(100 \text{ °C} - 0 \text{ °C})$$

$$\approx 5.9 \times 10^4 \text{ s}$$

Time in hours is: $(5.9 \times 10^4 \text{ s}) / (60 \text{ s/min})(60 \text{ min/h}) \approx 16$ h.

• While conduction occurs in liquids and gases, the more dominant form of heat transfer is convection.

Convection

• While conduction involves atoms and molecules transferring heat energy by intermolecular collisions, **convection** involves the atoms and molecules themselves moving from one place to another. Convection involves heat transfer by the movement of mass from one place to another as a heated fluid is caused to move away from a source of heat, carrying energy with it. It occurs in liquids and gases. Heat energy can be carried from one place to another in the currents of fluids and gases in the process of convection.

When most substances are heated, they expand so that their volumes increase while their densities decrease. When a liquid or gas is heated in a localized region, the heated fluid will experience a buoyant force and will rise through the cooler surrounding fluid. For example, when a pot full of water is heated from the bottom (heat is initially transferred by conduction), the warm water expands, becomes less dense (more buoyant) than the cooler water, and rises, beginning a convective process which transports energy. As the warm, less dense water rises, it is replaced with cooler water that is then heated, becoming less dense and rising. This forms a circulation of convection. This process of convection circulation within a fluid depends on the change in density that results from heating or cooling.

While convection can be induced by something external, such as a fan which causes a liquid or gas to move, it also occurs without an externally driven flow, such as when air next to a hot surface moves as a result of density differences or when we heat water in a pot. Convection is believed to be involved in the movements of the hot magma within the Earth and also in the transporting of energy from the center of the Sun to its surface.

To quantify convection with an equation, the particular situation needs to be taken into account, such as velocity of the fluid and distance traveled by fluid elements or bubbles. Nevertheless, convection is proportional to the temperature gradient and the surface area.

Radiation

• When you stand near a fire or place your hand near an incandescent light bulb, you feel heat. Some of that heat is due to conductive and convective processes, but most of it is due to radiation. **Radiation** is a type of heat or energy transfer in which heat energy is *not* transferred by molecules or mass but by electromagnetic waves. In essence, it *is heat transfer by the emission of electromagnetic waves which carry heat energy away from the emitting object.*

Heat transfer by way of electromagnetic waves (electromagnetic radiation) includes microwaves, sunlight, and infrared radiation. Radiation is the only way heat can be transferred through the relative emptiness of space. Other forms of heat transfer require motion of molecules. Energy arrives in the form of radiation from the Sun. Heat from the Sun reaches us as radiation in the form of invisible electromagnetic waves and visible light waves. A familiar example of energy transfer by radiation is a microwave oven, which transmits electromagnetic waves into food causing molecules in the food to vibrate faster.

Electromagnetic waves are generated by the atoms in any substance or material, which are always in motion. This motion increases with temperature so that collisions occur and some of the kinetic energy of the colliding atoms is absorbed by the atom's electrons. The atoms do not retain this extra energy and the electrons quickly return to their normal lower energy states in the atom as the energy that had been absorbed is emitted in the form of electromagnetic radiation. This process allows some of the kinetic energy of the rapidly moving atoms to be converted to radiant energy. Therefore, as atoms and molecules transform their kinetic heat energy into electromagnetic waves, that heat energy is then transmitted by the waves.

Forms of electromagnetic radiation or waves include radio waves, microwaves, infrared, visible light, ultraviolet, X-rays, and gamma rays. The only feature that distinguishes different types of electromagnetic radiation from one another is their frequencies or wavelengths. Otherwise they all have the same basic properties. For example, X-rays and ultraviolet radiation have higher frequencies than visible light, while radio waves and microwaves have lower frequencies.

Any object with a temperature above absolute zero emits radiation. The wavelength of the radiation is dependent on the temperature of the object. At room temperature radiation is mostly in the infrared region of the electromagnetic wave spectrum, with wavelengths longer than those of the visible spectrum. This means a human body or the walls of a room are glowing, but the glow can only be detected by special sensors. Hotter objects such as a stove burner or a star are visible because they are hot enough to emit radiation in a range of wavelengths that includes visible light. When a nonflammable object is heated, the radiation it emits increases in frequency as the temperature rises.

• Experiments have shown that the amount of energy radiated per second, or rate of energy emission, Q_e/t, by a glowing object is proportional to the fourth power of its temperature. Besides temperature, energy radiated per second has also been shown to be proportional to an object's surface area and to its *emissivity* e. This tells us that a heated object radiates energy in the form of electromagnetic radiation at a rate that is proportional to the fourth power of the temperature (in degrees K) of the object. Therefore, the **rate at which emitted energy Q_e leaves an object per second** is the **radiated power**:

$$\boxed{Q_e/t = e\sigma AT^4}$$

where e is the **emissivity**, a dimensionless number between 0 and 1 which represents the ability or efficiency of the object to emit radiation. The symbol σ is the Stefan-Boltzman constant (this equation is called the

Stefan-Boltzmann Law). Letter A represents the surface area of the radiating object, and letter T is the temperature of the radiating object on the Kelvin scale. The **Stefan-Boltzmann constant** σ is universal and is: $\sigma = 5.67 \times 10^{-8}$ J/s·m^2·K^4 = 1.36×10^{-12} cal/s·cm^2·K^4. Technically, the Stefan-Boltzmann Law describes the total energy being emitted at all wavelengths by a blackbody and describes the fourth power relationship between energy radiated and temperature. (A "**blackbody**" is a simple idealized body or object which is a perfect absorber, and therefore also a perfect emitter).

When a heated object is radiating energy to cooler surroundings which are at temperature T_c, the net loss in radiation from the emitting object is:

$$Q_e/t = e\sigma A(T^4 - T_c^{\,4})$$

Note that you will often see Q_e/t written as P for power, since power is the change in energy per time.

The **emissivity** of a substance varies with its shininess. A shiny polished metal has a nearly zero value of e and a black substance such as coal has a value near one. Dark objects usually have an emissivity of 0.9 or greater whereas light-colored or shiny objects have values of e = 0.2 or less. A perfect radiator would have e = 1. For example, carbon black soot has e = 0.95, and polished aluminum has e = 0.05. Objects that are good emitters of radiation are also good absorbers. For example, if you take two identical sheets of aluminum, except one is polished and the other is coated with carbon black, and place them in direct sunlight, after they reach equilibrium the temperature of the black sheet will be greater than the temperature of the shiny sheet. This is due to difference in emissivity. The black sheet will absorb energy and increase its temperature until the rate of absorption is balanced by the rate of emission. (The black sheet absorbs more and therefore must emit more radiation.) The polished sheet will reflect most (absorb less) of the incident energy and emit energy at a lower rate as it maintains equilibrium at a lower temperature.

• **Example**: If one object is twice as hot as another, how much more energy will it radiate?

$2^4 = 16$ times more energy per unit of surface area.

8.5. Key Concepts and Practice Problems

- Temperature reflects the average internal or kinetic energy of the atoms and molecules in a substance.
- Heat is a form of energy that flows due to temperature differences.
- $T_{Celsius} = (5/9)(T_F - 32)$; $T_{Fahrenheit} = (9/5)T_C + 32$; $T_{Kelvin} = T_C + 273$.
- Linear (length) expansion: $\Delta L = \alpha L_0 \Delta T$, with linear expansion coefficient α.
- Thermal stress: $F/A = \alpha \Delta T Y$, with expansion coefficient α, Young's modulus Y.
- Volume expansion: $\Delta V = \beta V_0 \Delta T$; with volume expansion coefficient β.
- For uniform solids volume expansion coefficient β: $\beta = 3\alpha$.
- Internal energy is increased by adding heat.
- First Law of Thermodynamics (energy conservation): $\Delta U = Q - W$, internal energy change ΔU is heat in Q minus work done.
- Specific heat capacity c is heat required to raise the temperature of a unit of mass by a unit change in temperature: $Q = mc\Delta T$.
- Latent heat of transformation $Q_L = mL$ is heat required to change phase.
- Heat energy flows spontaneously from a hot substance to a cold substance.
- Heat is transferred by: conduction via intermolecular collisions; convection via fluid motion; and radiation via electromagnetic waves.

Practice Problems

8.1 At what temperature are Fahrenheit and Kelvin degrees the same?

8.2 (a) A copper electrical cable is strung between two towers. Before being energized, the 20 °C cable is 400 m long. On a hot day it carries a large current and heats up to 90 °C. Assuming a coefficient of expansion of $1.6 \times 10^{-5}\,°C^{-1}$, how long does the cable become? **(b)** A round pond with a diameter of 1 km has a thick sheet of ice at a temperature of –5 °C. On a cold night the ice's temperature drops to –35 °C. What happens to the ice? Assume ice's coefficient of expansion is $5.1 \times 10^{-5}\,°C^{-1}$.

8.3 You and your friend are winter camping and need 1 liter (L) of water for freeze-dried meals and hot drinks. You propose taking snow with a temperature of –10 °C and heating it to 60 °C. Your friend proposes drawing 1 L of 5 °C water from an open stream and first heating it to 100 °C to make sure it is safe. You are running low on fuel. Whose plan uses the least fuel? Assume a specific heat of ice of 2,000 J/kg·°C.

8.4 (a) Your electric space heater has four Nichrome cylindrical heating elements with diameters of 1 cm and lengths of 25 cm. You turn it on "low" in a cold room at 10 °C and the elements take on a dull red glow, implying a temperature of 500 °C. At what rate is the heater radiating energy? Assume an emissivity coefficient of 0.7. (Remember to use Kelvin.) **(b)** You turn the heater to "high" and the elements begin to glow red-orange, implying a temperature of 1,000 °C. What is the new rate of energy radiation?

Answers to Chapter 8 Problems

8.1 $T_F = (9/5)(T_K - 273.15) + 32$ and $T_F = T_K = T$.
$T - (9/5)T = (9/5)(-273.15) + 32$, $(-4/5)T = -459.67$, $T \approx 574.6$ degrees.

8.2 (a) $L_{new} = L_0(1 + \alpha\Delta T)$, $L_{new} = 400(1 + (1.6 \times 10^{-5} \, °C^{-1})(70)) \approx$ 400.45 m. **(b)** $L_{new} = 1,000(1 + (5.1 \times 10^{-5} \, °C^{-1})(-30)) \approx 998.5$ m. The ice may crack in places to relieve stress.

8.3 1 L = 1 kg of water. The energy to heat the stream water is $Q = mc\Delta T = (1)(4,186)(100 - 5) = 397,670$ J. The energy to heat your snow up to 0° is $(1)(2,000)(10) = 20,000$ J. The energy to melt the snow is $Q = m_i L_i = (1)(3.33 \times 10^5 \, J/kg) = 333,000$ J. To heat the melted snow to 60° takes $Q = (1)(4,186)(60) = 251,160$ J. The total energy using snow is 604,160 J, or about 52% more than your friend's proposal. (To save more fuel, you could heat less than 1 L of stream water to 100 °C and then cool it back to 60 °C with snow!)

8.4 (a) The surface area of a cylinder is length × circumference, so the heating surface = $(4(0.25))(0.01\pi) = 0.0314$ m². $Q_e/t = e\sigma A(T^4 - T_c^4)$
= $(0.7)(5.67 \times 10^{-8})(0.0314)(773.15^4 - 283.15^4) \approx 437.3$ J/s.
(b) $Q_e/t = (0.7)(5.67 \times 10^{-8})(0.0314)(1,273.15^4 - 283.15^4) \approx 3,266.4$ J/s.

Chapter 9

GAS AND THERMODYNAMICS

9.1. Definitions—Mass, Moles, and Amu
9.2. The Gas Laws
9.3. Kinetic Theory
9.4. Diffusion and Osmosis
9.5. The Laws of Thermodynamics
9.6. Vaporization, Vapor Pressure, and Phase Change
9.7. Key Concepts and Practice Problems

> *"At quite uncertain times and places,*
> *The atoms left their heavenly path,*
> *And by fortuitous embraces,*
> *Engendered all that being hath"*
>
> Attributed to James Clerk Maxwell

9.1. Definitions—Mass, Moles, and Amu

Before learning about gases, let's review a few key definitions:

- The **Periodic Table of Elements** lists the elements and their isotopes, atomic numbers, and atomic masses.
- **Atomic number** is the total number of protons in an atomic nucleus.
- **Mass number** or **atomic mass number** is the total number of nucleons (protons and neutrons) in an atomic nucleus.
- **Atomic mass** is the rest mass of a neutral atom in its ground state measured in atomic mass units (amu). It is approximately equal to the sum of the number of protons and neutrons in an atomic nucleus.
- **Atomic mass unit (amu** or **u)** is the unit of mass defined such *that the atom ^{12}C or carbon-12 (carbon with 6 protons and 6 neutrons) has a mass of exactly 12 amu.* It is defined as 1/12 the mass of one ^{12}C atom in its ground state and is used to express the masses of atomic particles. Therefore: 1 ^{12}C atom has a mass of 12 amu. 1 amu also has the value of about 1.66×10^{-27} kg = 1.66×10^{-24} g, which is about the mass of

one nucleon (proton or neutron), which is about 1.67×10^{-27} kg. Also, 1 amu = 1 gram/mole.

- **Isotopes** of an element have the *same atomic number* (same number of protons in their nuclei) but *different mass numbers*. Atoms of a particular element by definition have the same number of protons but can have different numbers of neutrons. For example, ^{238}U and ^{235}U are isotopes of uranium, each having an atomic number of 92.
- The **atomic weight** of an element is the weighted average of the atomic masses of the different isotopes of an element.
- The **mass of any specified nucleus** is about equal to its **mass number** (total nucleons) multiplied by the **atomic mass unit** (1.66×10^{-24} g).
- **Avogadro's number** N_A is the number of atoms in one mole of an element, $N_A = 6.022 \times 10^{23}$ molecules/mole. One **mole** (mol) of any substance contains the same number of basic matter units (molecules) as does one mole of any other substance. Therefore, a mole of oxygen contains the same number of molecules as a mole of hydrogen. A mole of any substance contains Avogadro's number of molecules.
- One **mole** of a pure substance (such as ^{12}C) is a mass of that substance in grams equal to its **molecular mass** (total nucleons) in atomic mass units. Since the mass of one ^{12}C atom is 12.0 amu, then one *mole* of ^{12}C atoms has a mass of 12.0 grams.
- A mole is also defined as the number of ^{12}C atoms in exactly 12 grams of ^{12}C. A mole of any substance contains Avogadro's number of molecules, $N_A = 6.022 \times 10^{23}$. Therefore, in the case of ^{12}C:

$$6.022 \times 10^{23} \text{ atoms } ^{12}C = 12 \text{ g } ^{12}C$$
$$1 \text{ atom } ^{12}C = 12 \text{ amu}$$
$$6.022 \times 10^{23} \text{ atoms} = 1 \text{ mole of atoms}$$

The *molar mass* in g/mol of a single ^{12}C atom is:

$$(12 \text{ g } ^{12}C)/(6.022 \times 10^{23} \text{ atoms}) \times (6.022 \times 10^{23} \text{ atoms})/(\text{mole}) = 12 \text{ g/mol } ^{12}C$$

which is called the *molar mass* of ^{12}C. The **molar mass** is the mass of one mole of a substance. Note that if you weighed a 12-g sample of carbon, it would have 6.022×10^{23} atoms of carbon.

If you have two different elements with mass numbers 1 (for H) and 12 (for C), then 1 gram of H will contain the same number of atoms as 12 grams of C, which is true for all elements and is Avogadro's number $N_A = 6.022 \times 10^{23}$. Therefore, the number of atoms in 1 gram of hydrogen is 6.022×10^{23}, and the number of atoms in 12 grams of carbon is also 6.022×10^{23}. Similarly water, which has a combined mass number of 18, has 6.022×10^{23} molecules in a quantity whose mass is 18 grams.

9.2. The Gas Laws

• A **gas** can expand and contract in volume, depending on temperature and pressure. In the 17th and 18th centuries the behavior of gases was studied by a number of people including Amontons, Boyle, Charles, and Avogadro. Together they discovered important relationships that were combined into the **Ideal Gas Law**:

$$PV = nRT$$

where T is temperature and must be given in degrees Kelvin; P is pressure; V is volume; n is the number of moles of gas molecules with 1 mole = 6.022×10^{23} molecules; and R is the **Universal Gas Constant**, or 8.314 J/mol·K. By observing the equation you can see that P and V are each directly proportional to T, and P and V are inversely proportional to each other. This equation helps us see how T, P, or V change with respect to the others, particularly when one is held constant. Note: The characteristics of an **ideal gas** include assumptions that collisions between atoms or molecules are perfectly elastic, there are no inter-molecular attractive forces, internal energy exists as kinetic energy, and any change in internal energy is reflected in temperature. The Ideal Gas Law is useful in solving problems because it serves as a good model for the behavior of real gases at low densities.

Development of the Ideal Gas Law

• **Amontons' Law:** Amontons discovered that there is a linear increase in pressure with temperature. Amontons' Law, as it is called, states that the ***pressure of a gas is directly proportional to its temperature*** in Kelvin at ***constant volume***. (Subscripts denote initial and final.)

$$P \propto T \quad \text{when } \textbf{\textit{V is constant}}$$
$$P_i/P_f = T_i/T_f$$

• **Boyle's Law:** Boyle discovered that when the ***temperature*** of a confined gas is held ***constant***, the ***volume is inversely proportional to the pressure***. This means that when temperature is constant and pressure increases, the volume will decrease, and vice versa. Therefore:

$$P \propto 1/V \quad \text{or} \quad PV = \text{constant when } \textbf{\textit{T is constant}}$$
$$P_i V_i = P_f V_f \quad \text{when T is constant so } T_i = T_f$$

You can model this by submersing a gas-filled cylinder in an ice-water bath and decreasing its volume by pressing on a plunger in the cylinder. When you slowly decrease the volume of the gas, the heat generated by work being done on the gas will immediately escape to the environment so that the gas remains at its initial temperature T_i. This type of constant temperature system is called an **isothermal** (equal temperature) system.

• **Charles' Law**: Charles discovered that when the *pressure* of a contained gas is held *constant*, the *volume and temperature are directly proportional*:

$$V \propto T \quad \text{when } P \text{ is constant}$$
$$V_i/T_i = V_f/T_f \quad \text{when P is constant}$$

If you plot volume versus temperature at constant pressure, a linear relationship will result, which is Charles' Law. If you extrapolate the line to zero on the volume axis, it will also cross the temperature axis at $-273.15\,°C$. This was foundational to understanding absolute zero. Charles' Law explains the principle behind hot air balloons such that the air in the balloon expands when heated making it less dense than surrounding air and able to "float" (until cooled).

• **Avogadro's Law or Hypothesis**: This hypothesis states that *equal volumes of different gases at the same temperature and pressure contain equal numbers of molecules*. It also states that *at a constant temperature and pressure, the volume of a gas is directly proportional to the number of moles, n, of that gas*:

$$V \propto n \quad \text{or} \quad V/n = \text{constant} \quad \text{when } P \text{ and } T \text{ are constant}$$

Because of Avogadro's Law, the ideal gas constant is the same for all gases.

• The relationships or laws of Amontons, Boyle, and Charles, when combined with Avogadro's Law, form the relationship:

$$PV \propto T$$

The proportionality constant, **when one mole of a gas is present**, is the **Universal Gas Constant R**, which is $R = 8.314$ J/mol·K. This gives:

$$PV = RT \quad \text{for 1 mole}$$

For **n moles** of a gas the law is:

$$\boxed{PV = nRT}$$

which is the standard form of the **Ideal Gas Law**. It is also written in terms of the *number of molecules* N and the **Boltzmann Constant** k:

$$\boxed{PV = NkT}$$

where N = number of molecules, $k = R/N_A = 1.38 \times 10^{-23}$ J/K, and Avogadro's number $N_A = 6.022 \times 10^{23}$ molecules/mole. Therefore, we can express the Ideal Gas Law in terms of the number of molecules N in a sample or the number of moles n. We can equate the Boltzmann Constant k and the Universal Gas Constant R:

$$Nk = nR$$

Because the ratio N/n is the number of molecules per mole, which is Avogadro's Number:

$$N_A = N/n \quad \text{or} \quad N = nN_A$$

where N_A is Avogadro's number and n is the number of moles. The gas constant R is:

$$R = Nk/n = N_A k \quad \text{or} \quad k = R/N_A$$

- It is often useful to write the Ideal Gas Law as:

$$P_i V_i/n_i T_i = P_f V_f/n_f T_f \quad \text{or} \quad P_i/P_f \times V_i/V_f \times n_f/n_i = T_i/T_f$$

where the subscripts denote initial and final. We have seen that gases can be described in terms of the four variables: pressure P, volume V, temperature T, and the amount of gas n. When certain variables are held constant, the Ideal Gas Law reduces to one of the simpler laws we just discussed. If the number of moles remains constant, then:

$$P_i V_i/T_i = P_f V_f/T_f$$

If the temperature and number of moles remain constant, we have Boyle's Law:

$$P_i V_i = P_f V_f$$

If the pressure and number of moles remain constant, we have Charles' Law, which is useful for modeling gases that remain at constant atmospheric pressure:

$$V_i/T_i = V_f/T_f \quad \text{or} \quad V_f = V_i T_f/T_i$$

- Because gases depend on temperature and pressure, it is important to specify their values. Gases are often quoted or referenced according to conditions of **standard temperature and pressure** (STP). The standard temperature is the freezing point of water 0 °C = 273.15 K, and the standard pressure is one atmosphere or 1.013×10^5 Pa. Note that at STP, the **standard volume** that one mole of an ideal gas will occupy is 22.4 liters (L). Therefore, at the same temperature and pressure, any two gases will occupy the same volume, and at 0 °C and 1 atm pressure, that volume is 22.4 L.

- **Example**: Prove that the standard volume that one mole of an ideal gas will occupy is 22.4 L.

We can solve PV = nRT for V, and set n = 1 mole, T = 273.15 K, and P = 1 atm = 1.013×10^5 Pa:

$$V = nRT/P = (1 \text{ mol})(8.314 \text{ J/mol·K})(273.15 \text{ K})/(1.013 \times 10^5 \text{ Pa})$$
$$V \approx 0.0224 \text{ m}^3 = 22.4 \text{ L}$$

Note that 1 liter = 1,000 cm^3 = 0.001 m^3, or 1 m^3 = 1,000 L.

• **Example**: Suppose your friend asks you what the pressure of a gas will be if it is held in a cylinder at constant temperature while a piston compresses it to half its original volume.

You model this as an ideal gas and realize that at constant temperature (and no change in number of moles) the Ideal Gas Law, PV = nRT, can be written as Boyle's Law, $P_iV_i = P_fV_f$. You see that the pressure and volume are inversely proportional. If volume is reduced to half the initial volume, the pressure will double.

• **Example**: Satisfied with your last response, your friend comes up with another scenario for you to answer. If a gas in a sealed cylinder is heated from 0 °C to 273 °C, what will happen to its pressure?

You realize the volume is held constant due to the closed cylinder (and no change in number of moles), so you model this as an ideal gas using the Ideal Gas Law, PV = nRT, which becomes $P_i/T_i = P_f/T_f$. Since temperature needs to be measured in Kelvin, 0 °C to 273 °C is 273 K to 546 K, which you realize means that the temperature doubles. We can see according to the gas law equation that pressure and temperature are directly proportional, so if temperature is doubled, then the pressure will also double.

• **Example**: If you partially evacuate a 1-m^3 cylinder to a pressure of 0.01 atm, what is the mass of nitrogen gas, which has a molecular weight of 28, if the temperature is 10 °C?

You treat this as an ideal gas and solve PV = nRT for n: n = PV/RT. The temperature in Kelvin is 283 K.

$$n = (0.01 \text{ atm})(1.013 \times 10^5 \text{ Pa/atm})(1 \text{ m}^3)/(8.314 \text{ J/mol·K})(283 \text{ K}) \approx 0.43 \text{ mol}$$

The mass contained in this amount of gas is:

$$m = (0.43 \text{ mol})(28 \text{ g/mol}) \approx 12 \text{ g} = 0.012 \text{ kg}$$

• **Example**: If a helium balloon designed to survive high altitudes is filled with 300 m^3 of helium and released from sea level at 10 °C and rises to an altitude where the pressure is about 0.01 atm and the temperature is about −50 °C, what will be its volume? (Ignore pressure due to the elasticity of the balloon.)

The number of moles remain constant so if we model as an ideal gas, $P_iV_i/T_i = P_fV_f/T_f$, we can solve for V_f:

$$V_f = P_i V_i T_f / T_i P_f$$

Using: initial volume as 300 m³; initial and final temperatures as 283 K and 223 K; and initial and final pressures as 1 atm and 0.01 atm. Note that if we multiply pressure in atm by 1.013×10^5 Pa/atm to get pressure in Pa, the values of 1.013×10^5 will cancel, so we can leave this in atm.

$$V_f = (1 \text{ atm})(300 \text{ m}^3)(223 \text{ K}) / (283 \text{ K})(0.01 \text{ atm}) \approx 23{,}640 \text{ m}^3$$

Note: 23,640 m³ is almost 79 times greater gas volume than 300 m³. This illustrates why high altitude balloons are initially only partly filled.

9.3. Kinetic Theory

• The **kinetic theory of gases** describes gases as aggregates of individual particles. It relates the microscopic molecular properties of velocity and kinetic energy which obey Newton's Laws to the macroscopic properties of temperature and pressure which can be modeled using the Ideal Gas Law. The development of the kinetic theory makes the following assumptions about ideal gases:

- The number of molecules (or atoms) in a sample is large enough that statistical modeling of the average behavior of the particles is valid.
- The separation between molecules is great compared to molecular size so the molecules do not influence each other (except when colliding).
- The molecules impact the walls of a container, a surface, or each other in perfectly elastic collisions. There are no other forces exerted and no kinetic energy transferred or converted to potential energy or radiation.
- The molecules obey Newton's Laws of motion.
- The molecules are in random motion with a distribution of velocities that remains constant: $(v_x^2)_{Ave} = (v_y^2)_{Ave} = (v_z^2)_{Ave}$ (discussed below).

• We can now develop equations that relate the velocity and kinetic energy of gas molecules to the macroscopic properties using a sample of molecules. These equations will be valid for ideal gases or gases that can be modeled as ideal gases. We begin by imagining a sample of ideal gas molecules moving with random velocities confined within a cubical container having side length L. A single molecule has velocity vector **v** with components v_x, v_y, and v_z. The *momentum* p of the molecule in the x direction is $p_x = mv_x$, where m is its mass. If the molecule collides perpendicularly with the wall of its container in an elastic collision, the velocity components v_y and v_z do not change. The momentum of the molecule after collision is $-mv_x$, so the change in momentum is:

$$\Delta p = mv_x - (-mv_x) = 2mv_x$$

After the molecule collides with the right-hand wall, it bounces back to the left wall and then returns to the right wall. In this cycle it passes through a distance of 2L. Because the velocity is distance per time interval, $v_x = 2L/\Delta t$, the time it takes to complete one round trip is:

$$\Delta t = 2L/v_x$$

The average force ($F = ma = mv/t$) over time Δt that this one molecule exerts on the wall of the container is:

$$F_{Ave} = \Delta p_x/\Delta t = 2mv_x/(2L/v_x) = mv_x^2/L$$

Now we extend this to N molecules in the container to determine the **total force in the x direction on the container wall**:

$$\boxed{F_x = Nm(v_x^2)_{Ave}/L}$$

where $(v_x^2)_{Ave}$ is the **average of the square** of the x component of velocity, *not* the *square of the average*. This is an important distinction we can see by comparing $(v_{Ave})^2$ and $(v^2)_{Ave}$. For example, for velocities 1, 2, and 3:

$$(v_{Ave})^2 = [(1 + 2 + 3)/3]^2 = 4 \quad \text{and} \quad (v^2)_{Ave} = (1^2 + 2^2 + 3^2)/3 \approx 4.67$$

To find the total force exerted on any of the walls, we also consider the y- and z-components of velocity. *Since the gas molecules move **randomly**, the average velocity components in all three dimensions are equal:*

$$(v_x^2)_{Ave} = (v_y^2)_{Ave} = (v_z^2)_{Ave}$$

$$(v^2)_{Ave} = (v_x^2)_{Ave} + (v_y^2)_{Ave} + (v_z^2)_{Ave} = 3(v_x^2)_{Ave}$$

Substituting $(v_x^2)_{Ave} = (v^2)_{Ave}/3$ we find the **total force F exerted on any single wall by N gas molecules** is:

$$\boxed{F = Nm(v^2)_{Ave}/3L}$$

Since the area A of each wall is L^2, the **pressure P** is:

$$P = F/A = F/L^2 = Nm(v^2)_{Ave}/3L^3 = Nm(v^2)_{Ave}/3V \quad \text{or} \quad P = Nm(v^2)_{Ave}/3V$$

where V is the volume of the container. We can also write this as:

$$PV = \tfrac{1}{3}Nm(v^2)_{Ave}$$

Since the **average kinetic energy of a gas molecule** is:

$$(KE)_{Ave} = (1/2)(m)(v^2)_{Ave}$$

then:

$$PV = (2/3)(N)(KE)_{Ave}$$

We can now combine the Ideal Gas Law developed in the preceding subsection, $PV = NkT$, where N is number of molecules and $k = 1.38 \times 10^{-23}$ J/K, with the previous equation describing gas molecules to obtain an equation in terms of temperature:

$$PV = (2/3)N(KE)_{Ave} = NkT$$

$$\boxed{(KE)_{Ave} = (3/2)kT}$$

This equation for **average kinetic energy per molecule** shows that the *temperature of an ideal gas is proportional to the average kinetic energy per molecule and that the average kinetic energy does not depend on mass, just temperature.* Also, $k = R/N_A = 1.38 \times 10^{-23}$ J/K is the Boltzmann constant. We can also write **the average kinetic energy per mole** of an ideal gas using the Universal Gas constant, $R = Nk/n = N_A k = 8.314$ J/(mol·K):

$$\boxed{(KE)_{Ave} = (3/2)RT}$$

We can now solve for molecular velocity by taking the square root of $(v^2)_{Ave}$, which gives us the **root mean square velocity** v_{rms} (rather than the average velocity v_{Ave}). The *root mean square velocity* is the square root of the mean square velocity $(v^2)_{Ave}$. First solve

$$(KE)_{Ave} = (3/2)kT = (1/2)(m)(v^2)_{Ave}$$

for $(v^2)_{Ave}$:

$$(v^2)_{Ave} = 3kT/m$$

Taking the square root gives the **root mean square speed of the molecules** in a gas:

$$\boxed{v_{rms} = [3kT/m]^{\frac{1}{2}}}$$

where m is mass, T is temperature in Kelvin, and $k = 1.38 \times 10^{-23}$ J/K is the **Boltzmann constant**.

• **Example**: Your friend wants to know the average kinetic energy and root mean square velocity of the oxygen molecules in the 30 °C air he is breathing. (O_2 has a molecular mass of 32, and there are 1.66×10^{-27} kg per amu.)

You realize that the type of gas doesn't matter, just the temperature. You remember you need to use temperature in Kelvin and write the equation for the average kinetic energy of the molecules in a gas:

$$(KE)_{Ave} = (3/2)kT = (3/2)(1.38 \times 10^{-23} \text{ J/K})(303 \text{ K}) = 6.27 \times 10^{-21} \text{ J}$$

Your friend comments on what an unwieldy number this is and remembers that 1.602×10^{-19} J = 1 eV (electronvolt). Therefore:

$$(KE)_{Ave} = (6.27 \times 10^{-21} \text{ J})(1 \text{ eV}/1.602 \times 10^{-19} \text{ J}) \approx 0.039 \text{ eV}$$

Next you write the root mean square velocity as $v_{rms} = [3kT/m]^{\frac{1}{2}}$, or:

$$v_{rms} = [(3)(1.38 \times 10^{-23} \text{ J/K})(303 \text{ K})/(32 \text{ amu})(1.66 \times 10^{-27} \text{ kg/amu})]^{\frac{1}{2}} \approx 486 \text{ m/s}$$

"Wow! That's over a thousand miles per hour," your friend exclaims. You remind him there are a lot of collisions with other molecules so each molecule doesn't get too far.

• The velocities of the molecules within a gas vary and can be described as a distribution of velocities. The distribution of the velocities is given by **Maxwell velocity distribution** functions, which were developed by **James Clerk Maxwell**, a Scottish physicist, in the 1800s. Without considering the directions of the gas molecules, we can imagine the distribution of their speeds. If a gas is at a certain uniform temperature, the speeds of the molecules depend on the temperature of the gas. The general form of the **Maxwell distribution of speeds** for molecules in an ideal gas can be described by:

$$n(v) = 4\pi[m/2\pi kT]^{3/2} v^2 e^{-(m)(v \times v)/(2kT)}$$

where $n(v)$ is the relative number of molecules having speed near v, m is mass, $k = 1.38 \times 10^{-23}$ J/K is the Boltzmann constant, and T is the temperature in K. The **most probable speed** and the **mean speed** for the distribution are:

$$v_p = [2kT/m]^{\frac{1}{2}} \quad \text{and} \quad v_m = [8kT/\pi m]^{\frac{1}{2}}$$

As we described above, the **root mean square speed** of the molecules in the gas is:

$$v_{rms} = [3kT/m]^{\frac{1}{2}}$$

If you graph $n(v)$ the relative number of particles having speed v, you can locate the most probable speed v_p, mean speed v_m, and root mean square speed v_{rms}. This graph depicts a gas at two different temperatures T_1 and T_2 (where $T_1 < T_2$), with v_p, v_m, and v_{rms} shown for T_2.

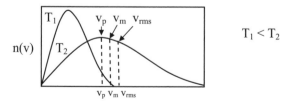

The Maxwell distributions for T_1 and T_2 show that *the range of velocities for the molecules is greater at higher temperatures*. We can also see that the distribution is not symmetric as there are more molecules at greater speeds. If you consider the gas around a planet, the velocities of gas molecules that significantly exceed v_{rms} may allow them to eventually escape from the atmosphere if they exceed the escape velocity for that planet. The Maxwell distributions can be used to determine certain characteristic speeds of the molecules in a gas such as the fraction of molecules having speeds over a particular value at various temperatures.

9.4. Diffusion and Osmosis

Diffusion

• Consistent with kinetic theory, the molecules in both gases and liquids are in constant random motion due to internal thermal kinetic energy. This motion causes the molecules to collide with each other as well as any nearby surface. The motion caused by their kinetic energy results in the **diffusion** of individual molecules from one location to another. The high velocities of the molecules, which are associated with their thermal energy, cause them to diffuse. The process of diffusion allows gas molecules to fill an evacuated chamber or allows two or more different types of gas molecules to intermingle.

For example, if you have two gases separated by a barrier, the molecules of both gases, which are in constant motion, will continually collide with each other and the barrier between them. If you remove the barrier, the random motion of the gases will cause them to intermingle. After some time the gasses will reach equilibrium and a uniform mixture of both types of gas molecules will occur.

  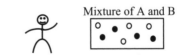

If you open a bottle of perfume in the corner a sealed room with no wind or temperature gradient, the perfume will eventually diffuse throughout the room until the concentration of perfume molecules everywhere in the room is the same.

• What drives diffusion? *Diffusion in a gas or liquid is driven by and therefore occurs in response to a concentration gradient.* As long as there is no obstruction, molecules will move from regions of higher concentration (or higher energy) to regions of lower concentration (or lower energy). Diffusion can be described using equations which reflect the driving force being a concentration gradient. If C_1 is the concentration (molecules per unit volume) of perfume molecules in the air at position 1, C_2 is the concentration at position 2, and Δx is the distance between, the molecules will tend to diffuse from the region of higher concentration (C_2) toward the region of lower concentration (C_1). The *rate of movement* will be proportional to the concentration gradient between C_2 and C_1. The **concentration gradient** is:

$$(C_2 - C_1)/\Delta x = \Delta C/\Delta x$$

If there is a concentration gradient perpendicular to a permeable surface the rate molecules diffuse across that surface is proportional to the surface area A. The **number of molecules ΔN that diffuse across the surface during time interval Δt** is $\Delta N/\Delta t$:

$$\Delta N/\Delta t = -DA\Delta C/\Delta x = -DA(C_2 - C_1)/\Delta x$$

This is called **Fick's Law**. The negative sign indicates that the direction of diffusion occurs from high to low concentration and is therefore opposite to increasing concentration. *D* is the **diffusion coefficient**, also called the **diffusivity**, of the particular substances and is in **units** of length^2/time, or m^2/s or cm^2/s. Just as the rate of diffusion depends on the speed of the diffusing molecules (which depend on mass and temperature), the diffusivity *D* depends on molecular mass of the diffusing gas or liquid and the temperature, as well as on the pressure and properties of the fluid in which diffusion takes place. **Diffusivity** is a measure of how quickly a substance diffuses through a medium. The diffusivity or diffusion coefficient mathematically is a proportionality constant between the molar flux due to molecular diffusion and the driving force for diffusion, which is the concentration gradient of the diffusing atoms or molecules. (Note: **Flux** is the rate of flow of particles, fluid, or energy through an area.) Sample diffusion coefficients for O_2 in air is 0.2 cm^2/s = 2.0×10^{-5} m^2/s, and O_2 in water is 2×10^{-5} cm^2/s = 2.0×10^{-9} m^2/s.

• **Fick's Law** can also be written in terms of density ρ, with the letter J used rather than $\Delta N/\Delta t$ to describe the mass flow rate in kg/s or mol/s:

$$J = DA\Delta\rho/\Delta x = DA(\rho_2 - \rho_1)/\Delta x$$

where J represents the mass flow rate more generally called the *diffusion flux,* D is the diffusivity or diffusion coefficient, and $\Delta\rho/\Delta x$ is the concentration gradient. The negative sign is not used since this models mass flow rate and has a time component. Note also that we are presenting Fick's Law in a simple form without using calculus and assuming one-dimensional steady state diffusion. Because J represents the mass flow rate or m/t, the **time required for diffusion** can be modeled by solving for t:

$$t = (m\Delta x) / (DA(\rho_2 - \rho_1))$$

To consider volume, we multiply the numerator and denominator by Δx and remember that $A\Delta x = $ Volume V:

$$t = (m(\Delta x)^2) / (DV(\rho_2 - \rho_1))$$

Since the average density is $\rho_{Ave} = (\rho_1 + \rho_2)/2 = m/V$:

$$\boxed{t = ((\Delta x)^2 \rho_{Ave}) / (D(\rho_2 - \rho_1)) = ((\Delta x)^2(\rho_1 + \rho_2)) / (2D(\rho_2 - \rho_1))}$$

These equations are true whether the concentrations are in kg/m³ or mol/m³. Note that as molecules move they collide with other molecules repeatedly, and randomly change direction in each collision proceeding in what is called a "random walk" rather than straight-line motion.

• **Example**: If you open a bottle of perfume at one end of a closed room with static air, how long will it take, on average, for a perfume molecule to diffuse 10 m? Assume the diffusion coefficient is 1×10^{-5} m²/s.

We begin with: $t = ((\Delta x)^2(\rho_1 + \rho_2)) / (2D(\rho_2 - \rho_1))$.
But since the initial density is 0, the equation reduces to:

$$t = (\Delta x)^2 / 2D = (10 \text{ m})^2 / 2(1 \times 10^{-5} \text{m}^2/\text{s}) = 5 \times 10^6 \text{ s}$$

Osmosis

• **Osmosis** involves selective diffusion through a porous barrier or semi-permeable membrane that is permeable to certain substances but not others. A semi-permeable membrane or barrier may allow or block atoms and molecules according to their size or some other property such as charge. Suppose you have two chambers separated by a semi-permeable membrane with pure water on one side and a sugar solution on the other side. Imagine that the semi-permeable membrane allows the smaller solvent (water) molecules to pass but not the larger solute (sugar) molecules. It turns out that the solvent will tend to diffuse across the membrane from the side with either zero or a lower concentration of solute to the side containing the solute or having a greater concentration of solute. This process is called **osmosis**.

The fine dots represent the solvent (water), the large dots the solute (sugar), and the dashed vertical line the semi-permeable membrane. The solvent particles can pass through the membrane while the solute cannot. The figure shows that the solvent (water) diffuses toward the side containing solute (or a higher concentration of solute). The solvent will flow until the solvent reaches equilibrium so that the partial pressure of the free-flowing solvent is the same on both sides of the membrane. At equilibrium for the solvent, there will be the additional partial pressure of the solute which makes the pressure on the solute side greater.

To visualize this using the figure, we see that initially before osmosis the total liquid pressure is the same in both chambers because the liquid levels are equal, but the pressure due to water molecules is greater in the pure water chamber. Since all the molecules are in thermal motion, the water molecules are hitting the membrane from both sides, and the sugar molecules are hitting the membrane from the right side. The semi-permeable membrane allows water molecules to pass freely but sugar molecules cannot pass. The rate that the water molecules reach the membrane surface is smaller on the side containing sugar molecules since the sugar molecules near the membrane block some of the water molecules. This allows for a greater rate of transport of water molecules through the membrane from the pure water side. Therefore, during osmosis the water flows from higher water pressure on the pure water side to lower water pressure on the sugar side. This net flow of water into the sugar side continues until the pressure due to water molecules hitting the membrane is the same on both sides.

After osmosis, the water pressure is the same on both sides of the membrane, although there is additional pressure due to the sugar that is all on the sugar side. The total pressure in any fluid (liquid or gas) is equal to the sum of the partial pressures of its various constituents. After osmosis, the side with the sugar has a greater total pressure than the side with only water, and the difference measures the **osmotic pressure** which is equal to the partial pressure of the sugar. This is depicted as the difference in liquid levels after osmosis, h (as shown in the above figure), and measures the pressure difference $\rho g h$, which also reflects the **osmotic pressure** of the sugar solution. The osmotic pressure of a solution therefore depends on the concentration of particles that will not pass through the membrane.

• Interestingly, osmosis allows the creation of a pressure difference or a greater pressure difference across a membrane. During osmosis, a semi-permeable membrane acts both as a filter by allowing some substances to pass but not others, and as a pump by allowing the creation of a pressure difference. Both functions are important in biological systems including the transport across cell membranes, the stiffness of plants and leaves, and the filtering function of waste removal by the kidneys.

• In 1887 J.H. van't Hoff discovered that **osmotic pressure** Π varies directly with both solute concentration C and absolute temperature T. Using R as the proportionality constant he found:

$$\Pi = R_o CT$$

where C = n/V = (number of moles of solute)/(unit volume of solution). Therefore:

$$\Pi V = nR_o T$$

This has the same form as the Ideal Gas Law equation PV = nRT, where the Universal Gas Constant is R = 8.314 J/mol·K. However, the constant R_o is determined from osmotic pressure measurements as:

$$R_o = 0.0827 \text{ atm·L/mol·K}$$

If we transform the conventional units of the gas constant R_o into R:

$$R = 0.0821 \text{ atm·L/mol·K}$$

where L is liters, R is the gas constant in units 0.0821 L atm/mol·K. This means $\Pi = R_o CT$ represents an approximation of the Ideal Gas Law equation for the solute molecules. Van't Hoff realized that the gas law PV = nRT applied to liquids by relating the molecules of solute (acting as a gas) moving though the solvent (acting as the medium or Universe). We can see that the Ideal Gas Law can be written as P = (n/V)RT, where (n/V) is moles divided by liters, which is molarity. The equation $\Pi V = nR_o T$ is only valid, however, for concentrations of solute in the solvent that are not too large (tenths of mol/l or less). For larger solute concentrations, the osmotic pressure is greater than that predicted by this simple equation and requires more advanced mathematics.

9.5. The Laws of Thermodynamics

• **Dynamics** studies objects in motion and, since "**thermo**" refers to heat, *thermodynamics* is the study of heat that is moving or flowing. There are four laws of thermodynamics you will likely encounter which are labeled Zero through Three. You may discover that the laws are presented a little differently depending on the discipline. In general, **thermodynamics** studies the effects of heat, energy, and work on a system at the macroscopic level. Its development and early focus involved the design and operation of steam engines. It is a complement to kinetic theory which involves the microscopic level.

The **Zeroth Law** involves thermodynamic equilibrium, temperature, and temperature measurement. The **First Law** involves heat, energy, work, and the conservation of energy. The **Second Law** involves entropy. The **Third Law** involves absolute zero temperatures and thermal motion.

Each law leads to definitions of thermodynamic properties which help us understand and model physical systems.

The Zeroth Law of Thermodynamics

• The **Zeroth Law of Thermodynamics** involves thermodynamic equilibrium and states that *when two objects are each in thermodynamic equilibrium with a third object, they are in thermal equilibrium with each other.* That means if two systems or objects have the same temperature, they are in thermal equilibrium. For example, if objects A and B have the same temperature and are therefore in thermal equilibrium, and if objects B and C have the same temperature and are therefore in thermal equilibrium, then objects A and C have the same temperature and are therefore in thermal equilibrium.

The Zeroth Law of Thermodynamics has been used as the basis for measurement of temperature and setting its scale by creating a thermometer. We can calibrate the change in a thermal property, such as the volume of mercury (or the length of a column of mercury), by putting a thermometer in thermal equilibrium with a known physical system at several reference points. For example, Celsius thermometers have their reference points fixed at the freezing and boiling points of pure water. If we then bring the thermometer into thermal equilibrium with any other system, such as a cup of coffee, we can determine the temperature of the coffee by measuring the change in the thermal property. Therefore, measuring temperature involves the use of a reference object or system that has some thermodynamic property such as volume which changes reliably with changing temperature and can be measured (such as the height of a column of mercury).

Because of the Zeroth Law, when two objects of different temperatures come into contact with each other, heat (which is a form of energy) will flow from the higher-temperature object to the lower-temperature object until they reach thermal equilibrium. When two objects are in thermal equilibrium, they have the same temperature.

The First Law of Thermodynamics

• Energy can exist in and be converted between various forms including internal (atomic vibrations), chemical (in chemical bonds), electrical, potential, and kinetic. **Conservation of energy**, which is a statement of the **First Law of Thermodynamics,** says that *energy cannot be created or destroyed but can be transformed from one form to another or moved from place to place.* (See Section 5.9 for a discussion of conservation of energy. Also as mentioned in Section 5.9, in light of relativity, conservation of energy may be thought of in a larger picture as the conservation of mass-energy.)

Heat and work are forms of energy or a means of transferring energy. Any heat or work that is absorbed or done to a system, or released or done by a system, will affect the *internal energy* of that system. **Internal energy** U arises from molecular motion and is mostly a function of temperature. It is associated with the random motion of the particles. When heat Q is added to (or released from) a system, and the system does work W on its surroundings (or the surroundings do work on the system), then the internal energy U in a system changes. Energy is related to heat and work by the **First Law of Thermodynamics**, which can be written:

$$\Delta U = Q - W$$

where energy, heat, and work are in the same units, Q is heat added to the system, and W is work done by the system. **The First Law of Thermodynamics** *defines the internal energy change of a system, ΔU, as equal to the heat transfer Q into a system minus the work W done by the system to its surroundings*. Note that we can also write the First Law on a per unit mass basis as: $\Delta u = q - w$, with units in J/kg.

• When heat is transferred *into* the system, Q is *positive*, and when heat is transferred *out* of the system, Q is *negative*. When the system does work on its surroundings, work W is *positive*, and when the surroundings do work on the system, W is *negative*. Note that work is subtracted from Q so there is a negative sign before W. You will therefore see the assignment of the positive and negative to work written both ways. Let's show some examples:

If a system does 3 Joules of work on its surroundings while 4 Joules of heat is added to the system:

$\Delta U = Q - W = [4\ \text{J heat in}] - [3\ \text{J work out}] = 1\ \text{J change in U}$

Alternatively, if a system does 3 Joules of work on its surroundings while 4 Joules of heat is released from the system:

$\Delta U = Q - W = [-4\ \text{J heat out}] - [3\ \text{J work out}] = -7\ \text{J change in U}$

If a system has 3 Joules of work done on it by its surroundings while 4 Joules of heat is added to the system:

$\Delta U = Q - W = [4\ \text{J heat in}] - [-3\ \text{J work in}] = 7\ \text{J change in U}$

If a system has 3 Joules of work done on it by its surroundings while 4 Joules of heat is released from the system:

$\Delta U = Q - W = [-4\ \text{J heat out}] - [-3\ \text{J work in}] = -1\ \text{J change in U}$

• Any thermodynamic system that is in a state of equilibrium possesses the so-called state variable internal energy ($\Delta U = Q - W$). If there is a change from one state of equilibrium to another, there will be a change in the internal energy. While work and heat depend not only on the initial and final state of a gas but also on process, the difference between the heat flow into a gas and the work done by a gas depends only on the initial and final states of the gas and *not* on the process or path which produces the final state. Therefore the change in internal energy ($Q - W$) of the gas depends only on the state of the gas and not on any process. Internal energy is a form of energy just like potential energy or kinetic energy, and they can be transformed from one to the other.

• **Example**: In Section 8.3 we introduced the First Law of Thermo-dynamics and illustrated a piston holding a contained gas under pressure. We said that if the gas is heated, it expands, thereby doing work on the piston by lifting it against gravity, and that if the piston is pushed down, the piston does work on the gas. Suppose heat is added to the piston system. Instead of letting the gas heat up (which increases its pressure thereby pushing up the piston), you slowly lift the piston and reduce the gas pressure so that the temperature remains constant. What happens to the heat energy that is added to the system?

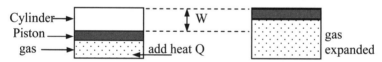

Because the temperature remains constant, the internal energy of the system remains constant ($\Delta U = 0$). The First Law becomes:
$$\Delta U = Q - W = 0 \quad \text{or} \quad Q = W$$
As the gas expands, it performs work W on the piston because the force of the gas's pressure acts on the piston through a vertical distance, helping you lift the piston. Because $\Delta U = 0$, the work done by the gas on the piston must be equal to the amount of heat Q added to the gas. Therefore, all of the heat energy is converted into work as the gas expands its volume, pushing up the piston. Note that the sum of the work done by the gas and the work done by you in lifting the piston equals the total work performed on the piston.

• Just as we described the Ideal Gas Law with certain conditions held constant, thermodynamic processes can be undertaken and described by restricting the change in certain variables. Work can be performed under the conditions: **adiabatic** with constant heat, **isothermal** with constant temperature, **isochoric** with constant volume, and **isobaric** with constant pressure. When we describe these processes we assume they proceed slowly enough that the temperature and pressure are the same throughout the gas or system. This slow change is referred to as being quasi-static.

Constant Heat: Adiabatic

• If a system is thermally insulated from the outside environment, it is possible to have a thermodynamic process and change of state in which **no heat is transferred into (or out of) the system**. This is called an **adiabatic process** and $Q = 0$. The change in internal energy:

$$\Delta U = Q - W \quad \text{becomes} \quad \Delta U = -W$$

The work done in compressing the gas goes into the internal energy. Work done in any adiabatic process is independent of path or process, so that the final state will be the same for a certain amount of work regardless of how it was performed.

• For an adiabatic process temperature and pressure can be described using simple equations. In Section 9.3 above we learned that the average kinetic energy per mole of an ideal gas is:

$$(KE)_{Ave} = (3/2)RT \quad \text{where} \quad R = 8.314 \text{ J/(mol·K)}$$

For an ideal gas the internal energy can therefore be written:

$$U = (3/2)nRT \quad \text{or} \quad \Delta U = (3/2)nR(T_f - T_i)$$

where the initial temperature is T_i and the final T_f. Since $\Delta U = -W$, the **work done in an adiabatic system** is:

$$\boxed{W = (3/2)nR(T_i - T_f)}$$

If the temperature decreases, the gas does work on its surroundings. The work done in an adiabatic process can be seen from the graph of pressure versus volume as the area under the "adiabat", which is the adiabatic curve plotted.

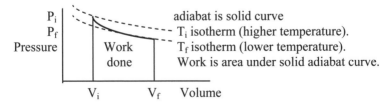

Constant Temperature: Isothermal

• A system in which a gas is held at constant temperature by allowing heat flow in or out is an **isothermal process**. The system usually has a large heat reservoir which is held at a constant temperature T to maintain the temperature. While the temperature remains constant in an isothermal system, other variables such as volume and pressure can change. In an isothermal system of an ideal gas where temperature is constant, the **internal energy** is $U = (3/2)nRT$ (which we discussed above in the adiabatic system) and will remain constant:

$$\Delta U = 0 = Q - W \quad \text{or} \quad Q = W$$

In an isothermal system having a contained ideal gas and a piston that maintains the temperature of the gas constant, if you add heat and the volume increases (raising the piston) the product of volume V and pressure P remain constant since temperature T is constant. This follows from the Ideal Gas Law: $PV = nRT$, where $R = 8.314$ J/(mol·K) and T is in degrees Kelvin. To derive an equation for work we need to use some simple calculus. Calculus is beyond the scope of this book, but I will show you this simple integral so you know how the equation for work in an isothermal system evolved. When a gas changes from an initial to a final state, the work done can be described by the integral:

$$W = {_{Vi}}\!\int^{Vf} P dV$$

From the Ideal Gas Law we can replace P with $P = nRT/V$:

$$W = nRT {_{Vi}}\!\int^{Vf} (1/V)dV$$

where nRT is a constant so it can be removed from the integral. Since the integral of $(1/x)$ is by definition the natural log of x, or $(\ln x)$, we have $(nRT)(\ln V)$ evaluated at V_f minus V_i:

$$W = nRT(\ln V_f - \ln V_i)$$

Using a property of logarithms that $(\log x - \log y = \log(x/y))$ **work done in an isothermal system** is given by:

$$\boxed{W = nRT \ \ln(V_f/V_i)}$$

As volume changes, **work** can be shown on a graph of pressure versus volume as the **_area under the isotherm curve_**. (Note that an integral equals the area under a curve.)

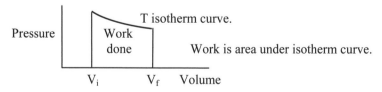

In this diagram we see that a contained ideal gas obeying $PV = nRT$ held at constant temperature T while pressure and volume are allowed to slowly vary, plot out a curve on the P vs. V plane, which is called an isotherm at temperature T.

• **Example**: If you have an isothermal system at 25 °C with 1 mole of a contained ideal gas having a piston and you add heat while it expands from a volume of 0.05 m^3 to 0.06 m^3, what is the internal energy change ΔU, work W, and heat Q added?

The internal energy change for an isothermal system is $\Delta U = 0$. Because $\Delta U = 0$, $Q = W$ so all the added heat becomes work done. We can calculate the work using (remember to use Kelvin):

$$W = nRT \ln(V_f/V_i) = (1.0)(8.314)(25 + 273.15)\ln(0.06/0.05) \approx 452 \text{ J}$$

Because $Q = W$, the heat added is 452 J.

Constant Volume: Isochoric

• An **isochoric process** means one in which volume remains constant while other system properties such as pressure and temperature vary. If you sketch a constant volume system on a pressure versus volume graph, you get a vertical line from P_i to P_f. You realize that if there was a curve, work would be the area under the curve, but since there is no curve, the **work in an isochoric system** must be zero:

$$\boxed{W = 0}$$

No work is done during an isochoric process, since no area accumulates under a vertical curve. Therefore, since $W = 0$:

$$\Delta U = Q - 0 \quad \text{or} \quad \Delta U = Q$$

We see that any energy transfer in the form of heat reflects the change in *internal energy* ΔU. In other words, Q provides the energy input and U will increase (and temperature increases).

Suppose a sealed container such as a can of gas has an increase in temperature and the pressure increases, it will maintain a constant volume ($PV = nRT$). If the pressure becomes too great, however, it can eventually burst. As long as the can maintains its integrity, no work is done.

Constant Pressure: Isobaric

• In an **isobaric process** a gas is held at constant pressure, while other system properties such as volume can vary. In an isobaric system:

$$\Delta U = Q - W$$

If you imagine a piston holding down on a contained gas and then the container is heated, the piston can rise to alleviate any pressure so that it remains constant. Meanwhile, as the floating piston allows the pressure to remain constant, the temperature and volume can increase. Let's look at the work on an isobaric system. Remember $W = Fd$, force times distance. We can make substitutions using area A to express it in terms of pressure P and volume V as follows:

$$W = Fd = PAd = PV$$

As the gas expands the volume V will increase, so we write **work done in an isobaric system** as:

$$W = P\Delta V$$

Providing the quantity of gas remains constant, we can substitute using the Ideal Gas Law, $PV = nRT$ (where n is the number of moles, R is the gas constant 8.314 J/mol·K, and T is temperature in degrees Kelvin), to write **work done in an isobaric system**:

$$\boxed{W = P\Delta V = nR\Delta T}$$

The graph of an **isobaric system** shows that at some fixed pressure P, as volume changes from V_i to V_f, the work can be shown on a graph of pressure versus volume as the *area under the curve*.

Pressure   Work done   Work is area under the curve.

V_i V_f Volume

• **Example**: You have an isobaric system having a piston with 1 mole of a contained ideal gas at 25 °C. If you add heat allowing it to expand from a volume of 0.05 m³ to 0.06 m³, what is the work and new temperature? Assume pressure is 1.013×10^5 Pa.

Work is: $W = P\Delta V = (101{,}300 \text{ Pa})(0.06 \text{ m}^3 - 0.05 \text{ m}^3) = 1{,}013$ J
Using $W = P\Delta V = nR\Delta T$:

$$\Delta T = W/nR = (1{,}013 \text{ J}) / (1 \text{ mol})(8.314 \text{ J/mol·K}) \approx 121.8 \text{ K}$$

The Celsius and Kelvin scales are the same so the new temperature is:

$$25\,°C + 121.8 = 146.8\,°C$$

Specific Heat Capacity

• In the last chapter we learned that the **specific heat,** or **specific heat capacity, c** is the heat required to raise the temperature of a unit of mass by a unit change in temperature. More specifically, the **specific heat** is the amount of heat energy required to raise the temperature of 1 kg of a substance by 1 degree Celsius or Kelvin. The specific heat c relates heat Q and temperature T: $Q = mc\Delta T$. The **specific heats of gasses** are affected by changes in pressure or volume as temperature changes, and are measured with either volume or pressure held constant while the other varies. The **specific heat at constant volume** c_v and the **specific heat at constant pressure** c_p are:

$$c_v = Q/m\Delta T_v \quad \text{and} \quad c_p = Q/m\Delta T_p$$

At constant pressure: $T_p < T_v$

This means for given values of Q and m, we find: $c_p > c_v$

If we divide c_p by c_v, the ratio is given the symbol γ: $\gamma = c_p / c_v$

• The **heat capacity** of a gas depends on whether it can expand in volume or is held at constant volume (raising its pressure) when heated. When volume is constant, the heat goes into the internal energy of the gas molecules, $\Delta U = Q$, and the temperature increases. If the gas can expand when heated, $\Delta U = Q - W$, and part of the energy goes into work. We see that for a given temperature increase more heat input is needed for an expanding gas. The difference in the required heat input to change the temperature for a constant pressure and a constant volume gas is the amount of work done (W) by the expanding gas. Work is given by: $W = F\Delta d$, where F is the force exerted on the gas and Δd is the distance over which the force is exerted. As we just showed above using substitutions for area A, we write W in terms of pressure P and volume V: $W = Fd = PAd = PV$. For an expanding gas, work is therefore: $W = P\Delta V$, which we recall is work in an isobaric system. Therefore, the heat required for a given temperature change is more than $P\Delta V$ if the gas expands (pressure is constant) than if volume is constant.

It is useful to describe this in terms of specific heat. Since more heat is required to raise the temperature of a unit quantity of a substance one degree (the specific heat) for a constant-pressure gas than a constant-volume gas, the specific heat at constant pressure c_p is larger than the specific heat at constant volume c_v. To find the difference between c_p and c_v we can convert the specific heat to the molar heat capacity, which is the amount of heat required to raise the temperature of a mole of gas by 1 °C. The **molar heat capacities** for constant pressure and constant volume are written C_p and C_v. Since the mass in kg of a mole of gas is numerically equal to the molecular weight m, we can relate the specific heat capacities to the molar heat capacities as:

$$C_p = mc_p \quad \text{and} \quad C_v = mc_v$$

To increase the temperature ΔT of n moles of gas under constant pressure requires $nC_p\Delta T$ heat and under constant volume requires $nC_v\Delta T$ heat. The difference between C_p and C_v is $P\Delta V$. Therefore:

$$nC_p\Delta T - nC_v\Delta T = P\Delta V$$

Substituting the Ideal Gas Law, $PV = nRT$, when P is constant is $P\Delta V = nRT$, which we can substitute into the above equation as:

$$nC_p\Delta T - nC_v\Delta T = nR\Delta T$$

or

$$\boxed{C_p - C_v = R}$$

When heat Q is added to a gas at constant volume, it causes a change in internal energy, $\Delta U = Q$. When a gas is made up of single atoms (**monatomic gas**), and heat is added, it goes into increased internal kinetic energy of the atoms, resulting in a rise in temperature. The internal energy can be written: $U = (3/2)nRT$, where R, the gas constant, is 8.314 J/mol·K. If the temperature of the gas changes by ΔT, the **internal energy** U changes by:

$$\Delta U = (3/2)nR\Delta T = nC_v\Delta T$$

We can obtain **equations for C_v and C_p** by first solving for C_v:

$$\boxed{C_v = (3/2)R}$$

Using $C_p - C_v = R$, we get, $C_p = C_v + R = (3/2)R + R$, or:

$$\boxed{C_p = (5/2)R}$$

For gases that are not monatomic the molar heat capacity is larger.

• Note that if you insulate a cylinder-plunger system, making it **adiabatic**, and then decrease its volume, heat cannot leave the gas so the work done on the gas is transformed into internal energy and the temperature will rise. If you know the initial pressure P_i, initial volume V_i, and final volume V_f, you can find the final pressure P_f using:

$$PV^\gamma = \text{constant} \quad \text{or} \quad P_iV_i^\gamma = P_fV_f^\gamma \quad \text{where } \gamma = c_p/c_v$$

where c_v is the specific heat at constant volume and c_p is the specific heat at constant pressure.

• Note also that working with the **First Law of Thermodynamics** for gases introduces the state variable, enthalpy. **Enthalpy** H is given by:

$$H = U + PV$$

We can understand enthalpy by considering that since a gas has "moved" its surroundings in order to occupy its space using pressure P, it did work equal to PV in order to make a place for itself. Therefore, enthalpy is a measure of the total energy H of a gas which includes the internal energy U plus the energy required to exist at a certain volume V and pressure P.

The Second Law of Thermodynamics

• The **Second Law of Thermodynamics** states that *heat does not spontaneously flow from a cooler region to a warmer region*. For heat to flow from a cooler region to a warmer region, work must be done from the outside. If you leave a hot cup of coffee in a cool room, it will not spontaneously get hotter, but will instead slowly cool to the temperature of the room (as the heat from the coffee enters the room). Similarly, if you leave a glass full of ice in a warm room, it will not spontaneously get colder but will warm to the temperature of the room (as the ice absorbs heat from the room).

• The same principle that specifies that heat spontaneously flows from hot to cold but not from cold to hot, also stipulates that ordered systems naturally move toward disorder while disordered systems do not spontaneously move toward order. The Second Law therefore also involves the definition of the state variable **entropy**, which is a measure of the disorder of a system. The **Second Law of Thermodynamics** also specifies *that the total entropy of the Universe (a system and its surroundings) does not decrease, rather, all natural changes increase entropy*. The natural tendency of an isolated system and the Universe as a whole is to move spontaneously toward a state of disorder.

If you place salt and pepper in separate sides of a divided container and carefully remove the divider, you have an ordered system. If you then shake the container, the salt and pepper will mix into a disordered state. No amount of shaking will separate and reorder the salt and pepper. You may be wondering about living organisms that grow and become more ordered. Living organisms release heat (and waste products) to their environments as they grow, creating net disorder. Thinking in terms of **time,** the Second Law suggests that over time the Universe moves toward increasing disorder. It has been suggested that the Second Law dictates the direction of time. Thinking about entropy on a molecular level, during a mechanical process some amount of mechanical energy is converted into heat. When this occurs internal energy increases as the *random motions of the molecules that make up the system increase*. In any mechanical process that produces heat (such as through friction), some of the energy will be irretrievably "lost" (as low-grade thermal energy). The randomness of the motions of the molecules is a measure of the entropy. By the Second Law energy flows from higher to lower: concentration, pressure, voltage, or temperature until equilibrium.

• While **entropy** S is related to energy, it is a different aspect of energy
since it explains spontaneous changes such as heat transfer from a hot
region to a cold one or the expansion of a gas into an empty space. In
such processes, entropy increases, ΔS. If a system absorbs heat energy Q
at a constant temperature T, the entropy increase ΔS can be written:
$\Delta S = \Delta Q/T$, where entropy is defined as the heat transfer ΔQ into a
system divided by temperature. More specifically, for a system going
from one state A to another state B as heat is input, the entropy increase
is $S_B - S_A = \Delta Q/T$. For a thermodynamic process that occurs over a range
of temperatures as heat is applied or removed from the system, the
differential form (from calculus) is used, $dS = dQ/T$, and integration is
required to evaluate (sum) the entropy at various temperatures.

Heat Engines and Heat Pumps

• The fact that heat will always flow from regions of high temperature to
regions of lower temperature is foundational to the operation of heat
engines. **Heat engines** are *machines which transform heat energy into
mechanical energy or work*. Examples of heat engines include steam
engines, diesel and gasoline-powered motors, and turbines in power
plants. A heat engine uses the flow of heat Q from a region of high
temperature T_{hot} to a region of lower temperature T_{cold} to do work. Often
heat produces an expansion of a gas which is used to do work. Heat
engines are comprised of three basic elements: a high temperature heat
source at temperature T_{hot}, a low temperature heat sink at temperature
T_{cold}, and a device that converts heat into work. Heat engines are often
operated to run in a cycle with flow directed between the heat source and
heat sink repeatedly as work is extracted during each cycle.

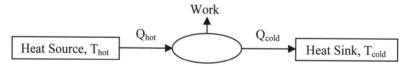

• A property of the heat engine is that its **maximum efficiency** is
determined by the operating temperatures, T_{hot} and T_{cold}. The maximum
efficiency is available only by *ideal*, *reversible engines* (that can operate
in either direction) in which no energy is converted into random motion
and no heat is lost through friction or to the environment. In a reversible
process entropy remains constant. In reality machines cannot work at
100% efficiency as they all generate some heat, and some of that heat
will be lost to the environment. Note that a **reversible process** is an ideal
process in which the system and surroundings (the Universe) can be
restored to their original state after a change in state. In a reversible
process the combined entropy of the system and the environment is

constant. In an **irreversible process** the system and surroundings cannot be restored after a change in state and the entropy increases.

Nineteenth Century scientist **N. L. S. Carnot** analyzed an idealized engine, called the **Carnot engine**, with a four-phase cycle consisting of two isothermal processes and two adiabatic processes. In each cycle the temperature and pressure of an ideal gas change as it expands and contracts. The basic Carnot cycle for a heat engine moves from a high-temperature isothermal expansion, to an adiabatic expansion cooling down, to a low-temperature isothermal compression, to an adiabatic compression warming up. Energy is removed from the high-temperature reservoir as heat, with some transformed to useful mechanical work, and the remaining heat energy going to a low-temperature reservoir (a sink).

As shown in the above diagram, *heat flow in* is Q_{hot} and *heat flow out* is Q_{cold}. If all the input heat Q_{hot} is converted to work W plus output heat Q_{cold}, then:

$$Q_{hot} = W + Q_{cold} \quad \text{or} \quad W = Q_{hot} - Q_{cold}$$

The **efficiency of the heat engine** is the ratio of work done W to the input heat Q_{hot}:

$$\text{Efficiency} = W / Q_{hot} = (Q_{hot} - Q_{cold}) / Q_{hot} = 1 - (Q_{cold}/Q_{hot})$$

Because of the idealized case of the Carnot cycle, **efficiency** becomes:

$$\boxed{\text{Efficiency} = (T_{hot} - T_{cold})/T_{hot} = 1 - (T_{cold}/T_{hot})}$$

Therefore, *the efficiency of the ideal heat engine can be determined from the two operating temperatures*, T_{hot} and T_{cold}. We can combine **efficiency** in terms of Q with that of T:

$$1 - (Q_{cold}/Q_{hot}) = 1 - (T_{cold}/T_{hot}) \quad \text{or} \quad (Q_{cold}/T_{cold}) = (Q_{hot}/T_{hot})$$

Calculating the efficiency of the idealized, reversible Carnot engine gives the theoretical maximum for the fraction of heat that can be used for work in a heat engine.

• **Example**: If you design a personal Carnot-style power plant for your home that provides an input heat of 200 °C and an output heat of 25 °C, what is its maximum efficiency?

First convert to Kelvin: $T_{hot} = 473.15$ K and $T_{cold} = 298.15$ K. Then:

$$\text{Efficiency} = 1 - (T_{cold}/T_{hot}) = 1 - (298.15 \text{ K}/473.15 \text{ K}) \approx 0.37$$

Your maximum efficiency is 37%.

• If this Carnot cycle is run backward, it is a *heat pump*, which requires work and must be driven by some type of external motor. **Heat pumps** are *heat engines that work in reverse by forcing heat to flow from a cool region to a hotter one*. Examples of heat pumps include *refrigerators* and *air conditioners*. Since heat does not flow spontaneously from a cool region toward a hotter region, work is required. Heat pumps consist of two operating temperatures and a device which transfers heat from a cold reservoir to a warmer reservoir. They operate as the reverse of the figure above for heat engines.

In a **heat pump**, the basic Carnot cycle moves from an adiabatic expansion cooling down, to a low-temperature isothermal expansion, to an adiabatic compression warming up, to a high temperature isothermal compression. In this direction, the cycle either removes heat from the low-temperature reservoir or adds heat to the high-temperature reservoir. Most heat pumps use a coolant fluid that experiences a phase change and undergoes both pressurization and expansion.

Third Law of Thermodynamics

• There are several statements of the **Third Law of Thermodynamics**. One is that *you cannot cool a substance to absolute zero (0 Kelvin) using any process that uses a finite number of steps*. Even though each step in a cooling process can move toward absolute zero, you cannot reach absolute zero unless you have an infinite number of steps, which is not practical. Therefore, it is not practically possible to cool all the way to absolute zero Kelvin.

Another statement of the **Third Law of Thermodynamics** says: *If you could cool to absolute zero Kelvin, ideally a pure substance would form a perfectly ordered and uniform crystal structure, all molecular motion would stop, and entropy would be zero*. In reality, substances are not made up entirely of identical molecules that are identically aligned with perfectly uniform motion. Therefore, even if absolute zero Kelvin was reached, entropy is unlikely to be zero. This allows us to state that *entropy cannot be negative*. Therefore, by the Third Law of Thermodynamics, theoretically, if you had an ideal pure substance with a perfectly ordered, perfectly uniform crystal structure and you somehow cooled it to absolute zero Kelvin and all molecular motion slowed to a stop, entropy could be zero *but never negative*.

Finally and similarly, the **Third Law** is sometimes written: *the entropy of a perfectly pure substance approaches zero as the temperature approaches zero Kelvin*. This would occur since at absolute zero (0 Kelvin) there is no thermal energy or heat and the atoms in a pure crystalline substance would perfectly align and have no motion.

9.6. Vaporization, Vapor Pressure, and Phase Change

• A common constituent of air is water vapor. Water vapor enters air from the surfaces of oceans, lakes, and rivers by way of evaporation as individual water molecules escape due to their own kinetic energy. **Evaporation** is the result of the molecular motion of liquid molecules. During **evaporation**, or **vaporization**, the relatively strong attractive intermolecular forces at the surface of a liquid are broken and a molecule escapes into a gaseous state. Only the molecules with very high energies can escape from a liquid, so that any vaporized molecule takes with it energy which is greater than the average energy of the remaining liquid molecules. Therefore, the vaporization process causes a lowering of the overall energy of the liquid molecules, leaving the liquid slightly cooler.

At the same time molecules are leaving the surface, water vapor molecules in the air hit the liquid surface and are absorbed. The process by which they enter the liquid is called **condensation**. Above the surface of liquid water, water vapor molecules move about and interact with air molecules. If the rate that molecules are being vaporized equals the rate they are being absorbed into the liquid, the liquid and vapor will be in **equilibrium**. At equilibrium the air just above the liquid will contain a constant number of water vapor molecules per volume and will be **saturated** with vapor.

Imagine a sealed container with liquid in the bottom and vapor at the top. The vapor in the confined volume above the liquid has a certain pressure defined as the number of molecules per unit volume (n/V) at a specified temperature T. The pressure obeys the Ideal Gas Law $PV = nRT$ (and is independent of the type of gas). Whether there is just water vapor present or vapor plus air and/or other gases, the water molecules in the vapor move independently, leaving and returning to the liquid surface.

If you have a larger sealed container with several cups containing different liquids sitting inside so they are all exposed to the same confined volume of air space, each liquid will have its own **vapor pressure**, or **partial pressure**, to the total pressure of the contained air space. In fact, the total pressure on the container walls is the sum of the individual partial pressures exerted by the different gases. The **partial pressure** is the pressure that each gas or vapor would exert if it were the only gas in the container. The partial pressure of each gas obeys the Ideal Gas Law for its molecular concentration. The **vapor pressure** is defined as the pressure of a vapor just above its condensed (liquid or solid) phase when it is in thermodynamic equilibrium with its condensed phase.

254 Master Math: Essential Physics

The vapor pressure of a liquid does not depend on the volume, or space, above its liquid surface. A liquid will establish its equilibrium with its vapor phase regardless of volume so that n/V is the same for a given temperature T. *Vapor pressure of a particular liquid is only dependent on temperature T.* This is because the temperature determines the energy of the molecules and the rate at which they escape the surface at equilibrium (thereby increasing the vapor pressure). The *evaporation rate for a particular liquid therefore depends on temperature.* As water evaporates, the partial pressure of water vapor will increase, resulting in an increase in the condensation rate. Eventually, *equilibrium* is reached as the rate of molecules leaving equals the rate they are absorbed so that *the condensation rate will equal the evaporation rate* (providing the liquid supply is not exhausted). This is when no further evaporation occurs and the air is saturated. The **partial pressure** of water vapor at saturation is called the **saturation pressure**, or simply the **vapor pressure**.

The strengths of the intermolecular forces are different for different liquids. Therefore *vapor pressure values are different for different liquids at the same temperature.* For example, for a more volatile substance like alcohol, molecules can escape the surface more readily, so alcohol has a greater vapor pressure than water at a given temperature.

When a small container of water is in a large room it will usually evaporate before equilibrium is achieved. However, if a small container of water is covered so that there is very little air space, the water will evaporate until a saturated vapor pressure is achieved at that temperature. The *ratio of the actual vapor pressure of water in air to the saturated vapor pressure at the same temperature* is called the **relative humidity**. Relative humidity at a specified temperature is the amount of water vapor in air expressed as a percentage of the maximum amount of water vapor that could be in the air at that temperature. When the relative humidity is 100%, the air holds all the water vapor it can.

When air is cooled, it will eventually become saturated with water and the partial pressure of water will equal the vapor pressure. As cooling continues there will be an excess amount of water vapor in the air. This vapor will return to a liquid state by **condensation**, forming water droplets in the air called **fog** or **clouds** or forming water droplets on any cool surface called **dew**. The temperature at which water vapor becomes saturated is called the **dew point**. If the temperature is lowered below 0 °C when saturation occurs, the excess water vapor will form **frost** or an **ice fog**.

• **Example**: How would you use the dew point to measure the relative humidity if the air temperature is 25 °C and, as a piece of metal is cooled, water droplets begin to condense on it when it reaches 15 °C?

Relative humidity is the ratio of the actual vapor pressure of water in air to the saturated vapor pressure at the same temperature. In this case the air that is in contact with the metal is cooled by conduction to the temperature of the metal. Because water vapor condenses on the metal when the temperature is 15 °C, the air must become saturated at 15 °C. Therefore the partial pressure at 25 °C is equal to the saturated pressure at 15 °C. Looking up the saturated pressures as 12.8 mmHg at 15 °C and 23.8 mmHg at 25 °C:

Relative humidity is $12.8/23.8 \approx 0.538$ or about 54%

Phase Change

• In Section 8.3, in the subsection *Phase Change and Latent Heat*, we learned that when heat is added to or removed from a substance, there can be a temperature change, a phase change, or both (sequentially). During the process of a phase change when the internal structure of the substance is changing, the temperature does not change until the substance has completed its phase change even while a certain amount of heat is being added or removed. The heat required to *change the phase* is the **latent heat of transformation**, Q_L. The **latent heat of fusion** is the quantity of heat energy removed or added when 1 kg of a liquid solidifies or when it changes from solid to liquid without its temperature changing. The **latent heat of vaporization** is the quantity of heat energy required to vaporize 1 kg of a liquid into a gas or vapor without its temperature changing. Phase change temperatures include the **melting point** (from solid to liquid), and the **boiling point** (from liquid to gas). **Sublimation** and **deposition** occurs when the phase change is directly from solid to gas or gas to solid.

• For each phase transition process there is an equilibrium state where the two phases coexist, but there is not a net conversion of one phase to the other due to saturation. Tables listing saturation vapor pressures have been constructed, and *phase diagrams* showing *equilibrium pressure versus temperature* are also available. **Phase diagrams** display where equilibrium exists between phases, and therefore the conditions for which transitions between the phases occur. A *phase diagram*, which plots pressure (P or ln P) vs. temperature (T), is a common way to represent the various phases of a substance and the conditions in which each phase exists. There are curves or lines on a phase diagram which show phase transitions and specific temperature and pressure conditions where a phase change is at equilibrium (where two phases can coexist). These diagrams show points where it is possible for two or three phases to coexist at equilibrium as well as other regions where only one phase can exist at equilibrium.

There is a **triple point** (of a certain temperature and pressure) where the three phases of solid, liquid, and gas can exist simultaneously at equilibrium. The triple point of water occurs for a pressure of 4.58 torr (where: 1 atm = 760 torr = 1.013×10^5 Pa = 760 mm Hg) and a temperature of 0.01°C. There is a **critical point** temperature and pressure above which it is no longer possible to distinguish between gas and liquid phases. The critical point shows the highest temperature at which the gas or vapor phase of a substance can be in equilibrium with its liquid phase, regardless of the pressure. Above the critical temperature and pressure, the substance is called a super-critical fluid. The critical point for water occurs at 374 °C and 218 atm. Liquid water does not exist for temperatures higher than critical point temperature regardless of the pressure. Note that different phases can also exist together for temperature and pressure combinations that are not lying on one of the solid curves in a phase diagram, but they are not in equilibrium so that one of the phases is gradually transitioning to the other.

9.7. Key Concepts and Practice Problems

- Ideal Gas Law: $PV = nRT$ or $PV = NkT$ (T in Kelvin, R = 8.31 J/mol·K Universal Gas Constant, $k = 1.38 \times 10^{-23}$ J/K Boltzmann Constant).
- Kinetic Theory of Gases relates molecular properties of velocity and KE to macroscopic properties of T and P that obey the Ideal Gas Law.
- Average KE per molecule $(KE)_{Ave} = (3/2)kT$ or per mole $(KE)_{Ave} = (3/2)RT$.
- Diffusion Fick's Law: $\Delta N/\Delta t = -DA\Delta C/\Delta x$ or $J = DA\Delta\rho/\Delta x$.
- Diffusion time: $t = ((\Delta x)^2(\rho_1 + \rho_2)) / (2D(\rho_2 - \rho_1))$.
- Osmosis is selective diffusion through a semi-permeable membrane.
- Zeroth Law of Thermodynamics: two objects in thermodynamic equilibrium with a third object are in thermal equilibrium with each other.
- First Law of Thermodynamics: Conservation of Energy, which says energy cannot be created or destroyed, but can be transformed from one form to another or moved from place to place. The First Law also defines the internal energy change ΔU as (heat transfer Q in) minus (work W done): $\Delta U = Q - W$.
- Second Law of Thermodynamics: heat does not spontaneously flow from cold to hot; and total entropy does not decrease.

- Third Law of Thermodynamics: you cannot cool a substance to zero Kelvin using any process that uses a finite number of steps. This law also says the entropy of a perfectly pure substance approaches zero as the temperature approaches zero Kelvin.

Practice Problems

9.1 (a) You have a 1 oz pure gold bar. How many moles of gold atoms does it contain? **(b)** How many atoms? Assume an atomic weight of 196.9665.

9.2 (a) Before a trip on a 10 °C morning you check your tires. You find the left rear tire low—only 20 psi—so you pump it up to 35 psi. Assuming no change in temperature or volume, by what percentage did you increase the amount of air in the tire? **(b)** After driving across a hot desert, you stop and check the tire's pressure. It is now 43 psi. Assuming no leakage and a constant volume, what is the tire's temperature in °C?

9.3 The escape velocity for a particle launched from Earth is about 11.2 km/s. At what temperature will the root mean square speed of oxygen molecules equal the escape velocity?

9.4 You and your friend take a long hike on an ocean beach. Your friend runs out of fresh water, becomes thirsty, and proposes to drink seawater. After reviewing Section 9.4, what would you advise him?

9.5 Using the First Law of Thermodynamics, explain why a physical machine can never be more than 100% efficient?

9.6 Explain why the side of a mountain range receiving prevailing sea breezes is typically cloudy and rainy while the other side is clear and dry.

Answers to Chapter 9 Problems

9.1 (a) 1 oz ≈ 28.35 grams. (28.35 g)/(196.9665 g/mole) ≈ 0.1439 moles. **(b)** (0.1439 mole)(6.022 × 10^{23} atoms/mole) ≈ 8.666 × 10^{22} atoms.

9.2 (a) Using $PV = nRT$, if V and T are constant, then $\Delta n = \Delta P$. $n_f/n_i = P_f/P_i = (35 \text{ psi})/(20 \text{ psi}) = 1.75$. Therefore, the amount of air increased 75%. **(b)** Using Amontons' Law, $T_f = (T_i)(P_f)/P_i = (283 \text{ K})(43 \text{ psi})/(35 \text{ psi}) = 348$ K or ≈ 75 °C.

9.3 $v_{rms} = [3kT/m]^{1/2}$; $T = (v_{rms})^2(m)/3k = (11,200 \text{ m/s})^2(1.66 \times 10^{-27} \text{ kg/amu})(32 \text{ amu}) / 3(1.38 \times 10^{-23} \text{ J/K}) = 1.6 \times 10^5 \text{ K}$.

9.4 The concentration of salts and other minerals is higher in sea water than in his body fluids. Drinking seawater will actually result in a net loss of water from his body. Water will leave his cells to equalize water pressure in his salty fluids, which will dehydrate his cells. He will urinate more to remove salt, overtax his kidneys, and become even more thirsty. Your advice is not to drink seawater.

9.5 Considering the machine as an isolated system, within that system energy can be transformed from one form to another but cannot be created or destroyed. Therefore, the work output cannot be greater than the work input. Since all physical machines incur frictional losses where energy is converted into low-grade internal thermal energy, and friction cannot be zero or negative, the useful work output will always be less than 100% of the input.

9.6 The saturated marine air is forced upward by the mountains. It loses pressure and also cools. It exceeds its saturation point, forming clouds and then rain. As the same air, after losing moisture in the form of rain, descends the lee side, it regains pressure and temperature, becoming much less saturated. The clouds disappear and rain is scarce.

Chapter 10

ELECTRIC FIELDS AND CURRENTS

10.1. Electric Field and Electric Potential
10.2. Capacitance
10.3. Electric Current
10.4. Electric Resistance
10.5. Electric Power
10.6. Circuits
10.7. Key Concepts and Practice Problems

"Nothing is too wonderful to be true, if it be consistent with the laws of nature..."
Attributed to Michael Faraday

In Section 4.2 *Electrostatic Force* we introduced electrostatics, which describes the effects of static or resting charges. In this chapter we first will expand on electrostatics and electric potential, and then will introduce charges in motion, or electric current, as well as capacitance, resistance, electric power, and circuits.

10.1. Electric Field and Electric Potential

Electric Fields

• In Section 4.2 we learned that every *charged particle or object exerts a force on every other charged particle or object*, and that *electric charge cannot be created or destroyed, only transferred or redistributed* (*Conservation of Charge*). We learned that atoms are made up of *protons*, *neutrons*, and *electrons*, and that the charge on an electron is -1.602×10^{-19} Coulombs, which is equal and opposite to the positive charge of a proton, $+1.602 \times 10^{-19}$ Coulombs. A **Coulomb** is the amount of charge that flows past a fixed point in one second in a current of one Ampere. The force that attracts an electron to the protons in a nucleus and holds it in "orbit" is the **electrostatic force**. Coulomb found that the electrostatic force between two charged objects varies as the *inverse square of the distance between them*. The electric force $\mathbf{F_E}$ acting on a point charge q as a result of the presence of a second point charge Q is described by the **Electrostatic Force Law** or **Coulomb's Law**:

$$F_E = KqQ/r^2$$

where F_E is the *force of electrostatic attraction or repulsion* in Newtons; q and Q are the charges in Coulombs C; r is their separation (in meters); and K is **Coulomb's Constant**, $K \approx 8.98755 \times 10^9$ N·m^2/C^2 or about 9.0×10^9 N·m^2/C^2. The distance between the charges (or charged objects) is considered to be the center-to-center distance. If the signs of q and Q are different (+ and −), the force is attractive and the direction of the force on each charge is toward the other. If the signs of q and Q are the same (+ and + *or* − and −), the force is repulsive and the direction of the force on each charge is away from the other.

Coulomb's Law of Electrostatic Force can also be written as:

$$F_E = qQ/4\pi\varepsilon_0 r^2 \quad \text{with } K = 1/4\pi\varepsilon_0 \text{ or } \varepsilon_0 = 1/4\pi K$$

where $\varepsilon_0 \approx 8.8542 \times 10^{-12}$ C^2/N·m^2 and is called the **permittivity of free space,** or the *electric constant.* This equation describes the force between charges q and Q when the distance r can be modeled as a vacuum.

If the distance r contains a substance, Coulomb's Law is written $F_E = qQ/4\pi\varepsilon r^2$, where ε is the **absolute permittivity** of the substance. **Permittivity** measures the effect of the substance on the electric field or the effect of the field on the substance, and whether the substance polarizes and reduces the electric field.

• **Electric (or electrostatic) fields** are mathematical representations of the effects of isolated charges on each other. Electric fields exist around charged particles or objects. An electric field can be shown to exist at any point in space when a test charge that is placed at that point experiences an electrical force. (A **test charge** q is a small charge used in measurements that has negligible effects on its environment.) The existence of electric fields is observable by the fact that charged objects can influence each other without making physical contact. The presence of an electric charge can be shown to produce a force on other nearby charged particles. (Note that electrostatics refers to static charges and fields around static charges whereas electrodynamics or electromagnetics refers to electric fields brought about by changing magnetic fields.)

• Using Coulomb's Law we can calculate the force exerted by charge Q on charge q. For example, if two charges, q and Q, are at a certain distance and at rest, and charge Q is moved closer to charge q, the force exerted by Q on q will increase. These charges exert a force on one another by disturbing their surrounding space, such that each electrically charged object generates an electric field which attracts or repels other nearby charged objects. The electric field **E** generated by a set of charges can be measured by placing a point charge q at a given location. This test charge q will experience an electric force F_E. The electric field at the location of the test charge is defined as the force **F** per charge q, so that **electric field equals electric force per charge**:

$$E = F_E/q$$

where **units** for **E** are in force per charge, or Newtons per Coulomb (N/C), or equivalently in volts/meter. If a charge q is placed in an electric field **E**, the **force F_E the charge experiences** will be:

$$F_E = qE$$

That means if you know the electric field **E** at a specific point, the force charge q experiences when it is placed at that point is F_E.

• The **strength of the electric field** can be described by an equation, which we find by rearranging Coulomb's Law, $F_E = KqQ/r^2$, as:

$$F_E/q = KQ/r^2$$

So that if *test charge* q is placed at distance r from charge Q (which we can think of as a source charge), the magnitude of the electric field, or **electric field strength**, can therefore be written:

$$E = F_E/q = KQ/r^2$$

Electric fields are vector fields, and therefore each point in an electric field has both magnitude and direction. The **direction** of the field is the direction of the force it exerts on a positive test charge, which is the direction the test charge will move at a given location.

• **Electric field lines** are used to sketch electric fields that exist around charged particles and run through a point in the same direction as the field at that point. The field lines are drawn so that, at a given point, the tangent of a line is in the direction of the electric field at that point. **Each field line begins at a positive point charge and ends at a negative point charge** (since positive charges repel a positive test charge and negative charges attract the positive test charge). These field lines are referred to as either *field lines* or *lines of force*. Because electric field strength **E** equals the electric force F_E divided by the magnitude of a test charge q, $E = F_E/q$, the field lines describe both the electric force and the electric field.

The electric field runs radially outward from a positive charge and radially inward toward a negative point charge. The field is largest where the field lines are closest together, so that the *density of lines is proportional to the magnitude of the field*. These fields reflect the force that acts between charged particles causing them to either attract or repel one another. A field diagram, therefore, reveals the direction of the electric force F_E and electric field E at any point and information about their magnitude. Note that the actual number of lines drawn to depict the field is arbitrary, the density and direction of the lines are what is used to visually represent the field.

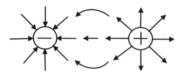

Superposition Principle

• The **Superposition Principle** states that the **total electric force F_{net}** on a particular charge q due to other charges is the vector sum of all the individual forces.

$$F_{net} = F_1 + F_2 + F_3 + \ldots$$

where each of the individual charges contributes a force as if it were the only charge and other charges were not present. Similarly, the **total electric field E_{net}** at a point is the vector sum of the individual contributing fields at that point:

$$\boxed{E_{net} = E_1 + E_2 + E_3 + \ldots}$$

• **Example**: If two charges are along the x-axis with a charge of $+1Q_1$ at the origin and a charge of $-2Q_2$ at x = 1.00 m, where along the x-axis is the electric field zero?

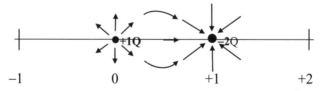

Since the field is the vector sum of the fields from the individual charges, we can add the fields from the two given charges and see where they add to zero. Because the field between a positive and negative charge begins at a positive and ends at a negative charge, the space between the charges cannot be zero. In this example the field points from x = 0 to the right toward x = 1. On the right side of the −2Q charge the field from the +1Q charge points to the right, but the field from the −2Q charge points left. Since the −2Q charge dominates the +1Q charge and the −2Q charge will always be closer to any point to the right of it, the fields cannot add to zero right of −2Q. On the left side of the +Q charge, we can imagine a point on the x-axis where the fields could sum to zero. We can set up an equation using:

$$\mathbf{E_{net}} = \mathbf{E}_1 + \mathbf{E}_2 \quad \text{and} \quad E = KQ/r^2$$

where $\mathbf{E}_1 = K1Q/r^2$ and $\mathbf{E}_2 = -K2Q/r^2$.

If the point where they sum to zero is a distance x to the left of +1Q:

$$0 = K1Q/x^2 - K2Q/(x + 1)^2$$

Divide the equation by K and Q:

$$0 = 1/x^2 - 2/(x + 1)^2 \quad \text{or} \quad 1/x^2 = 2/(x + 1)^2$$

$$\text{or} \quad 2x^2 = (x + 1)^2 \quad \text{or} \quad 2x^2 = x^2 + 2x + 1 \quad \text{or} \quad x^2 - 2x - 1 = 0$$

Using the quadratic formula, $x = (-b \pm [b^2 - 4ac]^{\frac{1}{2}}) / 2a$:

$$x = (-(-2) + [(-2)^2 - 4(1)(-1)]^{\frac{1}{2}}) / 2(1) = (2 + [4 + 4]^{\frac{1}{2}})/2 = 2.41$$

$$x = (-(-2) - [(-2)^2 - 4(1)(-1)]^{\frac{1}{2}}) / 2(1) = (2 - [4 + 4]^{\frac{1}{2}})/2 = -0.414$$

We defined x as the distance to the left of +1Q, so 2.41 m is the root and where the charges sum to zero. The root −0.414 is in the opposite direction, or between the charges, and is where the fields from +1Q and −2Q have the same magnitude but point in the same direction and cannot sum to zero. Thus, the point with zero field is 2.41 to the left of zero.

Electric Potential

• In Section 5.6 *Electrostatic Potential Energy* we learned that the electrostatic potential energy of charge q that is distance r from charge Q is given by:

$$PE_E = KqQ/r$$

If we divide potential energy by q, we obtain **potential energy per unit charge**, also called the **electric potential** V:

$$\boxed{V = PE_E/q = KQ/r}$$

where V has units of Joules per Coulomb (J/C). The **electric potential** V is the potential at a point that is distance r from charge Q, or $V = KQ/r$. Electric potential is a measure of the potential energy per unit charge, PE_E/q. In Section 5.7 *Introduction to the Electron Volt* we learned that **work** is required to move positive charge q from a negatively charged plate to a positively charged plate, and that the **potential difference** between two points is a measure of the work per unit charge required to move a charge from one location to the other:

Potential difference: $\Delta V = W/q$

In measured units: $1 \text{ V} = 1 \text{ J/C}$

where V is a scalar quantity. If the electric potential difference between two locations is 1 V, then 1 C of charge will gain 1 J of potential energy when moved between those two locations. The work required to move a charge from one location to another is a measure of the **potential energy difference** ΔPE_E between the two locations. The *work per unit charge* required to move this charge is also the change in potential, or ΔV, between the locations. Therefore:

$$\boxed{\Delta V = W/q = \Delta PE_E/q}$$

The quantity ΔV (often written V) is used to describe the **potential difference** or the **voltage** between the points. **Units** of electric potential and potential difference are energy per unit charge, or J/C, which is equal to a volt. The volt V is defined as: 1 Volt = 1 Joule/Coulomb, which means electric potential V can be measured in volts V, where the first V refers to the electric potential and the second V is the unit of measure.

• Because a volt is the unit of measure for the potential difference, you will find that a **potential difference is often called the voltage**. This can be a bit confusing. You may also find the terms *potential, electric potential, potential difference*, and *voltage* used interchangeably. Just to clarify, **electric potential** reflects the effect of an electric field at a specified point within the electric field and is the amount of potential energy per unit of charge at that specified location.

Electric Potential $= PE_E/q$

When a charge is moved against an electric field from an initial point to a final point, **work** must be done on the charge by an external force. The work done on the charge will increase its **potential energy** (PE), such that the *change in PE will be equal to the amount of work*. As potential energy changes, there is a difference in electric potential between the initial and final locations of the charge. This difference in electric potential is called the electric **potential difference** ΔV. In other words, the electric **potential difference** ΔV is the difference in electric potential V between the initial and final locations of the charge when work is done to the charge (to move it) which changes its potential energy. The **electric potential difference** is:

$$\Delta V = V_{final} - V_{initial} = \text{Work/Charge} = \Delta PE/\text{Charge}$$

Electric potential difference is often referred to as the **voltage**. When the word **potential** is used instead of **potential difference**, it is assumed that the given potential is measured with respect to a zero point, such as the potential of the ground (therefore, $V - V_0 = V - 0 = V$). Note that electrical appliances operate by extracting work from electrons accelerated by potential differences, so we often think of appliances as requiring a certain voltage.

We have learned that when a charged particle is moved against the force exerted by an electric field, **work** is done, and the particle's **electric potential energy** is changed by an amount equal to the work. Note that the **work** required to move like charges (+ and + *or* − and −) together or unlike charges (+ and −) apart increases the electric *potential energy*. Conversely, moving like charges apart or unlike charges together decreases potential energy.

• Electric potential also obeys the principle of **Superposition**:

$$\boxed{V_{net} = V_1 + V_2 + V_3 + \ldots}$$

• **Equation Summary**:

Force $\mathbf{F_E} = q\mathbf{E} = KqQ/r^2$, units in Newtons N.
Electric field strength $E = F_E/q = KQ/r^2$, units in N/C or V/m.
Electric potential $V = KQ/r$, units J/C or volts V.
Potential difference $\Delta V = W/q = \Delta PE_E/q$, units in volt V.
Charge Q is measured in Coulombs C, and $K = 9 \times 10^9 \, N \cdot m^2/C^2$.

• **Example**: In Section 4.2 we used Coulomb's Law, $F_E = Kq_eQ_p/r^2$, to calculate the electrostatic force F_E of -8.2×10^{-8} N between a proton Q_p and an electron q_e in a hydrogen atom, given their charges (q_e and Q_p) are plus and minus 1.6×10^{-19} C and the average distance between the proton and electron is 5.3×10^{-11} m. Now calculate the electric field and the potential at the electron.

The electric field strength felt at the electron is:

$$E = F_E/q = KQ_p/r^2 = (9 \times 10^9 \, N \cdot m^2/C^2)(1.6 \times 10^{-19} \, C)/(5.3 \times 10^{-11} \, m)^2$$
$$\approx 5.1 \times 10^{11} \, V/m$$

The potential at the electron is:

$$V = KQ_p/r = (9 \times 10^9 \, N \cdot m^2/C^2)(1.6 \times 10^{-19} \, C)/(5.3 \times 10^{-11} \, m) \approx 27 \, V$$

We see that electric field strength E depends on $1/r^2$, while potential V depends on $1/r$. In a hydrogen atom r is very small (5.3×10^{-11} m), so the inverse squaring of r reflects the large field strength compared to the small potential.

• **Example**: If two charges Q_1 and Q_2 are separated by 1 m along an x-axis, what is the electric field $\mathbf{E_{net}}$ and electric potential V_{total} at point x halfway between if the values of Q_1 and Q_2 are each $+2 \times 10^{-6}$ C?

The point x halfway between two identical charges will feel equal but oppositely pointing forces. ←•→ ☺ ←•→

Because $\mathbf{E_{net}}$ is a vector quantity, the net force $\mathbf{F_E}$, where $\mathbf{E} = \mathbf{F_E}/q$, between Q_1 and Q_2 is zero. The vector quantity:

$$\mathbf{E_{net}} = \mathbf{E_1} + \mathbf{E_2} = 0$$

The total potential V_{total} is a scalar and adds algebraically. The potentials for charges Q_1 and Q_2 are:

$$V_1 = KQ_1/r_1 = (9 \times 10^9 \, \text{N·m}^2/\text{C}^2)(+2 \times 10^{-6} \, \text{C})/(0.5 \, \text{m}) = 36,000 \, \text{V}$$

$$V_2 = KQ_2/r_2 = (9 \times 10^9 \, \text{N·m}^2/\text{C}^2)(+2 \times 10^{-6} \, \text{C})/(0.5 \, \text{m}) = 36,000 \, \text{V}$$

Therefore: $V_{total} = V_1 + V_2 = 72,000 \, \text{V}$ or 72 kV.

This example illustrates that the electric field can be zero with a nonzero potential. It is also true that the potential can be zero while the electric field is nonzero.

• **Example**: Your friend asks how fast he could accelerate an electron from rest in a particle accelerator using a potential difference of 2,500 V.

You suggest the total energy is constant, so PE = KE. The potential energy lost is:

$$PE = qV = (-1.6 \times 10^{-19})(2,500 \, \text{V}) = -4.0 \times 10^{-16} \, \text{J}$$

Next, since the kinetic energy increases by the amount that PE decreased:

$$KE = (1/2)mv^2 = 4.0 \times 10^{-16} \, \text{J}$$

You look up the **mass of an electron** and find it to be about 9.1×10^{-31} kg and solve for v:

$$v = [2(4.0 \times 10^{-16} \, \text{J}) / (9.1 \times 10^{-31} \, \text{kg})]^{\frac{1}{2}} \approx 3.0 \times 10^7 \, \text{m/s}$$

Electric Potential and Electric Fields

• We have learned that work is done as a charged particle is moved against an electric field which changes its potential energy. Suppose you have a pair of parallel plates separated by a distance d, and a voltage V is applied across d. The electric field between the parallel plates is uniform and is:

$$E = F_E/q = KQ/d^2$$

where q is what a test charge in the field would experience and Q is the source charge of the field. If you move the test charge q from the negative plate to the positive plate, the work required is:

$$W = F_E d$$

Since $F_E = qE$:

$$W = F_E d = qEd \text{ or rearrange to } E = W/qd$$

The work done per unit charge, W/q, is the **voltage** (or **potential difference**):

$$\Delta V = W/q = Ed$$

Combine $E = W/qd$ with $\Delta V = W/q$ to find the **electric field strength** in a **uniform electric field** between a pair of parallel plates:

$$E = \Delta V/d \quad \text{or just} \quad E = V/d \quad \text{(uniform field)}$$

where V is the voltage difference between the plates. This equation for the electric field strength is valid for uniform fields. The units are V/m, which are equivalent to N/C. While uniform electric fields are not common, they can exist. For example, two parallel plates having opposite charges +Q and –Q have uniformly spaced field lines that are perpendicular to the surfaces (except near the edges where they bow out). A uniform field forms between these plates, as discussed below.

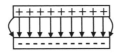

- **Example**: A uniform field can exist in a conducting wire where the field lines run parallel. What is the strength of a uniform electric field resulting from a 25 cm long wire attached to opposite terminals of a 12-V battery? Note that the field direction is along the wire from the positive to negative terminal.

We can use: $E = V/d = (12)/(0.25 \text{ m}) = 48$ V/m.

10.2. Capacitance

- **Capacitors** have several uses, including storing electrical (potential) energy. A **capacitor** is made up of a pair of conducting surfaces, called plates, separated by a gap. A capacitor can accumulate and store electric charge. We have learned that the **electric field strength** E in the region between a pair of parallel plates depends on the **voltage** V across the plates and the distance d that separates them:

$$E = V/d \quad \text{or} \quad V = Ed$$

A capacitor can be charged using a battery that has a potential difference across its terminals by connecting the positive and negative terminals of a battery to each capacitor plate using two wires.

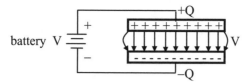

As each wire becomes an extension of its terminal, the positive and negative charges populate on each respective plate until a voltage is established across the capacitor plates (forming a potential difference) that is equal to the voltage across the battery terminals. The capacitor quickly reaches an equilibrium status in which the amount of positive charge on one plate equals the amount of negative charge on the other.

The resulting potential difference between the two plates is proportional to the magnitude of the equal but opposite charge Q on the plates. The resulting charge Q on the capacitor plates is also directly proportional to the applied voltage V, so that a higher voltage results in more charge at equilibrium. The proportionality constant between the applied voltage and resulting charge is the **capacitance** C of the capacitor:

$$Q = CV \quad \text{or} \quad C = Q/V$$

where Q is the **charge on the capacitor**, V is the potential difference between the plates, and C is the proportionality constant called **capacitance**. As we see, the **capacitance** is equal to the charge on the plates divided by the voltage between the plates. **Capacitance** is measured in **units** of Coulomb per volt (C/V), where 1 C/V = 1 **farad** F, (in honor of Michael Faraday, 1791–1867). Capacitors often operate in the range of picofarads (1 pF = 10^{-12} F) to microfarads (1 μF = 10^{-6} F).

• The **capacitance** of a pair of parallel plates depends on the area A of the conducting plates (C ∝ A), where larger plate areas allow more charge to accumulate for a given voltage V. The capacitance also varies inversely with the distance d between the plates (C ∝ 1/d) since increasing d results in less field strength for a particular voltage and decreasing d results in more field strength. (Remember, field strength E = V/d.) As we will discuss later, the capacitance also depends on the region between the plates and whether it is a vacuum or contains a dielectric material. **Capacitance between two plates** having area A separated by distance d is:

$$C = A/(d4\pi K)$$

where K ≈ 8.98755 × 10^9 N·m^2/C^2 or about 9 × 10^9 N·m^2/C^2 is the Coulomb's Law constant. **Capacitance** is also described using the **permittivity of free space** ε_0 as:

$$C = \varepsilon_0 A/d$$

where $\varepsilon_0 = 1/4\pi K \approx 8.842 \times 10^{-12}$ $C^2/N{\cdot}m^2 \approx 8.8542 \times 10^{-12}$ F/m.

• As opposing charges build up on parallel plate conductors, an electric field is established in the region between them. The **work** that was required to produce the charge separation becomes **electric potential energy** which is stored in the **electrical field** between the plates. For a parallel-plate capacitor, since $Q = CV$ and $C = \varepsilon_0 A/d$, the **potential between the plates** is:

$$V = Q/C = Qd/A\varepsilon_0$$

In this equation, Q is the charge on the capacitor, d is the plate separation, A is the area of each plate, and ε_0 is the permittivity of free space, which measures the effect of the substance (vacuum) on the electric field. The **capacitance** is therefore:

$$C = Q/V = \varepsilon_0 A/d$$

Energy Storage in a Capacitor

• We know that work is required to move a charge from one plate to another against the direction of the field, so moving a charge q through a potential difference ΔV requires an amount of work $W = q\Delta V$ or $W = qV$. If we move one electron e some distance d between uncharged plates, it requires very little work. If we move many electrons sequentially between plates, the final electron moved through the potential difference ΔV is $e\Delta V$. As electrons are transferred, the voltage between the plates steadily builds up so that the *average* potential difference through which all of the charge is moved is one-half the final voltage, or $(1/2)V$. That means the **work required to move the entire charge Q** is: $W = (1/2)QV$. Since $Q = CV$, the work becomes:

$$W = (1/2)QV = (1/2)CV^2 = Q^2/2C$$

The work that is done by moving charge Q from one plate to the other is **stored electrical potential energy** (residing between the plates in the electric field). This energy can be retrieved by connecting the two plates so that charge can flow freely between them. The flowing charges form current, which can do work. The **stored potential energy** is equal to the work that was done to place the charge on the capacitor:

$$W = PE = (1/2)CV^2$$

Since $C = \varepsilon_0 A/d$ (where $\varepsilon_0 = 1/4\pi K \approx 8.8542 \times 10^{-12}$ $C^2/N{\cdot}m^2 \approx 8.8542 \times 10^{-12}$ F/m, the permittivity of free space) and $V = Ed$, we have the **energy stored in a capacitor**:

$$PE = (1/2)CV^2 = (1/2)(\varepsilon_0 A/d)(Ed)^2 = (1/2)\varepsilon_0 AdE^2$$

Since Ad is the volume of the region between the plates, the **energy stored in the capacitor electric field per unit volume** can be written:

$$PE/Ad = (1/2)\varepsilon_0 E^2$$

where we see that the **energy stored** is proportional to the square of the electric field strength.

• **Example**: Your friend wants to test all this capacitor stuff and comes to you with his jury-rigged apparatus comprising two 0.5-m by 0.5-m smooth metal plates separated by a uniform 1-mm air gap. He wants to know how much energy he can store if he applies a voltage of 2,000 V across his plates. You also suggest calculating the electric field strength.

You begin with the capacitance of his plates:

$$C = \varepsilon_0 A/d = (8.85 \times 10^{-12}\ \text{F/m})(0.5\,\text{m} \times 0.5\,\text{m})/(10^{-3}\,\text{m}) \approx 2.2 \times 10^{-9}\ \text{F}$$

Then you calculate the energy his apparatus can store:

$$PE = (1/2)CV^2 = (1/2)(2.2 \times 10^{-9}\ \text{F})(2,000\ \text{V})^2 = 4.4 \times 10^{-3}\ \text{J}$$

The electric field strength is:

$$E = V/d = (2,000\ \text{V})/(10^{-3}\ \text{m}) = 2.0 \times 10^6\ \text{V/m}$$

Note that when the electric field across an air (dielectric) gap reaches a critical value in the range of 2 to 5 MV/m, which is near the value in this example, it can break down and spark.

• In practice, capacitors can be constructed with some flexibility. While some are made up of rigid parallel plates separated by an air gap, most are designed using metallic foils separated by an insulating dielectric sheet making them compact, inexpensive, and easily integrated into electronic devices and circuits. Note that in circuit diagrams, a capacitor is usually represented by two equal parallel lines.

Dielectrics

• A **dielectric** is the insulating substance or material allowed or placed into the region between the plates of a capacitor. Capacitance can be increased by filling the space between the plates with a **dielectric** material. A **dielectric** is a non-conductive substance or material which acts as an electrical insulator and is resistant to the flow of an electric current. Dielectric materials can be solids, liquids, or gases, or even a vacuum. The functioning of a capacitor and how well it can store electric energy, or charge, depends on the dielectric separating its plates. Solid materials such as plastic, glass, ceramic, Teflon, and porcelain are often used as dielectric materials, but air and other gases are also commonly used.

Dielectrics are used to enhance or increase the capacitance. The **dielectric constant** k is the ratio of the capacitance when a dielectric material is present, C_d, to the capacitance when a vacuum is between the plates, C_0. So $k = C_d/C_0$. To solve problems involving a dielectric we can combine $C = \varepsilon_o A/d$, or $d = \varepsilon_o A/C$ with $k = C_d/C_0$, giving useful equations:

$$d = \varepsilon_o A/C_o = \varepsilon_o A/(C_d/k) = \varepsilon_o Ak/C_d \quad \text{or} \quad C_d = \varepsilon_o Ak/d$$

We see that the capacitance is increased with higher k. The dielectric constant for a vacuum is 1.0 and it is near 1 for air, but it is larger for liquids and solids, such as 2.1 for Teflon and from 4 to 10 for glass.

Capacitors in Parallel and Series

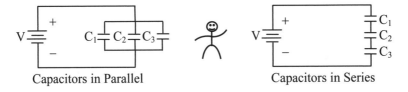

Capacitors in Parallel   Capacitors in Series

• Capacitors can be connected within electric circuits in different geometries. One common arrangement is to connect **capacitors in parallel**. In this arrangement the voltage V is the same across the capacitors. The charges, for example, on three capacitors are:

$$Q_1 = C_1 V \quad Q_2 = C_2 V \quad Q_3 = C_3 V$$

The total amount of charge Q accumulated on the three capacitors is:

$$Q = Q_1 + Q_2 + Q_3 = C_1 V + C_2 V + C_3 V = (C_1 + C_2 + C_3)V$$

The **total capacitance for parallel capacitors** is therefore the sum of the individual capacitances:

$$\boxed{C_{Total} = C_1 + C_2 + C_3 + \ldots}$$

• Another common arrangement is to connect **capacitors in series**. In this arrangement the voltage applied is not the same for each capacitor so for three capacitors the charges are:

$$Q_1 = C_1 V_1 \quad Q_2 = C_2 V_2 \quad Q_3 = C_3 V_3$$

Since a source voltage V, such as a battery, equals the sum of the individual voltages:

$$V = V_1 + V_2 + V_3$$

The individual charges are also equal:

$$Q = Q_1 = Q_2 = Q_3$$

A combination of capacitors in series has the same effect as a single capacitor C carrying a charge Q across which a voltage V is applied:

$$Q = C_T V \quad \text{or} \quad V = Q/C_T$$

So for three capacitors in series:

$$V = Q/C_T = V_1 + V_2 + V_3 = Q/C_1 + Q/C_2 + Q/C_3 = Q(1/C_1 + 1/C_2 + 1/C_3)$$

Dividing V by Q gives the **total capacitance for capacitors in series**, which is the reciprocal of the net capacitance and is the sum of the reciprocals of the individual capacitances:

$$1/C_{Total} = 1/C_1 + 1/C_2 + 1/C_3 + \dots$$

• In practice, capacitors are useful for regulating current or voltage. When a capacitor experiences an emf $\mathcal{E}$ source, charge accumulates until the potential across the capacitor equals the emf. After the capacitor is fully charged, current flow ceases.

10.3. Electric Current

• You are undoubtedly familiar with the terms **direct current** (**DC**) and **alternating current** (**AC**). A **DC** circuit operates from a steady voltage source such as a battery, and current flows in one direction. An **AC** circuit operates as current flow periodically reverses direction. While AC is commonly used in homes and buildings, DC is supplied by batteries. Studying DC provides a foundation for understanding electric circuits and machinery.

• **Current** is the movement of electric charge from one location to another. As electric current flows in a wire, positively charged atomic nuclei are fixed while some fraction of the electrons are relatively free to move within the conductor. In any substance or material (solid, liquid, or gas) that is able to conduct a current, some of the electrons are able to move freely within the material. In the absence of an applied electric field, these **free electrons** (referred to as **conduction electrons**) randomly move about with no net flow in any direction, resulting in zero electric current.

If the ends of a conductive material (such as a wire) are attached to the terminals of a battery, an electric field will be established within the wire. Within this wire the net motion of the electrons will be opposite to the electric field lines while the **current**, by convention, will flow in the same direction as the field lines.

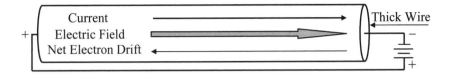

The direction that (negatively charged) electrons *drift* in an electric field is opposite to the direction of the field lines (since field lines begin at a positive point charge and end at a negative point charge, and electrons naturally move away from negative charge toward positive charge). Therefore, note that just as electric field lines are defined as having a direction of the force on a positive (test) charge, current flow is defined as having the direction that positive charges would move even though it is negatively charged electrons that actually move. This convention is valid since a negative electron current flow in one direction creates the same effects as a positive current in the opposite direction. The terms "current" or "current flow" are generally meant to describe current which flows in the same direction as the electric field lines. When the *flow of electrons* is specifically described, the terms "electron flow" or "electron current" may be used.

Electromotive Force Drives Current Flow

• **Current flow** is generated in a conductor when an established potential difference is present. A source of potential difference can be a device which is able to create and maintain a voltage, thereby allowing the flow of charge in the direction that would ultimately neutralize the potential difference. Such devices transform nonelectric energy into electric energy and include batteries which convert chemical potential energy to electric energy, generators which convert mechanical energy to electric energy, and other devices which convert mechanical, thermal, chemical, or other energy into electric energy. The potential difference created by a battery, generator, or other device is known as the **electromotive force (emf or $\mathcal{E}$)**. *The source of an emf can be any device that is able to cause the flow of current by converting nonelectric energy into electric energy,* such as the chemical reaction within a battery or the mechanical motion of a rotor in a generator that produces an emf.

An **emf $\mathcal{E}$ is measured in units** of **volts** (V), which are the same units used for potential difference. The emf, for example, of a small battery may be 9 V. (Note that, as conveyed by its units of measure not being units of force, electromotive force is *not* an actual force. Because of this the acronym emf is often used.)

The Definition of Current

• **Current** *can flow if a source of emf is present in a conductive material.*
If there is also a path through which charge can flow between points of
different voltage, the path can be referred to as a **circuit**. A simple
circuit can be formed when the potential difference between the terminals of a battery cause a current to flow through a conductive wire
connecting the two battery terminals.

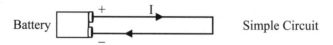

Electric current is represented by the letter I. In equation form, the
magnitude of an **electric current** I is equal to the net charge Q flowing
past a specified point (or surface) per second:

$$I = Q/t$$

Electric Current is therefore quantitatively written as the rate electric
charge flows past a given point or surface. Since electric current is the
net flow of electric charge past some point or surface along a conductive
material, it can be measured in charge Q per time, or Coulombs (C) per
second. Coulombs per second is defined as an Ampere (A) so that
1 A = 1 C/s. If 1 C of charge passes a certain point in 1 s, a current of
1 Ampere (A) is flowing. Electric current is therefore measured in
Amperes, and the **unit of current** is the **Ampere (A)**:

$$1 A = 1 C/s$$

Note that because the number of electron charges in one Coulomb is
about 6.24×10^{18}, a current of one Ampere reflects the flow of
6.24×10^{18} electrons per second. The electric current capacity that
household circuits generally carry is 15 to 20 A.

• **Example**: Suppose you have a copper wire with a cross-sectional area
of 1 mm^2, where each 1 mm length of the wire has about 8.5×10^{19} atoms.
You find the number n of atoms per unit volume using N_A, ρ, and mass:
n = $(6.02 \times 10^{23}$ atoms/mole)$(8.94 \times 10^3$ kg/m$^3)/(63.5 \times 10^{-3}$ kg/mol)
≈ 8.475×10^{28}/m$^3 \times 1/(10^9$ mm$^3/1$m$^3) ≈ 8.5 \times 10^{19}$ atoms/mm^3.
In copper metal each atom supplies one free electron. If the net (drift)
velocity of the electrons is 1 mm/s so that the free electrons in a 1-mm
length segment pass a point in the wire in 1 s, what is the current?

The current is:

$$I = Q/t = (8.5 \times 10^{19} \text{ electrons})(1.6 \times 10^{-19} \text{ C/electron}) / 1 \text{ s} ≈ 13.6 \text{ A}$$

Conductors possess a large numbers of free electrons, so relatively small drift velocities can produce substantial currents. Note, however, that while you may have a net drift velocity on the order of mm/s, the random velocities of individual electrons are on the order of 10^6 m/s.

10.4. Electric Resistance

• When electric current flows through a material, there is usually some amount of energy that is lost as conduction (free) electrons collide with atoms of the material. The conduction electrons flowing through a conductor are impeded somewhat by forces exerted by the atoms in the conducting material. Collisions between conduction electrons and atoms in the material result in a **resistance** of current flow. As electrons move through a conductor and collisions with atoms occur, energy is transferred to the atoms, which increases the conductor's internal energy (as the atoms vibrate more vigorously). This increase in internal energy raises the temperature of the conductive material causing a heating effect. The energy converted into heat when a current is flowing cannot be converted back into electric energy (without supplying even more energy) and is dissipated. The **resistance** to current flow in a conductive material, caused by collisions between conduction electrons and atoms in the material, therefore raises the internal energy of the material resulting in the conversion of electrical energy to thermal energy.

Most conductors resist the flow of current to some extent and can therefore also be called **resistors**. When current flows through a resistor, electrical energy is converted to heat. The degree to which a conductor (or resistor) resists current flow depends on its atomic and molecular properties. A material that acts as a good conductor does not strongly resist the flow of electric current. Silver and copper are good conductors.

Ohm's Law

• We have learned that if we connect a wire to a battery, a current which is proportional to the potential difference, or voltage, will flow. The current that flows through a conductive material is directly proportional to the voltage across the material. The **relationship between voltage and current** is called **Ohm's Law**:

$$\boxed{V = IR \quad \text{or} \quad I = V/R}$$

where the proportionality constant R between voltage and current is called the electrical **resistance** of the material. *Ohm's Law states that* *the current flowing through a given material is directly proportional to the voltage across the material.* For a certain voltage, a greater resistance corresponds to a smaller current. Ohm's Law, named in honor of George Simon Ohm (1787–1854), approximates the behavior of various conductive materials. Ohm's Law is useful for analyzing electric circuits made up of metallic materials. If the potential difference in a material, circuit, or section of a circuit is maintained by a source of emf ε, then $V = \varepsilon$, and the current I is also:

$$I = \varepsilon/R$$

The electrical resistance of a material is measured in **ohms** (Ω) where:

$$1\,\Omega = 1\,V/A$$

Therefore, a 1-V voltage causing a 1-A current to flow through a material has a resistance of 1 Ω.

• **Example**: If a 12-V voltage causes a 2-A current to flow through a material, what current will a 24-V voltage cause in the same material?

Because V = IR or R = V/I, this material's resistance is:

$$R = V/I = 12\,V\,/\,2\,A = 6\,\Omega$$

For a 24-V Voltage:

$$I = V/R = 24\,V\,/\,6\,\Omega = 4\,A$$

Therefore, a current of 4 A will flow.

Resistivity

• If you design a circuit, you may need to know the resistance of each component, including the wire and components which are not actual resistors. This will allow you to estimate the heat dissipated and the operating temperature of the circuit and to predict the potential difference across the circuit components, since they will be affected by the resistance of other circuit components.

Connecting a section of a conductive material to the terminals of a battery allows a current to flow as the conduction electrons in the material move in response to the potential difference supplied by the battery's emf. The movement of conduction electrons depends on their collision rate with atoms in the material—with higher collision rates corresponding to slower electron motion and higher electrical resistance. Resistance can be found using the potential difference or voltage and measuring the current using an ammeter (an instrument which measures electric current in Amperes), and then calculating resistance using Ohm's Law, V = IR.

• The **resistance of a material**, and therefore the flow of electrons, depend on several factors. One factor is the atomic and molecular properties of the material. If we look up the resistance of various materials we see that a section of either silver or copper wire, for example, will have lower resistance values than an identical section of tungsten wire.

Electrical resistance also depends on the size and shape of the material. The number of collisions between conduction electrons and atoms in a material are directly proportional to the length of a material (such as a wire) the electrons are traveling along, since the travel distance is greater and there is a greater opportunity for collisions. Therefore, the resistance R is proportional to the L, or R $\propto$ L. Alternatively, if the cross-sectional area A of the material (such as a wire) is increased, it would allow more area (for travel paths) for conduction electrons as they cross a specified point along the length of the wire per unit of time. This suggests that resistance R is inversely proportional to cross-sectional area A, or R $\propto$ 1/A. Combining the effects of length L and cross-sectional area:

$$R \propto L/A$$

Therefore, the **electrical resistance of a material is proportional to the ratio of its length to its area**. This suggests that current running along a longer, thinner conductor will experience a greater resistance than current running along a shorter, wider length of the same material. The dependence of electrical resistance on length and cross-sectional area explains why appliances such as space heaters and window-unit air conditioners have short, thick power cords. These appliances draw a lot of current, and a shorter, wider cord adds less electrical resistance while reducing resistance to power flow. We can write the proportionality for **resistance** in equation form using a proportionality constant ρ:

$$\boxed{R = \rho \, L/A}$$

where ρ (Greek letter rho) represents the **resistivity** property of the particular material, L is the length of the resistor, and A is the cross-sectional area. The units of ρ are ohm-meters (Ω·m), and it is an intrinsic property of the material. Examples of **resistivity values** ρ include 1.59×10^{-8} Ω·m for silver, 1.72×10^{-8} Ω·m for copper, 2.44×10^{-8} Ω·m for gold, 3.5×10^{-5} Ω·m for carbon, 3.0×10^{10} Ω·m for wood, and 1.0×10^{15} Ω·m for rubber. Silver, copper, and gold are good conductors, carbon is a poor conductor, and wood and rubber could be considered to be insulators. Insulators are used to prevent the full flow of current and keep it as small or negligible as possible.

• In addition to resistivity, the conductivity of a material is used to describe or evaluate the material for its electrical uses. The **conductivity**, σ, is equal to the reciprocal of its resistivity ρ:

$$\sigma = 1/\rho$$

• The electrical resistance of a material not only depends on its atomic and molecular properties and on the ratio of its length to its area, but also on **temperature**. Increasing a material's temperature causes an increase in the movement of atoms and molecules within the material, which increases the collision rates between conduction electrons and atoms of the material. This results in a correlation between temperature and electric resistance so that resistance increases with temperature. For most materials resistivity varies with temperature in a similar manner as thermal expansion of solids varies with temperature:

$$\Delta\rho/\rho = \alpha\Delta T$$

with $\Delta\rho/\rho$ as the fractional change in resistivity and α is the proportionality constant called the temperature *coefficient of resistance*. If we use the standard temperature as $0\,°C$ where resistivity is ρ_0, and at temperature $T\,°C$, the **resistivity** is ρ_T, we have:

$(\rho_T - \rho_0)/\rho_0 = \alpha(T - T_0)$ or $(\rho_T/\rho_0) - 1 = \alpha\Delta T$ or $\rho_T/\rho_0 = 1 + \alpha\Delta T$
or

$$\rho_T = \rho_0(1 + \alpha\Delta T)$$

where ρ_0 is the resistivity at a reference temperature such as $20\,°C$ and α is the **temperature coefficient of resistivity**. Approximate values of α for sample metals are: 0.0038 per deg C for aluminum, 0.0043 per deg C for copper, 0.0039 per deg C for platinum, 0.0045 per deg C for tungsten, and 0.0061 per deg C for silver. The correlation of resistivity ρ with temperature does not hold for most materials at extremely low temperatures approaching zero Kelvin, where resistivity does not continue to decrease to zero, and most materials still show some resistance. **Superconductor** materials, however, can achieve zero resistivity at very low temperatures and therefore conduct electricity without resistance. **Semiconductors** are a class of materials with moderate resistivities and electrical conductivity between that of a conductor and an insulator. These materials can be treated chemically to modify electric current. Semiconductors also typically have negative temperature coefficients of resistivity so the resistance decreases with increasing temperature.

Because resistance R is directly proportional to resistivity ρ, where $R = \rho\, L/A$, resistance varies with temperature just as does resistivity. So:

$$\Delta\rho/\rho = \alpha\Delta T \text{ can also be written } \Delta R/R = \alpha\Delta T$$

Therefore, using algebra as we did for $\rho_T = \rho_0(1 + \alpha\Delta T)$ above, we can write resistance as:

$$\boxed{R_T = R_0(1 + \alpha\Delta T)}$$

• **Example**: Estimate the change in resistance of a 1 mm diameter, 0.5 m length of silver wire if it is heated from 20 °C to 320 °C? Assume $\rho = 1.59 \times 10^{-8}$ Ω·m (at 20 °C) and α is 0.0061 per deg C for silver.

The initial resistance is:

$R = \rho L/A = \rho L/\pi r^2 = (1.59 \times 10^{-8}$ Ω·m$)(0.5$ m$)/(\pi)(0.5 \times 10^{-3}$ m$)^2 \approx 0.01$ Ω

After heating to 320 °C:

$$R_T = R_0(1 + \alpha\Delta T) = (0.01\ \Omega)[1 + (0.0061°C^{-1})(300\ °C)] \approx 0.03\ \Omega$$

There is an increase in resistance of about 0.02 Ω.

10.5. Electric Power

• We learned in Section 5.11 that **power** is the rate at which work is done or energy is expended or transformed. It is the amount of work per unit time, so for work W done in time t, the average **power** P expended is:

$$\boxed{\text{Power } P = \text{work/time} = W/t}$$

where the unit of power is work (or energy) per time or Joules per second, J/s, which is defined as a **Watt** W:

$$1\ \text{Watt} = 1\ \text{W} = 1\ \text{J/s}$$

We can combine the definitions of the volt (1 J/C) and the Ampere (1 C/s) to write an equation for electric power. Since a Watt W is a Joule per second:

$$1\ \text{Watt} = 1\ \text{J/s} = (1\ \text{J/C})(1\ \text{C/s}) = (1\ \text{V})(1\ \text{A})$$

This represents a current of 1 A driven by a 1-V potential difference, or voltage. If this 1-A current is flowing along a segment of wire, 1 Watt of power is expended in the wire and is equal to 1-V voltage times 1 A of current, so:

$$1\ \text{Watt} = 1\ \text{J/s} = (1\ \text{V})(1\ \text{A})$$

which represents the rate that electrical energy is expended in the wire. More generally, for a current I flowing along a segment of wire, the **power** P expended in the wire is the voltage V times the current I:

$$\boxed{P = VI}$$

We can substitute Ohm's Law, $V = IR$ to write **power** as:

$$P = VI = I^2R \quad \text{or} \quad P = V^2/R$$

which represents **power** in terms of the resistance R and the current I. The equation:

$$P = I^2R$$

is referred to as **Joule's Law**, and shows that heat produced per time, or energy dissipated per time, which is the power, is the square of the current times the resistance. This equation is useful in circuit analysis. Remember the units for electrical resistance are **Ohms** (Ω), where the relationship $R = P/I^2$ in units is:

$$1 \, \Omega = 1 \, \text{Watt} / A^2$$

Since power P is work per time, or $P = W/t$, **work** W is the energy (work) used in a section of an electric circuit and equals the power, or rate of energy use, times total time:

$$W = Pt = I^2Rt = V^2t/R$$

As we see, work can also be expressed in terms of resistance, current, and voltage by substituting $P = I^2R$ or $P = V^2/R$.

• **Example**: What is the energy use (in Joules and kilowatt-hours) of a circuit with a resistance of 3 Ω if a 3-A current flows for 30 min?

Energy (work) used is power times total time:
$$W = Pt = I^2Rt = (3 \, A)^2(3 \, \Omega)(30 \, \text{min})(60 \, \text{s/min}) = 48,600 \, J$$

The kilowatt-hour, kWh, is the unit normally used by power companies, and is 1,000 Watts times the time in hours. In our circuit:
$$W = (3 \, A)^2(3 \, \Omega)(0.5 \, h) = 13.5 \, \text{Watt-hours, Wh, or } 0.0135 \, \text{kWh}$$

• **Example**: Your parrot, Gerard, tells you he's too cold. You calculate the resistance when running your electric space heater on high to be about 9 Ω. If you run it continuously for 24 hours and your power company bills you at a rate of 12 cents per kWh, what will you spend to keep the bird happy? (Assume your household voltage is 120 V.)

The energy used in one day is
$$W = Pt = I^2Rt = V^2t/R = (120 \, V)^2(24 \, h)/(9 \, \Omega) = 38,400 \, \text{Wh or } 38.4 \, \text{kWh}$$
The cost is:
$$\text{cost} = (38.4 \, \text{kWh})(\$0.12/\text{kWh}) \approx \$4.61 \, \text{per day}$$

Note, you can equivalently use: $I = V/R = (120 \, V)/(9 \, \Omega) \approx 13.33 \, A$

So: $W = Pt = I^2Rt = (13.33 \, A)^2(9 \, \Omega)(24 \, h) \approx 38.4 \, \text{kWh}$.

10.6. Circuits

• **Electric circuits** provide specific amounts of electric energy at designated times. In electric circuits, electrical energy is converted to other forms, such as heat in a resistance heater, light in an incandescent bulb or light-emitting diode (LED), or mechanical energy in an electric motor. Electric circuits operate under rules and require **elements** such as capacitors, resistors, or emf sources. As we have seen in this chapter, certain symbols are used to represent circuit elements. Following are sample element symbols used in circuit drawings:

Junction	Ground	emf source	Capacitor	Resistor	Rheostat
	(either symbol)		(either symbol)		(either arrow)

Junctions are points between wires where electrical contact is made and current can flow, **ground** is where a circuit is grounded and the potential is defined as zero, and a **rheostat** is a variable resistor. In a circuit drawing, elements are located in the sequence they are encountered when current is flowing. The position and spacing in a drawing is not to scale.

Kirchhoff's Rules

• The laws of conservation, including conservation of energy and conservation of charge, apply to electricity and to circuits. Gustav Kirchhoff, a German scientist working in the mid-nineteenth century, developed rules for the flow of current in circuits based on conservation laws which are useful for analyzing circuits. **Kirchhoff's First Rule** is based on the conservation of charge and is the:

Junction Rule: The algebraic sum of the currents entering any junction point in a circuit is zero. This means the current or charge entering any point in a circuit equals the current or charge exiting that point, so that charge cannot accumulate at any point. The **Junction Rule** is written:

$$\sum I = 0 \quad \text{at a circuit point}$$

where $\sum$ represents the sum and I is the current.

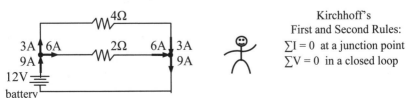

Kirchhoff's
First and Second Rules:
$\sum I = 0$ at a junction point
$\sum V = 0$ in a closed loop

Kirchhoff's Second Rule, based on the conservation of potential energy, refers to the total change in potential and is:

Loop Rule: The algebraic sum of the changes in potential around any closed loop is zero. This means that for any closed loop of a circuit, the sum of the rises in potential (from a battery or emf $\mathcal{E}$ source) equals the sum of the drops in potential (from a resistor, etc.). This is written:

$$\sum V = 0 \quad \text{or} \quad \sum \mathcal{E} - \sum IR = 0 \quad \text{in a closed circuit loop}$$

where $\sum V$ represents the sum of the potential differences or voltages (including both potential gains and drops), $\sum \mathcal{E}$ represents the sum of the potential gains from emf sources, and $\sum IR$ is the sum of the potential drops due to current I flow through resistances. The potential within a circuit is changed by emf sources, such as a battery, and by circuit elements that have some amount of resistance, thereby causing a voltage change of $\Delta V = IR$. The current adjusts so that the total change in potential is zero.

When analyzing a circuit, the source of emf determines the direction of the flow of current and therefore the direction of all changes in potential. Resistors decrease the potential in the direction of current flow, or equivalently, increase the potential in the direction opposite current flow.

• **Example**: In a simple circuit, such as the one above, if you only know the emf and resistances, show how to determine the currents.

There are three loops in the circuit, the top loop, the bottom loop, and the large loop.

1. The bottom loop has the 12-V battery and the 2-Ω resistor. The potential gain across the battery (+12 V) must be the same as the potential drop across the segment containing the 2-Ω resistor, or −12 V. We can calculate current in the bottom loop as: I = V/R = −12V/2Ω = −6A.
2. The large loop has the 12-V battery and the 4-Ω resistor. The potential gain across the battery (+12 V) must be the same as the potential drop across the segment containing the 4-Ω resistor, or −12 V. We can calculate current in the large loop as: I = V/R = −12V/4Ω = −3A.
3. The top loop has the 4-Ω resistor in the direction of current flow and the 2-Ω resistor opposite current flow. The fall in potential in the segment containing the 4-Ω resistor must equal the gain in potential in the segment containing the 2-Ω resistor. We see that this occurs since as the current flows in a loop, the potential drops as we move in the same direction as current flow (segment with 4-Ω resistor) and rises as we move opposite current flow (segment with 2-Ω resistor). The total change in potential around this top loop is therefore zero.

When more complex circuits are analyzed, algebraic equations are written and simultaneously solved for the different loops. In complex circuits there can be multiple resistors in the loops, and it is more difficult to determine current. In some cases combinations of resistors can be represented as one collective resistor.

Series and Parallel Resistor Circuits

• Circuits can have resistors in series (**series circuit**) or in parallel (**parallel circuit**). In a circuit diagram a **resistor** is designated by a sawtooth pattern. The figure shows parallel and series circuits with three resistors connected to a battery, which supplies the emf.

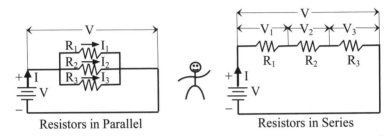

Resistors in Parallel   Resistors in Series

Note that when you are analyzing small simple electric circuits, it is often acceptable to ignore any resistance in the wires that connect various parts of the circuit since they have relatively small resistances.

• For **resistors in series** (as depicted for three resistors in series), the **current** I flowing through each resistor is the same.

$$I = I_1 = I_2 = I_3$$

The **voltage** across all of the resistors together is V and the voltage in each resistor is:

$$V_1 = IR_1 \quad V_2 = IR_2 \quad V_3 = IR_3$$

where the total potential drop V is the sum of the individual potential drops, or voltages, at each resistor ($V = V_1 + V_2 + V_3$). The total potential drop must equal the value of the emf source, which, in this circuit, is the voltage V across the battery:

$$V = V_1 + V_2 + V_3 \ = \ IR_1 + IR_2 + IR_3 \ = \ I(R_1 + R_2 + R_3)$$

The total voltage V for the circuit equals the total resistance R_T times the current I flowing through the circuit:

$$V = IR_T$$

The **total resistance for a series circuit** is therefore the sum of the individual resistances and is greater than any of the individual resistances:

$$R_T = R_1 + R_2 + R_3 + \ldots$$

• For **resistors in parallel** (as depicted for three resistors in parallel), the potential drop, or **voltage** V, across each of the parallel resistors is the same.

$$V = V_1 = V_2 = V_3$$

The total **current** I flowing through the circuit divides between the resistors (in the figure it divides between the three resistors). The current I for each resistor is:

$$I_1 = V/R_1 \qquad I_2 = V/R_2 \qquad I_3 = V/R_3$$

The total current I is:

$$I = I_1 + I_2 + I_3 = V/R_1 + V/R_2 + V/R_3 = V(1/R_1 + 1/R_2 + 1/R_3) = V(1/R_T)$$

where total current I is voltage V divided by total resistance R_T, or $I = V/R_T$. Rearranging:

$$I/V = (1/R_1 + 1/R_2 + 1/R_3) = (1/R_T)$$

Or the **total resistance for a parallel circuit** is:

$$1/R_T = 1/R_1 + 1/R_2 + 1/R_3 + \ldots$$

which is the sum of the reciprocals of the individual resistances. The equivalent resistance in a parallel circuit is less than any of the individual resistances.

• We can compare resistors and capacitors in series and parallel in a circuit and see their opposing forms:

For resistors in series: $R_T = R_1 + R_2 + R_3 + \ldots$
For resistors in parallel: $1/R_T = 1/R_1 + 1/R_2 + 1/R_3 + \ldots$
For capacitors in series: $1/C_{Total} = 1/C_1 + 1/C_2 + 1/C_3 + \ldots$
For capacitors in parallel: $C_{Total} = C_1 + C_2 + C_3 + \ldots$

• **Example**: Suppose you have two circuits that are each connected to a 12 V source. If one circuit has two resistors of 3 Ω and 4 Ω in series and the other circuit has two resistors of 3 Ω and 4 Ω in parallel, which circuit has a greater current flow?

For the resistors in series: $R_T = R_1 + R_2 = 3 + 4 = 7 \, \Omega$
$$I = V/R_T = (12 \text{ V}) / (7 \, \Omega) \approx 1.7 \text{ A}$$

For the resistors in parallel: $1/R_T = 1/R_1 + 1/R_2$
$$1/R_T = 1/3 + 1/4 = (1/3)(4/4) + (1/4)(3/3) = (4 + 3)/(12) = (7)/(12 \, \Omega)$$
$$R_T = 12 \, \Omega / 7 \approx 1.7 \, \Omega$$
$$I = V/R_T = (12 \text{ V}) / (1.7 \, \Omega) \approx 7 \text{ A}$$

Note in the parallel circuit you can also calculate the current through each resistor and then take the sum:

$$I_{3\Omega} = 12\,V / 3\,\Omega = 4\,A \quad \text{and} \quad I_{4\Omega} = 12\,V / 4\,\Omega = 3\,A$$

$$\text{Total } I = 4\,A + 3\,A = 7\,A$$

The current in the parallel circuit is greater.

Multiple Sources of emf

• What if you have more than one source of emf such as several different batteries? The emf sources may be in series or parallel.

For **emf sources in series**, the net potential difference, or voltage, across the series is equal to the sum of the potential differences of each of the emf sources. If you have four 12-V batteries in series the total emf ε is: $12\,V + 12\,V + 12\,V + 12\,V = 48\,V$.

For **emf sources in parallel**, if each of the emf sources has the same potential difference value, the potential difference (or voltage) across the combined parallel sources of emf has the same value as any of the individual emf sources.

Batteries have some amount of internal resistance, which decreases the potential difference V between its terminals. This potential difference can be described by:

$$V = \varepsilon - Ir$$

where r is the internal resistance and is usually a very small value, resulting in the battery's potential difference being close to its emf ε. When currents are large or the battery ages, however, Ir may be noticeable. Connecting batteries in parallel allows current to be divided among them, thereby reducing Ir for each battery. This allows a parallel group of batteries to maintain a potential difference nearly equal to the emf of the batteries, especially when currents are large.

More Simple Circuits

• If you have a circuit with both series and parallel resistors, you determine total resistance by first calculating each of the sections of the circuit that have parallel resistors to obtain the resistances for the parallel sets. Then, combine the sets of parallel resistors with other series resistors as a series calculation. In the figure below you first calculate the parallel resistors R_1, R_2, and R_3 and also R_5, R_6, and R_7 and then find the total resistance by adding the parallel resistance values to resistor R_4 which is in series with the parallel sets. Let's do an example for clarity.

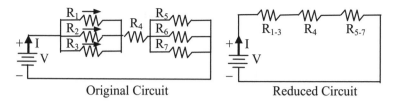

Original Circuit   Reduced Circuit

• **Example**: In the circuit shown, if $R_1 = R_2 = R_3 = 2\ \Omega$, $R_4 = 5\ \Omega$, $R_5 = R_6 = R_7 = 3\ \Omega$, and voltage $V = 12$ V, what is current I?

First find resistance in parallel $(1/R_T = 1/R_1 + 1/R_2 + 1/R_3 + ...)$:

$$1/R_{1\text{-}3} = 1/R_1 + 1/R_2 + 1/R_3 = 1/2 + 1/2 + 1/2 = 3/2, \text{ so } R_{1\text{-}3} = 2/3\ \Omega$$

$$1/R_{5\text{-}7} = 1/R_5 + 1/R_6 + 1/R_7 = 1/3 + 1/3 + 1/3 = 3/3, \text{ so } R_{5\text{-}7} = 1\ \Omega$$

Add resistance in series $(R_T = R_1 + R_2 + R_3 + ...)$:

$$R_T = R_{1\text{-}3} + R_4 + R_{5\text{-}7} = 2/3 + 5 + 1 = 2/3 + 15/3 + 3/3 = 20/3 \approx 6.7\ \Omega$$

For current, use $V = IR_T$, so:

$$I = V/R_T = (12\text{ V})/(20/3) = 1.8\text{ A}$$

Therefore, the current is $I = 1.8$ A.

• What if you had a simple circuit with two resistors (or two sets of parallel resistors) and two batteries? The total current in a simple circuit does not depend on the location of the batteries or resistors. The batteries will either add to or subtract from each other depending on how the battery terminals are connected with the circuit. Imagine the above drawn circuit but with two batteries, a 12 V and a 6 V. If the batteries are connected to oppose, the net voltage in the circuit will subtract. In this case, $12 - 6 = 6$ V. If the batteries are connected to enhance, the net voltage in the circuit will add. In this case, $12 + 6 = 18$ V. These two cases are depicted in the following summary diagram.

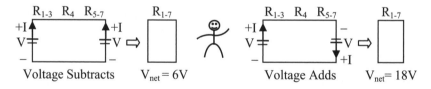

Voltage Subtracts $V_{net} = 6V$ Voltage Adds $V_{net} = 18V$

• **Example**: In the circuit in the last example, but with two batteries—a 12 V and a 6 V—determine the current if the batteries are connected to oppose or enhance voltage.

The total resistance is the same as the last example, $R_T = 20/3\ \Omega$. When the batteries oppose voltage is $12 - 6 = 6$ V. The current is:

$$I = V/R_T = (6\text{ V})/(20/3) = 0.9\text{ A}$$

When the batteries add voltage is $12 + 6 = 18$ V. The current is:

$$I = V/R_T = (18\text{ V})/(20/3) = 2.7\text{ A}$$

• There are circuits that do not lend themselves to the simple reductions we have just discussed. For such circuits Kirchhoff's Rules (introduced in this section) may be helpful.

Measuring Circuits

• How can you actually measure what is occurring in a circuit? Following is a brief introduction to some basic circuit measuring instruments, or **meters**. *Current* can be measured using an **ammeter**; *potential difference* (voltage) between two points can be measured using a **voltmeter**; *resistance* can be measured using an **ohmmeter** or a **Wheatstone bridge**; and emf can be measured using a **potentiometer**. A **multimeter** combines a number of measuring functions into a single instrument.

• An **ammeter** measures the electric **current** flowing between two points in a circuit. It can measure either direct or alternating electric current in **Amperes**. An ammeter is connected in series so the current flowing in the circuit flows through the ammeter. Since a meter should not significantly change the operation of the circuit, it is designed to have very low resistance. The central part of a classic ammeter is a current-carrying coil of wire called a **galvanometer**, whose deflection (in a magnetic field) is proportional to the flowing current. When attached to a pointer on a scale, it displays current in Amperes. In order to read a large range of currents, resistors are placed in parallel with the galvanometer to reduce the current flowing through the galvanometer. Galvanometers are able to directly read very small currents. To read larger currents a parallel resistor (shunt resistor) diverts a known fraction of the current, which runs parallel to the galvanometer (which has some internal resistance), and the galvanometer reads the current running through it. Digital ammeters have no moving parts and use a circuit to convert continuous analog current to an equivalent digital current value.

• A **voltmeter** measures the **potential difference**, or **voltage**, between two points in an electric circuit. Classic voltmeters have been referred to as galvanometers since they measure current. The current can be translated from Amperes to volts using Ohm's Law. A galvanometer can be converted into a voltmeter by adding a resistor in series with the galvanometer. (Remember, a galvanometer contains a coil of wire in a magnetic field which experiences a torque when current passes through the coiled wire. The coil is attached to a pointer and a spring which deflects the pointer proportionally to the current in the coiled wire.)

Unlike an ammeter, which is inserted into the circuit of interest in series so that all the current flows through it, a voltmeter is connected in parallel with the circuit. By connecting two points in a circuit through a voltmeter in parallel, a small amount of circuit current can be routed through the voltmeter. Resistance is inserted in series with the meter to keep the current flow through it small. Different resistors can be selected using a switch to allow measurement of a range of voltages. The current through the meter is equal to $I_{meter} = V/R_{meter}$, where V is the potential difference between the two points.

• An **ohmmeter** measures **resistance** by applying a known emf to a circuit (or circuit element) containing the unknown resistance. The unknown resistance is connected in series with an ammeter, which measures the current flowing through the resistor. Since the voltage is known, resistance can be calculated using $R = V/I$.

A more accurate measurement of **resistance** can be done using a Wheatstone bridge circuit. A **Wheatstone bridge** is a circuit that determines the unknown resistance by balancing it against known resistances. Once the balance is achieved, no current is drawn and the primary circuit being measured is not altered. A typical Wheatstone bridge circuit consists of a constant supply voltage (such as a battery); an ammeter, galvanometer, or potentiometer; and four resistors in a diamond shape so that current from a battery is divided and flows through the resistors before recombining into a single wire or conductor. The values of three of the resistors are known, with one of the three being adjustable (called a rheostat). The fourth resistor has the unknown resistance value, which is determined by manipulating the current flow through the bridge circuit. The diverging currents, which are measured using an ammeter, must become balanced with the help of the rheostat. The unknown resistance R_u is found using $R_u = (R_{k1})(R_{k2})/(R_{k3})$, where R_{k1}, R_{k2}, and R_{k3} represent the three known resistance values.

• A **potentiometer** has several functions, including the measurement of **emf** or **potential difference** by comparing with a known voltage, or as a mechanical **variable resistor** which can be used to control variables such as volume or brightness in electrical equipment.

When measuring potential differences, a known emf can be supplied to a circuit having an unknown emf until the two are balanced, revealing the value of the unknown emf. The unknown emf drives the current through a loop having one segment as a slide wire, which acts as a variable resistor. As the contact moves along the slide wire, the potential supplied to the circuit by a battery is changed until the loop current (and therefore the loop potential difference) measures zero (using a galvanometer) and the potential drop along the slide wire equals the unknown potential difference. This results in the known emf supplied by the battery being balanced against the unknown emf.

Like a resistor, a potentiometer has three terminals, two of which are connected to a resistance wire and the third to a sliding contact. As a variable resistor, one terminal is connected to a power source, another is connected to ground, and the third terminal can move across a resistive material. The resistive material can vary from low resistance at one end to high resistance at the other end. The third terminal connects the power source and ground, providing a means to adjust the position of the third terminal along the resistive strip, thereby manipulating resistance. By controlling resistance, a potentiometer can control current flow through a circuit. A potentiometer can also be designed to control the potential difference, or voltage, across a circuit.

• In practice, electronic measurement is commonly done with **multimeters**, which combine several measurement functions in one instrument. A typical multimeter will be able to measure current, resistance, and voltage, and perhaps other variables such as capacitance or inductance. Multimeters can be divided into two categories—analog which use analog circuits and digital which use digital circuits. Analog multimeters can be varied to employ the functionality of an ammeter, voltmeter, and ohmmeter, and have a scale with a pointer. Digital multimeters are solid state and employ greater functionality along with LCD screens displaying digits or bars. Some multimeters may be able to measure current and voltage in the two different modes, alternating current AC and direct current DC.

10.7. Key Concepts and Practice Problems

• Electrostatic Force Law or Coulomb's Law: $\mathbf{F_E} = KqQ/r^2$.
• Electric field equals electric force $\mathbf{F_E}$ per charge q: $\mathbf{E} = \mathbf{F_E}/q$.
• Charge q in electric field E will experience force $\mathbf{F_E}$: $\mathbf{F_E} = q\mathbf{E}$.
• Electric field felt by charge q at distance r from Q: $\mathbf{E} = \mathbf{F_E}/q = KQ/r^2$.
• Electric field lines, or lines of force, sketch direction and strength of electric fields. Field line begins at a positive charge and ends at a negative charge. Line density is proportional to field magnitude.
• Electric potential V is: $V = PE_E/q = KQ/r$.
• Potential difference or voltage between points: $\Delta V = W/q = \Delta PE_E/q$.
• Electric field strength in a uniform field: $E = \Delta V/d$ or $E = V/d$.

- Capacitors store electric charge Q. Capacitor charge is: $Q = CV$.
- Capacitance is: $C = Q/V = A/(d4\pi K) = \varepsilon_o A/d$, where A is plate area, d is distance between, K is Coulomb's constant, and ε_o is permittivity.
- Work to move entire charge Q is: $W = (1/2)QV = (1/2)CV^2 = Q^2/2C$.
- Energy stored in a capacitor: $PE = (1/2)CV^2 = (1/2)\varepsilon_o AdE^2$.
- Capacitance for parallel capacitors: $C_{Total} = C_1 + C_2 + C_3 + \ldots$
- Capacitance for series capacitors: $1/C_{Total} = 1/C_1 + 1/C_2 + 1/C_3 + \ldots$
- DC circuits: steady voltage source (battery). Current in one direction.
- AC circuit: current flow periodically reverses. Used in homes, etc.
- Current flows when a potential difference or emf source is present.
- Current I is charge Q flow past a point per time: $I = Q/t$.
- Ohm's Law: $V = IR$ or $I = V/R$, current is proportional to voltage.
- Resistance R: $R = \rho L/A$, where ρ is resistivity, L length, A area.
- Power: $P = VI = I^2R$ or $P = V^2/R$, where $P = I^2R$ is Joule's Law.
- Kirchhoff's First Rule: Junction Rule: $\sum I = 0$ at a circuit junction.
- Kirchhoff's Second Rule: Loop Rule: $\sum V = 0$ in a closed circuit loop.
- Total resistance for a series circuit: $R_T = R_1 + R_2 + R_3 + \ldots$
- Total resistance for a parallel circuit: $1/R_T = 1/R_1 + 1/R_2 + 1/R_3 + \ldots$

Practice Problems

10.1 In the example in Section 10.1 about the two charges Q_1 and Q_2 separated by 1 m, if Q_2 had been -2×10^{-6} C (rather than both being $+2 \times 10^{-6}$ C) what would be the electric field and potential at the midpoint between Q_1 and Q_2?

10.2 (a) In the example in Section 10.2, your friend's jury-rigged capacitor plates attached to a 2,000-V source stored $PE = 4.4 \times 10^{-3}$ J. If you linked 5 identical capacitors in parallel and applied the same 2,000 V, how much total energy would the capacitors store? **(b)** If instead you linked them in series, how much total energy would they store?

10.3 Compare the speed of the electron flow in two wires, one with a diameter of 1 mm and the other with a diameter of 1 cm, both carrying the same current.

10.4 A copper wire with a resistance of 1 Ω is drawn through an extruder so that its new length is 3.48 times longer. What is its new resistance?

10.5 You just left on a 2-week driving trip and remember leaving on a 100-W light in the garage. Your electric rate is 10¢/kWh. You are 10 mi from home, your car gets 20 mpg, and gas is $4/gal. Assuming your time is worth nothing and ignoring other factors, should you return home to turn out the light?

10.6 You take the pair of parallel resistors and the pair of series resistors in the example in Section 10.6, subsection *Series and Parallel Resistor Circuits,* and put the two pairs into a single loop circuit one after the other in series. Using the same 12-V battery, what will be the new total resistance and the new current?

Answers to Chapter 10 Problems

10.1 V_1 would remain 36,000 V, but V_2 would be –36,000 V. Therefore $V_{Total} = (36,000$ V$) + (-36,000$ V$) = 0$ V. Since $E_{Net} = E_1 + E_2$, find E_1 and E_2:

$E_1 = KQ_1/r^2 = (9 \times 10^9$ N·m$^2)$/C$^2)(+2 \times 10^{-6}$ C$)/(0.5)^2 = 72,000$ V/m
$E_2 = KQ_2/r^2 = (9 \times 10^9$ N·m$^2)$/C$^2)(-2 \times 10^{-6}$ C$)/(0.5)^2 = -72,000$ V/m

Because the negative charge vector E_2 is on the opposite side of the midpoint, its field lines point in the same direction as those of E_1 and they are additive: ←·→ ☺ →·← Therefore $E_{Net} = 144,000$ V/m.

10.2 (a) When in parallel, they all experience the same 2,000 V, so each capacitor will store 4.4×10^{-3} J, and together will store 5 times as much, or 2.2×10^{-2} J. **(b)** When in series, the sum of the voltages across all 5 capacitors must be 2,000 V, so the total stored energy is 4.4×10^{-3} J, the same quantity that the one capacitor stored in the original example.

10.3 The cross sectional area of the 1-mm wire is 1/100 that of the 1-cm wire, so the rate of electron flow is 100 times faster. This is similar to comparing the speed of an equal volume of water flowing through pipes of differing diameters.

10.4 A wire is shaped like a cylinder where volume = base × height. Since the volume of the copper is fixed, the base becomes $1/3.48 \approx 0.287$ times as large. Since $R \propto L/A$, $\Delta R = \Delta L/\Delta A = 3.48/0.287 \approx 12.1$ Ω.

10.5 Will the savings on your electric bill exceed the $4 of gas to return 10 mi home and again drive 10 mi outside of town? 14 days = 336 hrs. The light will use 0.1 kWh each hour, or 33.6 kWh. The cost of leaving the light on is 33.6 kWh × $0.10/kWh = $3.36. You keep driving.

10.6 Since the 2 pairs of resistors are in series, $R_{Total} = R_{Series} + R_{Parallel} = 1.7 + 7 = 8.7$ Ω. I = V/R_{Total} = 12 V/8.7 Ω $\approx$ 1.4 A.

Chapter 11

ELECTROMAGNETISM

11.1. Magnetism
11.2. Magnetic Forces on Moving Charges
11.3. Charged Particles Moving in Circular Motion in Magnetic Fields
11.4. Electric Currents Generate Magnetic Fields
11.5. Changing Fields Induce Current and Magnetic Flux
11.6. Key Concepts and Practice Problems

> *"Nature is our kindest friend and best critic in experimental science if we only allow her intimations to fall unbiassed on our minds."*
> Attributed to Michael Faraday

11.1. Magnetism

Magnets

• In Asia Minor, in a region of modern Turkey, more than 2,000 years ago people realized that a natural stone called *lodestone* attracted other lodestones as well as small pieces of iron. These naturally magnetized stones were found in a region called *Magnesia*, and therefore became known as **magnets**. Lodestone is made up of the mineral magnetite. It was found that when lodestone was either suspended freely or floated in quiet water on wood, it oriented itself along the Earth's north-south direction, forming a magnetic **compass**. Understanding magnets helps us to comprehend magnetic fields and magnetic poles.

Magnets have a north-pointing north N-pole and a south-pointing south S-pole. *The N-pole of a compass magnet points toward the Earth's S magnetic pole near the Earth's geographic North Pole.* (The S-pole of the Earth is near the geographic North Pole.) A magnet that is freely-suspended will orient itself along magnetic field lines of Earth from S-pole to N-pole. A compass identifies the direction of the Earth's magnetic field at any point.

The like **magnetic poles** of a magnet (N-N and S-S) repel each other, while the unlike magnetic poles (N-S and S-N) attract each other. The two **poles**, N and S, of a magnet always exist together. If a magnet is cut in half, the two resulting shorter magnets will each be magnets with N- and S-poles which point in the same direction as the original magnet. If the magnets are repeatedly cut in two, each smaller section will be a smaller replica of the original magnet.

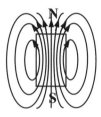

Magnetic Fields

• A magnet produces a **magnetic field**, which is the region where a *force* is exerted on another magnet (or on a current-carrying conductor). The direction of magnetic field lines are, by convention, the direction in which the N-pole of a compass magnet points. The magnetic field lines generated by a typical bar magnet run through its center pointing from the S-pole to the N-pole, and the same field lines run around the outside of the magnet point from its N-pole toward its S-pole. The Earth produces a magnetic field which is similar to that of a bar magnet. (Remember, the S-pole of the Earth is near the geographic North Pole.)

Magnetic fields possess both *magnitude* and *direction*, and therefore can be represented as a *vector*. The **magnetic field** is represented using the letter **B**, which can be thought of as describing *magnetic lines of force*. Magnetic field vector **B** points in the *direction* of the magnetic lines of force and has a *magnitude* or *strength* which corresponds to the density of the lines. Magnetic field lines are continuous, having no defined beginning or end.

Electromagnetism

• Electricity and magnetism are interrelated. *Moving charges produce magnetic fields, and moving magnets can induce electric currents.* The interrelated electric and magnetic effects are known as **electromagnetism**. When an electric charge moves, it creates a magnetic field. Even a single moving charge creates a field. In fact, a charge's rotational motion, or spin, can create a magnetic field. In 1820 Danish physicist Hans Christian Oersted discovered that a current-carrying wire influenced the orientation of a compass magnet. This occurred because current running through a wire produced a magnetic field which influenced the nearby compass magnet.

• **Current** flowing along a **straight wire** produces a magnetic field whose lines form circles around the center of the wire. The direction of the field lines can be detected using a compass magnet. *The magnetic field direction is perpendicular to the direction of the flow of electric current*. The direction of the magnetic field lines surrounding a current-carrying **straight wire** can be revealed using the **right-hand rule**. If you imagine tightly clutching a straight wire in your right hand with your right *thumb* pointing up along the wire shaft in the *direction of current flow*, your curled *fingers* encircling the wire will be in the *direction of the magnetic field* **B**.

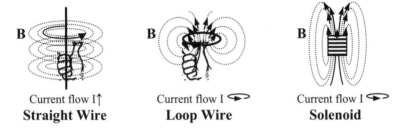

Current flow I↑	Current flow I ↩⟩	Current flow I ↩⟩
Straight Wire	**Loop Wire**	**Solenoid**

• If a **loop of wire** carrying a current forms a circular loop, moving charges in the wire generate a magnetic field similar to a very short bar magnet aligned perpendicular to the loop. The direction of the magnetic field lines surrounding a current-carrying loop of wire can also be revealed using the right-hand rule. If you bend your *fingers* in the *direction of current flow* along the loop, with your thumb pointing through the loop's center, your thumb (in this case) will point in the *direction of the magnetic field* **B**. The magnetic field lines that run through the loop's center are parallel to each other, whereas the lines outside the current loop curve and join forming closed loops themselves. Even an electron moving around its atomic nucleus forms a tiny loop of electric current and produces a small magnetic field.

• **Solenoids** are composed of a number of loops next to each other in the shape of a tube. They create a uniform magnetic field that points through the center of the solenoid. Solenoids can be constructed by coiling a single *current-carrying* wire. The *magnetic field lines along the length of a solenoid are parallel* and obey the same right-hand rule as a single loop of wire. The region of the generated magnetic field running along the inside of a long solenoid is uniform. If an iron rod is inserted in a solenoid or a current-carrying wire is wrapped around an iron rod, the iron becomes magnetized. The resulting magnetic field is the sum of the fields generated by the wire and the iron and can be strong. This combination of a current-carrying wire wrapped around a ferromagnetic core is referred to as an **electromagnet**. (**Ferromagnetic** materials are strongly attracted to magnetic fields and can retain their magnetic properties after the field is removed.)

11.2. Magnetic Forces on Moving Charges

Magnetic Force on a Moving Charge

• A charge in motion relative to a magnetic field experiences a force due to the magnetic field. The *direction of the magnetic force* F_M on a moving *positive test charge* is perpendicular to the velocity **v** of the moving charge and also perpendicular to the field **B**. The **magnitude of magnetic force** F_M has been shown to be directly proportional to the charge q and its velocity v, as well as to the magnetic field strength B. In fact:

$$F_M = qvB \quad \text{for } \mathbf{v} \text{ perpendicular to } \mathbf{B}$$

where F_M, v, and B are magnitudes of vectors F_M, **v**, and **B**. **Units** of **magnetic field strength** are Tesla T or Gauss G, where:

$$1\,G = 10^{-4}\,T \quad \text{or} \quad 1\,T = 10^4\,G$$

Magnetic field strengths are also measured in Webers (Wb) per square meter Wb/m^2, where 1 T = 1 Wb/m^2 = N/Amp·m = N·s/Coulomb·m.

When we studied the electric force vector F_E, we learned that force vector F_E has the same *direction* as the electric field vector **E**. In the case of magnetic forces and fields, it has been shown in experiments that the **direction of a magnetic force vector** F_M is perpendicular to both the velocity vector **v** of a moving charge q and the magnetic field vector **B**.

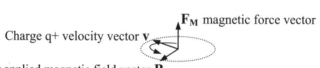

F_M magnetic force vector

Charge q+ velocity vector **v**

Externally applied magnetic field vector **B**

The perpendicular vectors can be modeled using a right-hand rule. For a positive test charge in an externally created magnetic field **B**, if you point the fingers of your right hand in the direction of the velocity **v** of the test charge and then curl your fingers toward the direction of field **B** (curved arrow in diagram), your thumb will point in the direction of the magnetic force vector F_M.

The **maximum force exerted on a moving charge q** or charged particle by a magnetic field occurs when the velocity vector **v** of the particle is perpendicular to the field vector **B**: $F_M = qvB$, for **v** perpendicular to **B**. There will be many cases when a charge (or charges) is not moving in a direction perpendicular to a magnetic field and **v** and **B** are not perpendicular (such as current flow through a wire within a magnetic field). The more general equation for the **magnetic force vector F_M** is:

$$F_M = qvB \sin \theta$$

where F_M, v, and B are magnitudes of vectors **F_M**, **v**, and **B**, and θ is the angle between vectors **v** and **B** (see previous figure). When $\theta = 90°$, **v** and **B** are perpendicular, $\sin \theta = 1$, and $F_M = qvB$, which is maximum force. When $\theta = 0°$, **v** and **B** are parallel, $\sin \theta = 0$, and $F_M = 0$. When θ is not zero, **v** and **B** fall on a plane which is always perpendicular to magnetic force vector **F_M**. The magnetic force vector **F_M** therefore depends only on the component of velocity vector **v** that is perpendicular to **B**:

$$v_{perpendicular} = v \sin \theta$$

In fact, multiplying vectors **v** and **B** is called a **vector product**, or **cross product**, written $A \times B = |A||B| \sin \theta$ (see Section 2.6). For the magnetic force vector **F_M**:

$$F_M = qv \times B = q|v||B| \sin \theta$$

When written as the size or magnitudes of the vectors we have:

$$F_M = qvB \sin \theta$$

The **magnetic field** B can therefore be written for force F_M on a positive test charge q moving at velocity v with the angle θ between the velocity and the field as:

$$B = F_M / qv \sin \theta$$

Magnetic Force on Current-Carrying Wire

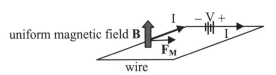

uniform magnetic field **B**

wire

• Consider a current-carrying wire placed into a uniform magnetic field **B**. Flowing current in a wire experiences a force **F_M** due to a magnetic field. If a straight section of the wire having length L is set in a direction so that it makes an angle θ with field **B**, then each electron moving along the wire will experience a magnetic force equivalent to:

$$F_M = qvB \sin \theta$$

If we imagine there are n number of electrons e in the length L of wire experiencing the magnetic field, we have nL total electrons (nLe) acted on by the field in that section of wire. The force on that wire section is:

$$F_M = nLevB \sin \theta \quad \text{where} \quad nLe = q$$

The charge q that passes a particular point in the wire per time t is the current I, or:

$$I = q/t = nLe/t$$

The distance charge will travel is L = vt or v = L/t, so current I becomes:

$$I = q/t = neL/t = nev$$

Substituting nev = I, we can write **magnetic force in terms of current in a wire** as:

$$\boxed{F_M = ILB \sin \theta}$$

where θ is the angle between the direction of the current I and the direction of field B, and L is the length of the section of wire. By rearranging, the **magnetic field** in a wire is:

$$B = F_M / IL \sin \theta$$

The **force is maximum** when the field lines are perpendicular to the current-carrying wire ($\theta = 90°$ and $F_M = ILB$), and the force is zero when they are parallel ($\theta = 0°$ and $F_M = 0$). We can also write the magnetic force as the **vector product** for current I at a particular point in terms of length **L** at that point within field **B**:

$$F_M = I\mathbf{L} \, \mathbf{X} \, \mathbf{B} = I \, |\mathbf{L}||\mathbf{B}| \sin \theta$$

• **Example**: Earth has a magnetic north and south pole and connecting magnetic field lines that form its *magnetosphere*. Since a magnetic field can be produced by circulating electrical charges, the Earth's magneto-sphere is formed, according to the **magnetic dynamo** theory, by the swirling motions of liquefied conducting material inside the planet which create the magnetic field. Mercury, Jupiter, Saturn, Uranus, and Neptune also have magnetospheres, while Mars and Venus have only traces of magnetic field lines. If you are at a location on Earth where the magnetic field measures 0.5 Gauss, or 5×10^{-5} Tesla, and you have a straight 0.5-km wire stretched perpendicular to the field with a 12-A current flowing through it, what force does Earth's magnetic field exert on the wire?

The magnetic force on the wire can be found using: $F_M = ILB \sin \theta$. Since $\sin 90° = 1$:

$$F_M = ILB = (12 \text{ A})(500 \text{ m})(5 \times 10^{-5} \text{ T}) = 0.3 \text{ N}$$

Force on a Coil Producing Torque

• Imagine you have a current-carrying coiled wire, and you place it into a uniform magnetic field **B**. The field exerts a force on the components of current perpendicular to the field. Remember from Section 2.6 *Torque*, if a force is applied to an object at a point other than its central axis of rotation, that force may create a **torque**. Torque is the product of an applied force **F** times the length of a lever arm **r**, which is the shortest radial *perpendicular* distance from the axis of rotation to a line drawn along the direction of force. Thus, torque $T = rF \sin\theta$, where **r** is the radial distance from the axis of rotation to the point at which the force is applied, and θ is the angle between **F** and the radial line **r** which is the acute angle between the lines of vector **r** and force vector **F**.

In this case, the force on your electric coil generates a torque, which is the force F = ILB times the moment arm. To develop an equation for torque, imagine a wire loop or coil is square shaped rather than round, with a moment arm of (1/2)d, where d is the length of a side.

Since there are two sides of a loop, torque can be written:

$$T = ILB(2)(1/2)d \sin\theta = ILBd \sin\theta$$

Because dL is the area of the loop, for a loop of area A and angle θ, the **torque for one loop of wire** is:

$$T = IAB \sin\theta$$

Since coils are made of multiple wires or loops, a wire coil having N loops has **torque T on a flat coil having N loops**, where each carries current I. In a uniform external magnetic field torque T is:

$$T = NIAB \sin\theta$$

where A is the area of the coil, and θ is the angle between the field lines and a perpendicular to the plane of coil. The **maximum torque** occurs when $\theta = 90°$ where $\sin 90° = 1$ and:

$$T = NIAB$$

To visualize the direction of rotation of the coil, you can use the following right-hand rule: With your right thumb perpendicular to the plane of the coil, so that your fingers run in the direction of the current flow, the torque acts to rotate the thumb into alignment with the external field (at which direction the torque will be zero).

Electromagnetic Forces

• When both an electric field **E** and a magnetic field **B** are present, a moving charged particle q will experience forces F_E and F_M from each of the fields:

$$F_E = qE \quad \text{and} \quad F_M = qv \times B$$

where **X** represents the cross product. The vector sum of the electric and magnetic forces provides a model of the total of the two types of forces on a hypothetical test charge q:

$$F_L = F_E + F_M = qE + qv \times B$$

where F_L is the electromagnetic force, also called the **Lorentz force**, after Dutch physicist Hendrik Antoon Lorentz (1853–1928).

The magnetic force is not a fundamental force of nature like the gravitational or electric force, but rather it is a manifestation of the electric force. Magnetic fields are created by the motion of electric charge. These charges may be free charges flowing in a wire or circulating or spinning charges that are bound in an atom.

11.3. Charged Particles Moving in Circular Motion in Magnetic Fields

• When a freely moving charged particle moves through a uniform magnetic field **B**, the force exerted by the field continually changes the particle's direction of motion. The force F_M and acceleration remain perpendicular to the field and to the direction of motion, and the speed of the charged particle remains constant. The change in the particle's direction results in **circular** or **spiral motion**. If the charged particle is initially moving perpendicular to the direction of **B**, the resulting motion will be a circle in a plane perpendicular to the field. If the charged particle has a velocity component parallel to **B**, it will continue to move along the field's magnetic field lines, but will develop circular motion in the perpendicular plane which creates a spiral path.

A magnetic field, therefore, changes the direction of motion of a moving charge. Because the force of the magnetic field on the moving charge is perpendicular to the charge's motion, there is a centripetal force which causes the moving charge to move in a circular motion. Suppose you have a charged particle of mass m moving with velocity **v** in uniform magnetic field **B**, and the velocity **v** is perpendicular to field **B**. The magnitude of the force F_M exerted on the charged particle q by the magnetic field **B** is:

$$F_M = qvB \quad \text{for } v \text{ perpendicular to } \mathbf{B}$$

The direction of the magnetic force vector F_M is always perpendicular to the instantaneous direction of the charged particle's motion v. In other words, the force F_M exerted on the particle by the magnetic field is always perpendicular to the charged particle's velocity. The magnetic force causes a constant **centripetal acceleration** that stays perpendicular to v and keeps the charged particle moving in a circular orbit. The **centripetal acceleration** is:

$$a_c = F_M/m = qvB/m$$

To find the **radius r** of the charged particle's circular motion and its frequency f, we first write the **centripetal acceleration** in terms of velocity and the radius of the orbit:

$$a_c = v^2/r$$

Combining the two equations for a_c:

$$qvB/m = v^2/r$$

and solving for r gives the **radius r of the charged particle's circular motion**:

$$\boxed{r = mv/qB}$$

This shows that the radius of a charged particle's circular motion in a uniform magnetic field is directly proportional to its momentum, mv, and inversely proportional to the field strength B.

To develop an equation for the **frequency** f of the charged particle, remember in Section 6.3 we learned that frequency is related to period τ, as $f = 1/\tau$. We also learned in Section 1.11 that the time required for one revolution or period τ is $\tau = 2\pi/\omega$, and that the speed of an object moving in a circular path of radius r is $v = r\omega$, so $\omega = v/r$. The period can therefore be written $\tau = 2\pi r/v$. We can rearrange the radius equation, $r = mv/qB$, as $r/v = m/qB$ and combine these as:

$$\tau = 2\pi r/v = 2\pi m/qB$$

Since frequency is $f = 1/\tau$, the **frequency of the charged particle's circular orbit** is:

$$\boxed{f = 1/\tau = qB/2\pi m}$$

The **angular frequency** ω can be found by combining $\tau = 2\pi/\omega$, or $\omega = 2\pi/\tau$, with $f = 1/\tau = qB/2\pi m$, or $\tau = 2\pi m/qB$:

$$\omega = 2\pi/\tau = qB2\pi/2\pi m = qB/m$$

or:

$$\omega = qB/m$$

These equations for radius and frequency model the *cyclotron radius* and *cyclotron frequency* of particles moving in circular paths in particle accelerators called *cyclotrons*. Cyclotrons accelerate charged particles to high velocities. In cyclotrons particles are accelerated using electric fields while they are confined by bending their path using a magnetic field. An applied electric field accelerates charged particles between D-shaped regions of a magnetic field. A magnetic force is used to bend moving charges into a semicircular path between the accelerations that are applied by the electric field.

• **Example**: What is the radius of motion for an electron and a proton each traveling 2×10^6 m/s in a 1-Tesla and in a 2-Tesla magnetic field?

The radius of a charged particle's circular motion is: $r = mv/qB$. We look up $q = \pm 1.6 \times 10^{-19}$ C, $m_e = 9.1 \times 10^{-31}$ kg, and $m_p = 1.67 \times 10^{-27}$ kg.

$r_{electron} = (9.1 \times 10^{-31}$ kg$)(2 \times 10^6$ m/s$)/(1.6 \times 10^{-19}$ C$)(1$ T$) \approx 1.14 \times 10^{-5}$ m

$r_{proton} = (1.67 \times 10^{-27}$ kg$)(2 \times 10^6$ m/s$)/(1.6 \times 10^{-19}$ C$)(1$ T$) \approx 2.09 \times 10^{-2}$ m

$r_{electron} = (9.1 \times 10^{-31}$ kg$)(2 \times 10^6$ m/s$)/(1.6 \times 10^{-19}$ C$)(2$ T$) \approx 5.7 \times 10^{-6}$ m

$r_{proton} = (1.67 \times 10^{-27}$ kg$)(2 \times 10^6$ m/s$)/(1.6 \times 10^{-19}$ C$)(2$ T$) \approx 1.04 \times 10^{-2}$ m

Note: Greater magnetic field strength corresponds with tighter radius.

• Notice that since there is no component of force $\mathbf{F_M}$ exerted in the direction of a particle's motion $\mathbf{v}$, there is no **work** done on a particle by a magnetic field. That means the work done by a magnetic field on a moving charge is zero. Remember, work is $W = Fs \cos \theta$ and θ is always 90° for charges in magnetic fields. Since $\cos 90° = 0$, $W = 0$. Because a magnetic field does no work on a moving charge, even though the direction of motion is continually changing as a result of the magnetic force, the speed of the particle v is constant and the charge's kinetic energy remains constant.

11.4. Electric Currents Generate Magnetic Fields

• **Magnetic fields** are created by moving free electric charges as well as by current flowing in a wire, which is made up of moving charges. The *strength* of the magnetic field B near the wire can be measured. We can develop a relationship between current I and magnetic field **B** by first considering a straight current-carrying wire. The magnetic field **B** around the wire is cylindrical with field lines forming circles centered on the wire. It has been observed in experiments that the **magnetic field decreases linearly with distance** r away from the center of a wire, and that the field strength B at a distance r from the wire is proportional to the current I and inversely proportional to r:

$$B \propto I/r$$

The proportionality constant between B and I/r for a straight wire has been shown to be in the form $\mu_0/2\pi$, where $\mu_0 = 4\pi \times 10^{-7}$ T·m/A is called the **permeability of free space**. The **magnetic field for a long straight current-carrying wire** can therefore be expressed as:

$$\boxed{B = \mu_0 I/2\pi r}$$

This equation has been shown to describe the magnetic field in a long straight wire.

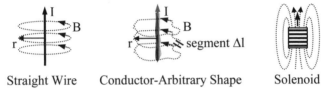

| Straight Wire | Conductor-Arbitrary Shape | Solenoid |

• A more general equation describing the magnetic field **B** in a current-carrying conductor of any shape was developed by French scientist A.M. **Ampere**. Ampere's equation describes the field that is not a perfect cylinder but some arbitrary shape surrounding a current-carrying conductor. Ampere summed the products $B_\parallel \Delta l$ around the field (since they vary with the shape of the conductor and therefore the field), where Δl represents a segment along B, and $B_\parallel$ is the component of **B** that is parallel to each Δl segment. Ampere found the relationship between current I and the sum of the product $B_\parallel \Delta l$ for the segments is:

$$\boxed{\sum B_\parallel \Delta l = \mu_0 I}$$

This equation is called **Ampere's Law** and is difficult to evaluate for arbitrary shapes without using calculus. *Ampere's Law* states that for any closed loop path of the field, the sum of the length sections times the magnetic field in the direction of the length section ($\sum B_\parallel \Delta l$) equals the permeability μ_0 times the electric current I enclosed in the loop.

As we saw with the straight wire producing field lines, which formed circles centered on the wire, there are special cases where the magnetic field can be described by a simple equation. We can develop the equation for the long straight wire from Ampere's Law. In this case B is the same for any **r**, so we draw a circle of radius r around the wire. The sum in Ampere's Law, $\sum B_\parallel \Delta l = \mu_0 I$, can be rewritten: $B_\parallel \sum \Delta l = \mu_0 I$. The sum of

all the segments around a circle is the circumference of a circle, or $2\pi r$.
So we can write Ampere's Law for the **long straight wire** as:
$B(2\pi r) = \mu_0 I$. Rearranging, we again get the equation for the *magnetic field for a long straight wire*:

$$B = \mu_0 I / 2\pi r$$

Above, we said the proportionality constant between B and I/r for the straight wire has been shown to be in the form $\mu_0/2\pi$, but we can see that the *more general proportionality constant* is $\mu_0 = 4\pi \times 10^{-7}$ T·m/A, the *permeability of free space*.

• A **solenoid** is another configuration in which Ampere's Law, $\sum B_{\parallel}\Delta l = \mu_0 I$, simplifies to an equation for magnetic field B. Remember a solenoid is a current-carrying coil. The magnetic field created *inside* the coil runs parallel to the axis of the coil and is uniform, except near the ends where the field lines diverge. The value of B inside a solenoid is strong and constant, so $\sum B_{\parallel}\Delta l$ becomes BL. The value of B at the ends and outside is negligible by comparison, so an approximation to $\sum B_{\parallel}\Delta l$ in Ampere's Law becomes simply BL. Along the length L of a solenoid there are N loops through which the current I flows. This means for N loops of wire within length L, the total current flowing along the length of the solenoid is NI. That means for a solenoid the proportionality for current and magnetic field is: BL $\propto$ NI. Using the proportionality constant μ_0, we have for a solenoid:

$$BL = N\mu_0 I \quad \text{or} \quad B = (N/L)\mu_0 I$$

which approximates the **magnetic field for the solenoid**. We can instead use n as the number of current loops per unit length of a solenoid, so n = (N/L), giving:

$$\boxed{B = (N/L)\mu_0 I = n\mu_0 I}$$

Because B does not depend on the location within the solenoid, solenoids can create uniform magnetic fields.

• **Example**: Your friend informs you that you live in an area where the Earth's magnetic field is about 5×10^{-5} Tesla. He wants to know how much current he needs to deliver to his 10-cm solenoid to produce the Earth's magnetic field strength providing it has 1,000 wire loops.

You use the magnetic field for solenoids equation $B = (N/L)\mu_0 I = n\mu_0 I$, with: n = 1,000/0.1 m = 10,000/m and $\mu_0 = 4\pi \times 10^{-7}$ T·m/A. Solve for I:

$$I = B/n\mu_0 = (5 \times 10^{-5} \text{ T})/(10,000/\text{m})(4\pi \times 10^{-7} \text{ T·m/A}) \approx 4 \times 10^{-3} \text{ A}$$

"Not much current..." he muses.

The Ampere

• The **Ampere** can be defined in terms of the force between two long straight parallel current-carrying wires. We have used the Ampere as a unit of measure of current, but there is a definition of the Ampere we can understand in the context of current and magnetic fields. Imagine two parallel wires of length L, which are separated by distance R. If wire 1 carries current I_1 and wire 2 carries current I_2, the magnetic field from each wire exerts a force on the other. The magnetic field at wire 2 due to the current I_1 is:

$$B_1 = \mu_0 I_1 / 2\pi R$$

The force wire 1 exerts on wire 2 is:

$$F = I_2 L B_1 = I_2 L \mu_0 I_1 / 2\pi R$$

(Remember F_M = ILB is the magnetic force when the field lines are perpendicular to the current-carrying wire.) To write the force F per length L, divide by L:

$$F/L = \mu_0 I_1 I_2 / 2\pi R$$

When two currents are in the same direction, it can be shown that the force is attractive, and when the currents are in opposite directions, the force is repulsive. (The right-hand rule illustrates the directions of the magnetic field and force vectors.) If each wire has a current flow of 1 A, and they are separated by 1 m, then the force per unit length is:

$$F/L = \mu_0 I_1 I_2 / 2\pi R = (4\pi \times 10^{-7}\,\text{T·m/A})(1\,\text{A})(1\,\text{A})/(2\pi)(1\,\text{m}) = 2 \times 10^{-7}\,\text{N/m}$$

where 1 T = 1 N/A·m. This is the definition of the Ampere, which says, for two long parallel wires spaced one meter apart that exert attractive forces on each other, the Ampere is the current flowing in the wires, which produces a force on a unit length of either wire of exactly $2 \times 10^{-7}\,\text{N/m}$. In other words, one Ampere is the current needed in each of two parallel wires separated by 1 m to create a force of $2 \times 10^{-7}\,\text{N/m}$.

Magnetic Fields and Relative Motion

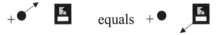

• Imagine a charged particle q and a meter that measures magnetic fields, where the charge and meter are near each other at rest in a coordinate system (which is an inertial reference frame). While there is an electric field from the charged particle, the magnetic-field-reading meter will measure **B** = 0 since the magnetic field is zero. Now imagine charge q is in motion while the meter is still stationary. Since *moving charges produce magnetic fields*, the magnetic-field-reading meter will detect a field and **B** will not be zero. It is the *relative motion between the charge and the meter that the meter detects*. Now suppose the charged particle q remains stationary and the meter moves relative to the charged particle. (This is equivalent to the scenario in which the charged particle moves while the meter is stationary.) When the charge is stationary and the meter moves, the meter will measure the presence of a magnetic field. A *magnetic field is produced by a changing or moving electric field* which, in this case, is a result of the relative motion between the charge q and the magnetic-field-measuring meter.

11.5. Changing Fields Induce Current and Magnetic Flux

• We have learned that a static magnetic field exerts a force on a moving charge or a current-carrying wire. If you have a wire that is not connected to a battery or other source of current, it has no current and no magnetic field. *If you move that current-free wire through a magnetic field, the motion of the wire in the magnetic field will induce a current to flow in the wire.* Moving the wire produces a magnetic force F_M on the charges in the wire, and they will move along the wire as long as the wire is moving in the field. Therefore, moving a wire in a field causes free charges to flow so that a *current is induced in the wire.* This is referred to as **electromagnetic induction**, and was discovered in 1831 by English physicist, Michael **Faraday** (1791–1867).

Changing Magnetic Fields Induce Current

• Before we learn a few useful relationships pertaining to inducing currents in conductors, we should be aware that *current and magnetic fields don't just appear or disappear, but grow or shrink over small time intervals.* Suppose you have a wire loop linked to a battery via a switch such that closing the switch allows current flow and opening the switch stops current flow. When the switch is moved from open to closed, current begins to flow and builds up to its final steady value over some time interval—it does not reach its final steady value instantly. After the switch is changed from open to closed, the magnetic field that is induced by current flow in the wire also increases (and spreads out) with time as it builds up from zero to its final value (when current is flowing at its final steady value).

• Now suppose you have two wire loops near each other—wire 1 connected to a battery via a switch and wire 2 connected to an ammeter. If you close the switch on wire 1 so that current begins to flow and the magnetic field begins to grow, that growing magnetic field (with its moving field lines) acts as a moving magnetic field and can induce a current in nearby wire 2 (which you can read on wire 2's ammeter) as the field grows and crosses wire 2. Once the current reaches its final steady value, the magnetic field no longer grows (and moves), and the induced current in wire 2 decreases to zero. Note that if the switch is then set to "open" so that current in wire 1 decreases to zero, its magnetic field will also decrease to zero. As the field decreases (and shrinks), another current will be induced in wire 2, but in the opposite direction as the current from the expanding field.

A changing magnetic field produces a voltage in a wire or coil, causing a current to flow. This voltage is known as the induced emf.

Inducing emf

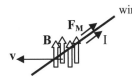

Current I induced by F_M as wire moves at velocity v perpendicular to **B**.

- As a conductor (such as a wire) moves in a magnetic field, the force that is exerted on the free electrons in the conductor creates an **induced emf** (or voltage). The *value of the emf* can be found by considering a straight section of wire having length L that is moving at velocity **v** within magnetic field **B** (with the wire oriented perpendicular to both **v** and **B**). We learned earlier in this chapter that the magnitude of force F_M a charge in a wire experiences is:

$$F_M = qvB \sin \theta$$

where q is the charge, v is velocity of the wire, B is field strength, and θ is angle between **v** and **B**. The direction of F_M is parallel to the wire (by the right-hand rule) and the charge flows along the wire. An emf, or voltage, is generated, and the force is sufficient to move charges. As charges move, **work** is done, which is given by:

$$W = F_M L = qvBL \sin \theta$$

where L is the distance the charge moves along the conductor. The work per charge done causes a *potential difference* or *voltage*, which is the **induced emf** $\mathcal{E}$:

$$\mathcal{E} = W/q = BvL \sin \theta$$

or just:

$$\boxed{\mathcal{E} = BvL \sin \theta}$$

When velocity **v** is perpendicular to the direction of **B**, $\sin \theta = 1$ and **induced emf** $\mathcal{E}$ is **maximum**:

$$\boxed{\mathcal{E} = BvL \quad \text{for } \mathbf{v} \perp \mathbf{B}}$$

Note that since voltage is V = IR and the emf $\mathcal{E}$ equals the potential difference, or voltage V, the **current I induced in the wire** is:

$$I = V/R = \mathcal{E}/R = BvL/R$$

where R is the resistance.

• A magnetic field does not need to be stationary for a current to be induced in a moving wire. Either the field or the wire or both need to move. It is the *relative motion* of a wire and a magnetic field that is needed for current flow. In fact a current will flow in the same direction within a wire whether a magnetic field moves, for example, to the right across a stationary wire or the wire moves to the left through a stationary magnetic field.

Magnetic Flux and Faraday's Law

• As we have seen, relative motion of a conductor (such as a wire) and a magnetic field results in an induced emf, which causes a current to flow. Is there a way to quantify the *amount of magnetic field in a particular region*? Yes, the **magnetic flux** Φ_M. The magnetic flux can be thought of as a measure of the number of magnetic field lines passing through a certain surface *perpendicularly*. The greater the number of field lines passing through a particular surface, the greater is the magnetic flux through that surface.

Imagine a uniform magnetic field **B** (or a small and essentially uniform region of a field). If you have a flat surface with area A that is *perpendicular* ($\perp$) to field **B**, the **magnetic flux** Φ_M that passes through this surface is defined as the *product of the perpendicular field strength $B_\perp$ and the surface area A*:

$$\Phi_M = B_\perp A$$

If the surface is not perpendicular to field **B**, the magnetic flux corresponds to the *perpendicular component of B, or B cos θ*, where θ is the angle between the direction of magnetic field **B** and the perpendicular to the area A. The **magnetic flux** Φ_M becomes:

$$\boxed{\Phi_M = B_\perp A = BA\cos\theta}$$

Note that A can be the area inside a circuit loop, and $B_\perp$ would be the component of the magnetic field that is perpendicular to the plane of that circuit loop with area A. In this case $B_\perp$ is equal to B cos θ, where θ is the angle between the direction of **B** and the *perpendicular to the plane* of the circuit loop. Therefore, if a loop of wire encircling area A is in a magnetic field B, the magnetic flux is: $\Phi_M = B_\perp A = BA \cos \theta$.

Magnetic flux can be imagined as the intensity of a magnetic field over a certain area A within a current loop. The density of the magnetic field lines within the area represents the intensity of the field, so that the **magnetic flux** is proportional to the number of lines enclosed by the current loop. Flux conveys the amount of a magnetic field going through a certain area. If the magnetic field is increased, the flux will be increased by that same amount.

• **Magnetic flux** is measured in **units** of $T \cdot m^2$ in the MKS (meters-kilogram-seconds) system, which reflects the magnetic field intensity in **Tesla times area** (m^2). It is also measured in the **Weber** (Wb), named for German physicist W.E. Weber (1804–1891), where $1 \text{ Wb} = 1 \text{ T} \cdot m^2$.

• As we have just seen, induced emf $\mathcal{E}$ is $\mathcal{E} = BvL \sin \theta = B_\perp vL$, and the **magnetic flux** Φ_M is $\Phi_M = B_\perp A = BA \cos \theta$. Let's consider the effect of a *change in area* on induced emf. If a conductor (such as wire) moves a distance x in time t through a magnetic field B, then $v = x/t$. The emf $\mathcal{E}$ becomes:

$$\mathcal{E} = BvL = B(x/t)L$$

The distance the wire moves times its length is xL, which is the area A traced out by the wire as it moves during time interval Δt. Therefore, the change in enclosed area ΔA during time Δt is the change in xL during Δt, and emf $\mathcal{E}$ can be written:

$$\boxed{\mathcal{E} = B\Delta A/\Delta t}$$

Since the **magnetic flux** Φ_M that passes through area A is the perpendicular field strength $B_\perp$ times the area A, or $\Phi_M = B_\perp A$, then the product BA in equation $\mathcal{E} = B\Delta A/\Delta t$ corresponds to the magnetic flux. Therefore, we can write for emf $\mathcal{E}$:

$$\mathcal{E} = B\Delta A/\Delta t = \Delta \Phi_M/\Delta t$$

This shows that the *induced emf* $\mathcal{E}$ equals the *change in magnetic flux per second,* or *rate of change of magnetic flux.* **Faraday** observed that *the emf generated in a circuit equals the rate of change of the magnetic flux through that circuit.* In equation form this is referred to as Faraday's Law:

$$\mathcal{E} = \Delta \Phi_M/\Delta t$$

where $\Delta \Phi_M$ is the change in the magnetic flux over time interval Δt. This equation is often written with a *negative sign* to show that the direction of the induced emf $\mathcal{E}$, which creates a current, opposes the flux change

(which is in accord with *Lenz's Law*). This equation which describes emf
ε as the change in flux per time is called **Faraday's Law**:

$$\varepsilon = -\Delta\Phi_M/\Delta t$$

*Faraday's Law shows that a change in magnetic flux creates an induced
emf.*

Faraday's Law often is used to describe a circuit that is a *coil* in a varying
magnetic field. You can increase the magnitude of the induced emf if a
circuit is designed as a coil having many wire loops rather than a single
loop. For a coil having N loops and a changing magnetic flux through the
coil, **Faraday's Law** becomes:

$$\varepsilon = -N\Delta\Phi_M/\Delta t$$

where the *induced emf* ε in the coil is measured in **volts**.

If flux changes $\Delta\Phi_M$, an emf ε will be induced. An emf can be induced in
a loop if the magnetic field changes ΔB, if the area changes ΔA, or if
there is a change in the angle between the field and the loop:

$$\varepsilon = -N\Delta\Phi_M/\Delta t = -\Delta(BA\cos\theta)/\Delta t$$

• **Units** for measuring **emf** ε are in volts providing $\Delta\Phi_M/\Delta t$ is in Wb/s.

• Faraday observed that a current can be induced in a conductor (such as
a wire) that is placed into a changing or varying magnetic field even if
there is no physical displacement of the conductor or magnetic field. We
discussed this phenomenon above in our example of the two nearby
wires, wire 1 and wire 2, as the actively increasing or decreasing
magnetic field in wire 1 induced a current in wire 2.

• **Example**: Your friend walks in with a 10-cm by 20-cm circuit he just
wired which has two loops. Given that the resistance in the circuit is 3 Ω,
he wants to know what the induced current flow will be if he allows the
magnetic field strength to steadily decrease from 6,000 G to 1,000 G
during a 2-s interval.

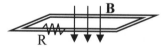

First you remind him that his current will only flow during the time that
the magnetic field is changing. Then you jot down Faraday's Law:

$$\varepsilon = -N\Delta\Phi_M/\Delta t = -\Delta(BA\cos\theta)/\Delta t$$

where in this case $\Delta\Phi_M$ is the changing field times area $A\Delta B$, so:

$$\mathcal{E} = NA\Delta B / \Delta t$$

where N = 2, Δt = 2 s, ΔB = 0.6 T – 0.1 T = 0.5 T (since 1 G = 10^{-4} T), and A is 10 cm by 20 cm or 0.1 m × 0.2 m = 0.02 m^2. Therefore:

$$\mathcal{E} = NA\Delta B / \Delta t = (2)(0.02\ m^2)(0.5\ T)/(2\ s) = 0.01\ V$$

Because V = IR and the voltage across the resistor is also 0.01 V, the current is:

$$I = V/R = (0.01\ V)/(3\ \Omega) \approx 3.3 \times 10^{-3}\ A \ \ \text{or} \ \ 3.3 \ \text{milliamps}$$

Note that the magnetic field strength is decreasing so current flow is in the opposite direction it would be if field strength was increasing.

• According to Faraday's Law the effect of changing the magnetic flux in a circuit can be synonymous with placing a battery into the circuit. Since $\mathcal{E} = B\Delta A / \Delta t = \Delta \Phi_M / \Delta t$, the current flow due to the flux $\Phi_M = BA$ through the circuit would be as if a battery with a voltage $V = \Delta \Phi_M / \Delta t$ was inserted into the circuit.

Lenz's Law

• For a wire moving relative to a magnetic field, by the principle of **Conservation of Energy**, the direction of the induced current is oriented so that the force on that current opposes the wire's movement. Therefore work is required to move the wire through the magnetic field. Stated another way, an induced emf has the direction which ōpposes the change in magnetic flux that produced it. If there is a flux increasing within a coil, the current produced by the induced emf creates a flux which attempts to cancel the increasing flux. Conversely, if there is a flux decreasing within a coil, the current creates a flux which attempts to rebuild the decreasing flux. This principle was discovered by German physicist Heinrich Lenz (1804–1865) and is known as **Lenz's Law**. **Lenz's Law** tells us that *an induced emf produces a current having an action which opposes the change that induced it. This means that an induced emf acts in a way that the resulting current produces an induced magnetic field that opposes the change in flux.* If this did not occur, an induced current would enhance the flux change that produced it, which would create an endlessly building process. The concept of Lenz's Law more generally states that if an electromagnetic change produces an effect, then that effect will induce a reaction which tends to oppose the original change.

11.6. Key Concepts and Practice Problems

- Magnets have two poles, N and S, which exist together.
- Magnetic fields **B** possess both magnitude and direction, and are therefore vectors.
- Moving charges produce magnetic fields, and moving magnets induce electric current.
- Interrelated electric and magnetic effects are called electromagnetism.
- The magnetic field direction is perpendicular to the direction of electric current flow.
- The right-hand rule shows the direction of the magnetic field lines for a current-carrying straight wire, a loop of wire, or a solenoid.
- A moving charge experiences a force due to a magnetic field.
- The magnitude of magnetic force is: $F_M = qvB \sin \theta$ or $F_M = qvB$ if $\mathbf{v} \perp \mathbf{B}$ or $F_M = ILB \sin \theta$, where L is wire length, θ is between I and **B**.
- Torque on an N-loop flat coil in a uniform magnetic field is: $T = NIAB \sin \theta$.
- A charged particle moving in uniform magnetic field **B** has its direction continually changed by force $\mathbf{F_M}$ exerted by the field causing circular or spiral motion. Radius r, frequency f, and angular frequency ω of the particle's circular motion are: $r = mv/qB$; $f = 1/\tau = qB/2\pi m$; $\omega = qB/m$.
- Magnetic fields are created by moving charges or flowing current.
- Ampere's Law for current enclosed in a loop of a magnetic field is: $\sum B_{\parallel} \Delta l = \mu_0 I$, with **B** parallel to section Δl and $\mu_0 = 4\pi \times 10^{-7}$ T·m/A.
- Magnetic field for a straight current-carrying wire: $B = \mu_0 I/2\pi r$.
- Magnetic field for a solenoid: $B = (N/L)\mu_0 I = n\mu_0 I$, n is number N of current loops per length L.
- Moving a current-free wire in a magnetic field induces current flow.
- Current and magnetic fields grow or shrink over small time intervals.
- A conductor moving in a magnetic field has force exerted on its free electrons inducing an emf or voltage: $\mathcal{E} = BvL \sin \theta = B_{\perp}vL$.
- Magnetic flux Φ_M is: $\Phi_M = B_{\perp}A = BA \cos \theta$.
- Faraday's Law: induced emf equals the change in magnetic flux per second, or rate of change of magnetic flux $\Delta\Phi_M$: $\mathcal{E} = -\Delta\Phi_M/\Delta t$.
- A change in flux $\Delta\Phi_M$ induces emf $\mathcal{E}$, which occurs in a loop if the magnetic field changes ΔB, the area changes ΔA, or the angle between **B** and the loop changes: $\mathcal{E} = -N\Delta\Phi_M/\Delta t = -\Delta(BA \cos \theta)/\Delta t$.
- Lenz's Law: an induced emf produces a current having a magnetic field that opposes the change in flux that induced it.

Practice Problems

11.1 Which end of a compass needle points north? Why?

11.2 In the example in Section 11.2, subsection *Magnetic Force on Current-Carrying Wire*, if your wire lies along the equator and the current runs from East to West, what will be the direction of the magnetic force?

11.3 What strength magnetic field is required to keep a proton moving around a cyclotron with a radius of 0.75 m at a speed of 2.3×10^7 m/s?

11.4 (a) Why is the magnetic field uniform within a solenoid? **(b)** Why is the magnetic field within a solenoid not a function of its diameter?

11.5 (a) What is the magnetic flux across a cross section of a solenoid having a length of 10 cm, a diameter of 4 cm, 1,000 loops of wire, and a current of 50 Amps? **(b)** What current would be induced in a stationary 10 cm wire pointing down the center of the solenoid?

Answers to Chapter 11 Problems

11.1 The N-pole of the magnetic compass needle points north. Since opposite poles attract, the Earth's "north" magnetic pole is in fact the S-pole of the Earth's magnetic field, but is called north by convention.

11.2 Using the right hand rule, pointing your fingers west and rotating them to the south (the direction of the Earth's magnetic field lines) indicates the magnetic force would be in the direction of your thumb, or upward, away from the center of the Earth.

11.3 $r = mv/qB$ or $B = mv/qr$
$= (1.67 \times 10^{-27}$ kg$)(2.3 \times 10^7$ m/s$) / (1.6 \times 10^{-19}$ C$)(0.75$ m$) \approx 0.32$ T.

11.4 (a) The strength of a magnetic field is inversely proportional to the distance from the wires (not the square of the distance). Beginning at the center of the cylindrical solenoid, moving away from wires on one side results in moving an equal distance toward wires on the other side, so the net change in B is zero. **(b)** Comparing two solenoids carrying equal current which are identical except the second has twice the diameter of the first, a point inside the second will be twice as far from the wires, but those wires are twice as long and conduct twice as many electrons per loop, so the net difference on B is zero.

11.5 (a) First, $B = (N/L)\mu_0 I = (1,000/0.1$ m$)(4\pi \times 10^{-7})(50$ A$) \approx 0.628$ T. $\Omega_M = B_\perp A = (0.628$ T$)(\pi(0.02$ m$)^2) \approx 7.9 \times 10^{-4}$ Wb.
(b) Since the wire is stationary with respect to the magnetic field, no magnetic force is being created and no current is induced.

Chapter 12

ALTERNATING CURRENT

12.1. Alternating Current
12.2. Capacitance in an AC Circuit
12.3. Inductance in an AC Circuit
12.4. RCL Circuits
12.5. Key Concepts and Practice Problems

> *"The seeds of great discoveries are constantly floating around us,*
> *but they only take root in minds well prepared to receive them."*
> Attributed to Joseph Henry, first director of The Smithsonian Institution

12.1. Alternating Current

● In a simple **generator** a loop-shaped conductor (which can be a wire circuit loop or a wire-wrapped iron core) called an armature is rotated in a magnetic field, thereby inducing a current. A generator uses the principle of inductance and a changing magnetic flux to create electric current. A generator is most efficient when the circuit loop is perpendicular to the magnetic field and to its direction of motion. The total induced current can be further increased by increasing the number of wires in the armature, thereby adding to the total length of wire moving within the magnetic field.

Work is supplied to the system in order to rotate the armature. In electricity generation plants, work is often supplied by steam-powered turbines where the steam is generated using coal, oil, gas, nuclear reactions, etc. The current flow in the armature loop reverses direction with each half revolution. This switching of current flow produces an **alternating current**. The *alternating current* produced by an armature steadily rotating in a constant magnetic field graphs as a *sinusoid*.

Direct current can be obtained from an armature using a switching device called a **commutator** (having fixed contacts which act as sliding switches such that as the armature rotates, the coil comes into contact with them alternately and the direction of the current does not change). A modern alternative is a **diode**, which acts as a **rectifier**, filtering out current in one direction allowing only the flow in the other direction.

Alternating Current

• In direct current (DC) circuits, current flows in one direction as a constant voltage is supplied (e.g., by a battery). In **alternating current (AC) circuits**, the **voltage oscillates** in a **sine wave pattern** which varies with time. In AC circuits both current and voltage reverse direction at regular intervals. Alternating currents are generated by **alternating voltages**:

$$V(t) = V_{max} \sin \omega t$$

where V_{max} is the maximum or peak voltage measured in volts, ω is angular frequency in radians per second, t is time in seconds. The **angular frequency** ω is related to the **frequency** f by $\omega = 2\pi f$, where f is measured in **Hertz** and is the number of cycles per second.

The equations describing AC circuits reflect the rotation of a simple armature loop or coil having area A between two magnets in a simple generator. Using **Faraday's Law**, $\varepsilon = \Delta\Phi_M/\Delta t$, with flux, $\Phi_M = B_\perp A$, the component of the magnetic field $B_\perp$ perpendicular to the plane of the coil varies in a sinusoid pattern with time Δt, giving the graph of emf (voltage) versus time (or ωt) as a sine wave. (See Section 6.3 for a discussion on periodic sinusoid functions.) Equations for **emf ε, or equivalently voltage V, in AC circuits**:

$$\varepsilon = \varepsilon_{max} \sin \omega t = \varepsilon_{max} \sin 2\pi t/\tau = \varepsilon_{max} \sin 2\pi ft$$
$$V = V_{max} \sin \omega t = V_{max} \sin 2\pi t/\tau = V_{max} \sin 2\pi ft$$

where $\tau = 2\pi/\omega$ or $\omega = 2\pi/\tau = 2\pi f$ and $f = 1/\tau$. The time required for one revolution or period is: $\tau = 2\pi/\omega$. Angular frequency, or angular velocity, ω is measured in radians per second and describes the number of radians an angle ϕ in circular motion completes per second, where each complete cycle corresponds to a change in ϕ of 360° or 2π radians, so that $\omega = 2\pi$ radians. Frequency f is the number of complete oscillations or cycles per second, and is measured in Hertz Hz. Household current in the U.S. has a frequency of 60 Hz ($\tau = 1/60s$).

• The flow of current in an AC circuit through a resistor obeys $V = IR$. A resistor does not react to a changing voltage (unlike capacitors and inductors which we will discuss). Therefore, the relationship $V = IR$ applies for resistors in an AC circuit, even though the voltage is changing. The **current through a resistor in an AC circuit** is:

$$I = V/R = (V_{max}/R) \sin \omega t = I_{max} \sin \omega t = I_{max} \sin 2\pi t/\tau = I_{max} \sin 2\pi ft$$

Root Mean Square Voltage and Current

• We have seen that voltage, or emf, in an AC circuit is described by $\mathcal{E} = \mathcal{E}_{max} \sin \omega t$, where $\mathcal{E}_{max}$ is the maximum emf during a cycle of the oscillation. The voltage or current that is measured using a voltmeter, ammeter, or multimeter is generally not the maximum values ($\mathcal{E}_{max}$, V_{max}, or I_{max}), but rather the **rms (root mean square)** value. Voltages and currents for AC circuits are generally expressed as rms values.

Remember in Section 9.3 we used *root mean square velocity* to describe the molecular velocity v by taking the square root of the mean square velocity $(v^2)_{Ave}$. The rms value is a statistical measure of a varying quantity x such that, $x_{rms} = [(x^2)_{Ave}]^{1/2}$. To calculate rms, square the quantities, take the average of the squares, then take the square root. For a quantity such as voltage, which varies sinusoidally with time, there are several ways to derive the rms value. Here, we find V_{rms} by beginning with $V = V_{max} \sin \omega t$ and squaring and averaging, $[(\sin^2 \omega t)_{Ave}]$. Then since $\sin \omega t$ has the same shape as $\cos \omega t$, just shifted along t by 1/4 period, we substitute into the trigonometric identity $\sin^2 \omega t + \cos^2 \omega t = 1$, and average it to show $2(\sin^2 \omega t)_{Ave} = 1$ or $(\sin^2 \omega t)_{Ave} = 1/2$. Therefore, $V_{rms} = V_{max}[(\sin^2 \omega t)_{Ave}]^{1/2} = V_{max}[1/2]^{1/2}$, or the **root mean square voltage** is:

$$V_{rms} = V_{max} / [2]^{1/2} \approx V_{max} (0.707)$$

Therefore, the **root mean square voltage** V_{rms} is the maximum voltage V_{max} divided by the square root of two, and the **root mean square current** I_{rms} is the maximum current I_{max} divided by the square root of two:

$$I_{rms} = I_{max} / [2]^{1/2} \approx I_{max} (0.707)$$

V_{rms} would be what a voltmeter measuring AC would read across the terminals of a simple generator. While the maximum voltage in a household circuit in North America is about 170 volts, we refer to a typical household voltage of 120 volts, which is V_{rms}. The following graph is of a sine wave over one 360° cycle with the *dashed line* showing the **root mean square, rms,** value near 0.707.

• **Example**: Your friend examines his standard household wall plug from which he can measure a 120 V voltage at 60 Hz AC using his new multimeter. He notes that the smaller of the two rectangles is what is called the "hot" wire. He asks you what the voltage equation is that best describes his circuit and how he could find the maximum voltage.

First you find the maximum voltage using $V_{rms} = V_{max} / [2]^{1/2}$, where the measured voltage is the rms voltage $V_{rms} = 120$ V, so that:

$$V_{max} = V_{rms}[2]^{1/2} = (120 \text{ V})([2]^{1/2}) \approx 170 \text{ V}$$

The equation describing voltage is:

$$V = V_{max} \sin 2\pi ft = 170 \text{ V} \sin 2\pi (60 \text{ Hz})t = (170 \text{ V}) \sin 120\pi t$$

Root Mean Square for Power

• **Power** is dissipated by a resistor as heat. While we remember that power is voltage times current, $P = VI$, since the current and voltage in an AC circuit vary with time, we need to use the root mean square values of the sinusoidal voltage and current for power:

$$P = V_{rms} I_{rms} = (V_{max}/[2]^{1/2})(I_{max}/[2]^{1/2}) = (1/2)V_{max} I_{max}$$

Using the V_{rms} and I_{rms} values, **power in an AC circuit** can be written as:

$$\boxed{P = V_{rms} I_{rms} = I_{rms}^2 R = V_{rms}^2 / R}$$

Graph of Current and Voltage in AC Circuit with Resistor

• Graphs of current and voltage in an AC circuit commonly form sine wave patterns. We know that when the voltage across a resistor in an AC circuit is $V = V_{max} \sin 2\pi ft$, then the current through that resistor is $I = V/R = (V_{max}/R) \sin 2\pi ft = (V_{max}/R) \sin \omega t = I_{max} \sin \omega t$. In a **graph of an AC circuit which has only resistors, the current and voltage will be *in phase* with each other**. This means that the maximum or peak voltage is reached at the same time the maximum or peak current is reached. An **AC circuit** is designated by a circle around what looks like a backward horizontal "S":

AC
Circuit
with
Resistor

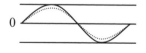

Voltage is solid curve.
Current is dashed curve.
Horizontal axis is time.

If an AC circuit has capacitors and inductors, the phase relationships between current and voltage will differ.

12.2. Capacitance in an AC Circuit

• In Section 10.2 *Capacitance,* we learned that a **capacitor** is a pair of conducting surfaces or plates separated by a gap that can accumulate and store electric charge. Imagine a circuit with only a capacitor and an AC power source (such as a wall outlet). A capacitor in an AC circuit exhibits a resistance-like effect called **capacitive reactance** X_C, which is measured in **units** of **Ohms** Ω (just as resistance is measured for a resistor). *Capacitive reactance* acts as a resistance or counter emf that is induced as current begins to flow into a capacitor. It reflects the degree to which a capacitor behaves like a resistor as the frequency in an AC circuit changes. **Capacitive reactance** depends on the frequency f of the AC voltage and has been experimentally measured as:

$$X_C = 1/\omega C = 1/(2\pi fC)$$

where X_C is **capacitive reactance** of the capacitor, C is capacitance, $f = 1/\tau$ is frequency, and $\omega = 2\pi/\tau = 2\pi f$ is angular frequency. We see that capacitive reactance is *inversely* proportional to *capacitance*, $X_C \propto 1/C$.

• Because capacitive reactance behaves as a resistance, we can calculate **voltage across a capacitor** and correlate the voltage V across a capacitor with the current I flowing through it using an equation similar to the resistance equation V = IR. We learned that for resistors in an A/C circuit $V_{rms} = I_{rms}R$, so for a capacitor we substitute the *capacitive reactance* X_C of the capacitor for *resistance* R, giving **voltage across a capacitor**:

$$V_{rms} = I_{rms}X_C$$

where rms values of voltage V_{rms} and current I_{rms} are generally used.

• If you know the applied voltage, you can calculate the **current through a capacitor**. Since we just learned that $V_{rms} = I_{rms}X_C$ and $X_C = 1/2\pi fC$, we combine these as $I_{rms} = V_{rms}/X_C = V_{rms}2\pi fC$, or just:

$$I_{rms} = V_{rms}2\pi fC$$

• **Example**: For a 1.5×10^{-6} F capacitor in an AC circuit with a 9 V rms voltage source, what would be the current when the frequency is 500 Hz, and is 1,000 Hz?

For 500 Hz: $I_{rms} = 2\pi fCV_{rms} = 2\pi(500)(1.5 \times 10^{-6})(9) \approx 0.0424$ A.

For 1,000 Hz: $I_{rms} = 2\pi fCV_{rms} = 2\pi(1,000)(1.5 \times 10^{-6})(9) \approx 0.0848$ A.

We see the linear dependence of current flow on frequency for a given capacitor in an AC circuit.

• The sinusoidal graphs of voltage and current across a *resistor* ($V_{rms} = I_{rms}R$) in an AC circuit are in phase. While **graphs of current and voltage for an AC circuit with a capacitor** have the same sinusoidal shape, they are offset along the horizontal time axis and are **out of phase**. (The amplitudes of the voltage and current are usually different.) In a capacitive circuit, the current reaches its maximum, or peak, one-quarter cycle (which is $\pi/2$, or 90°) before the voltage reaches its maximum peak. Therefore, in a capacitive circuit **current leads the voltage** by 90°, and they are out of **phase** by 90° or $\pi/2$. **Voltage and current for a capacitor** can be expressed with the following equations:

$$V = V_{max} \sin(\omega t)$$
$$I = I_{max} \sin(\omega t + \pi/2)$$

where $\omega = 2\pi f$. In trigonometry, we learn that $\sin(\omega t + \pi/2) = \cos(\omega t)$, which means we can also write **current for a capacitor** as:

$$I = I_{max} \cos(\omega t)$$

This form of the current equation reflects that a sine wave and a cosine wave are 90° out of phase with each other, and voltage and current in a capacitor are out of phase by 90°. We can also express current in terms of **capacitive reactance** X_C and peak voltage V_{max}:

$$I = (V_{max}/X_C) \cos(\omega t)$$

Let's look at the graph:

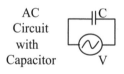

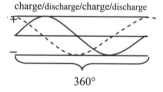

AC Circuit with Capacitor

charge/discharge/charge/discharge

Voltage: solid curve.
Current: dashed.
Horizontal axis: time.

360°

In capacitive AC circuits (like that depicted), the charge on a capacitor varies directly with the voltage. When the voltage reaches its positive peak, the capacitor reaches its maximum charge and current flow is briefly zero. Then the voltage declines from its peak, the capacitor discharges, and current reverses direction. As voltage passes through zero becoming negative, current is large and negative, and charge again builds up on the capacitor with an opposite polarity. When voltage reaches its negative peak, the capacitor again reaches its maximum charge but with opposite polarity. Voltage across the capacitor is $V = Q/C$, where Q is charge in a capacitor, C is capacitance, and the current is $I = \Delta Q/\Delta t$.

• Unlike resistors, which lose energy as heat, a capacitor in an AC circuit repeatedly absorbs and discharges power and uses no net energy.

12.3. Inductance in an AC Circuit

• We have learned that electromagnetic induction involves the generation of emf and current by a changing magnetic field. *Faraday* noted that a changing magnetic field in a circuit induced a current in a nearby circuit. American *Joseph Henry* made a similar discovery at about the same time. The creation of emf, or voltage, in a circuit or in a nearby circuit by a changing magnetic field is called **inductance**. There is also a property within electric circuits that opposes changes in electric current, which is also called **inductance**.

Self-Inductance and Mutual Inductance

• Remember that current flow in a conductor (such as a wire or coil) generates a magnetic field around that conductor. As the current flow in the conductor increases, decreases, or changes direction, the generated magnetic field will grow, shrink, or change direction. A change in the generated magnetic field in turn induces a new emf, or voltage, within the conductor. The induction of this new emf, or voltage, within the conductor is called **self-induction**. The *direction* of this new self-induced emf is in the *opposite direction of the current flow that produced it*. This is consistent with Lenz's Law, where an induced emf (voltage) in any circuit occurs in a direction opposite to the current that produced it. This opposing emf is sometimes referred to as back emf.

Self-induction in a circuit has the effect of *opposing any change in current flow in that circuit*. For example, when current begins to flow in a *coiled wire* and a magnetic field begins to build, the field buildup around one loop, or turn, of the wire affects adjacent turns of the coil. As a magnetic field builds around one loop of the coil, adjacent turns of the coil are subject to that changing field, and an emf is developed in the adjacent loops. According to Lenz's Law, the emf is in the direction that opposes the original flow of current in the circuit. This opposing, or back, emf causes the current in the coil to increase slowly to its maximum value, $I = V/R$.

Let's think of this in terms of voltage. In a circuit, when voltage is applied current begins to flow, which generates a magnetic field. As the field expands, a counter, or back, voltage is generated in the circuit. This back voltage causes a current flow in the opposite direction of the original current flow. This self-inductance opposes the buildup of current until the magnetic field becomes steady and the induced back voltage stops. If the current in the circuit is turned off, the induced magnetic field begins to decline. This declining magnet field generates another voltage in the direction that temporarily prolongs the original flow of current. The field induced by turning off the circuit eventually declines to zero, and the voltage it induced stops along with current flow. Self-induction from turning off the circuit also opposes changes in current flow. The result of self-induction is to oppose the buildup and delay the decline of current flow. Self-inductance in a coil acts to *resist sudden changes in the current or flux* through it.

• **Mutual inductance** was revealed in an experiment conducted by Faraday in which he placed two coils on a conducting iron ring. He showed that a changing magnetic field in one coil induced an emf, or voltage, in the second coil. It is called *mutual inductance when a changing magnetic field in one circuit induces voltage in a nearby circuit.* As expected by *Lenz's Law*, the direction of the induced emf, or voltage, is in the opposite direction of the current flow that generated it.

Inductors

• **Inductors** are components that are designed to provide inductance in a circuit. Inductors are made up of a *coil of conducting material* such as wire. While a small self-inductance can occur in a circuit even when the conducting material is straight, inductance is much greater when conductors are coiled since the magnetic field of each turn of a coil more directly impacts nearby turns. Inductance is enhanced by placing a material such as iron within the core of the coil. An iron core increases the strength of an inductor's magnetic field. Core materials with a higher permeability (which is a measure of the material's ability to magnetize) increase the magnetic field and therefore the inductance. Inductance of a coil is affected by: the number of turns in the coil (the more turns, the more magnetic field interactions, the more inductance); the cross-sectional area of the core; spacing between turns (a magnetic field gets weaker further away); and core permeability.

• Like capacitors, inductors react to alternating voltage. Current and voltage are out of phase in the opposite sequence as they are for capacitors. In the *graph of voltage V and current I for an AC circuit with an inductor L, the current lags the voltage* by a 90° or $\pi/2$ phase difference, so current reaches its peak one-quarter cycle (90°) after the voltage peaks.

AC
Circuit
with an
Inductor

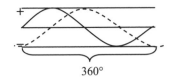

Voltage: solid curve.
Current: dashed.
Horizontal axis: time.

360°

As voltage from the power source increases from zero, the voltage on the inductor matches it with voltage from the changing flux through the coil (or changing current which changes the magnetic field in the coil). When the current is at its plus or minus peak maximum values and temporarily levels off, the induced voltage, which acts against changes in the flux in the inductor, falls or rises through its zero point. This is reflected in *current and voltage being out of phase as we see in the graph*. Since the applied **voltage** in the A/C circuit is:

$$V = V_{max} \sin(\omega t)$$

as the current lags the voltage by a 90° (or $\pi/2$) phase difference, the **current through the inductor** circuit is described by:

$$I = I_{max} \sin(\omega t - \pi/2)$$

Using trigonometry, **current** is equivalently written:

$$I = -I_{max} \cos(\omega t)$$

• Like a capacitor, an inductor alternately absorbs and discharges **power**, so it uses no net energy in an A/C circuit.

• Since inductors oppose any changes in current, they are useful for various applications, including as surge protectors, to stabilize direct current, or to control or restrict alternating current above certain frequencies. Inductor coils are also used in conjunction with capacitors in the tuning circuits of radios.

Inductors in Series and Parallel

• In a series circuit, all of the current passes through each of the components in the circuit. If the inductors are shielded from each other, or spaced to prevent mutual inductance, *the total inductance of the circuit is cumulative*. The **total inductance L_T of a series circuit** is the sum of all the inductors in the circuit:

$$L_T = L_1 + L_2 + L_3 \ldots$$

• In a parallel circuit, components are arranged such that the current path is divided. Placing inductors in parallel *decreases the total inductance* L_T of the circuit. If the inductors are shielded from each other or spaced enough to prevent mutual inductance, the **total inductance L_T of a parallel circuit** is:

$$1/L_T = 1/L_1 + 1/L_2 + 1/L_3 \ldots$$

Induced emf and Determining Inductance

• We have learned using *Faraday's Law* that total **induced emf, or voltage**, in a closely wound coil is proportional to the change in flux or current flow through the coil:

$$\mathcal{E} = -N\Delta\Phi_M/\Delta t$$

where $\mathcal{E}$ is the induced emf in volts, N is the number of turns in the coil, $\Delta\Phi$ is the change in magnetic flux in Webers (1 Wb = 1 T·m^2), and Δt is time in seconds. To describe inductance in equation form, remember that it is the property of a wire coil that determines its effectiveness in producing a **back emf** and therefore opposing changes in magnetic flux. It has been shown that if a back emf of 1 V is induced in a coil when the current through that coil is changing at a rate of $\Delta I/\Delta t$ equal to 1 **Ampere** per second, the coil will have an inductance of L which is 1 **Henry** (H). In other words, 1 H of **inductance** exists if 1 V of emf is induced when the current is changing at the rate of 1 A/s. The inductance can be written:

$$\boxed{L = \mathcal{E} / (\Delta I/\Delta t)}$$

where L (in honor of *Heinrich Lenz*) is *self*-**inductance** measured in Henries H (named after Joseph Henry), $\mathcal{E}$ is induced emf measured in volts V, ΔI is the change in current in Amperes A, and Δt is the amount of time in seconds for the change in current. Rearranging, we get an equation for *self*-**inductance** similar to Faraday's Law:

$$\boxed{\mathcal{E} = -L\Delta I/\Delta t \quad \text{or} \quad V = -L\Delta I/\Delta t}$$

where induced emf is also written V, the negative sign is included to represent *Lenz's Law*, and $\Delta I/\Delta t$ is the rate of current change. We can see that the induced flux through the N loops of coil, $N\Delta\Phi_M$ in $\mathcal{E} = -N\Delta\Phi_M/\Delta t$, is proportional to the current I flow through the coil. The inductance L is the constant of proportionality and: $\mathcal{E}\Delta t = N\Delta\Phi_M = L\Delta I$. This again gives: $\mathcal{E} = -L\Delta I/\Delta t$.

Inductive Reactance

• We have learned that **self-induction** in a coil has the effect of opposing any change in the flow of current by creating a "back" voltage which opposes the original flow of current. This scenario plays out as voltage is applied to a coil, current begins to flow, which causes a magnetic field to begin expanding, which induces a back voltage that is opposed to the original flow of current in the coil. The opposition to current flow is called **inductive reactance** and is measured in *Ohms* Ω (like resistance).

Inductive reactance in a circuit depends on the alternating current and its frequency and the amount of inductance. A circuit's **inductive reactance** X_L, also called effective resistance, can be shown to be:

$$X_L = \omega L = 2\pi f L$$

where X_L is measured in ohms, ω is angular frequency $\omega = 2\pi/\tau = 2\pi f$, $f = 1/\tau$ is the frequency of the applied alternating current in *Hertz*, and L is the inductance of the coil (inductor) measured in *Henries*. Inductance of the coil depends on its geometry and whether its core is ferromagnetic.

Similar to capacitors and capacitive reactance, in an AC circuit the **voltage across an inductor** depends on inductive reactance X_L and is:

$$V_{rms} = I_{rms} X_L$$

where the root mean square values for voltage V_{rms} and current I_{rms} are used. The inductive reactance of an inductor reflects the degree to which the inductor will resist the voltage across it. Unlike capacitive reactance which is *inversely* proportional to the *capacitance*, $X_C \propto 1/C$, the inductive reactance is *directly* proportional to the *inductance*, $X_L \propto L$.

• **Example**: Your friend has been occupying himself building small circuits and running them off his household 120-V voltage, 60-Hz AC electricity (which makes you nervous). He strung three 2 H inductors in series and wants you to calculate what the current should be. If he strings the inductors in parallel what would the current be?

For series: $X_L = 2\pi f L = (2)(\pi)(60/s)(2H + 2H + 2H) \approx 2{,}262 \ \Omega$, (where $L_T = L_1 + L_2 + L_3$).
Then current is: $I_{rms} = V_{rms}/X_L = (120 \ V)/(2{,}262 \ \Omega) \approx 0.053$ A.
For parallel L_T: $1/L_T = 1/L_1 + 1/L_2 + 1/L_3 = 1/2 + 1/2 + 1/2 = 3/2$.
So $L_T = 2/3$. Next: $X_L = 2\pi f L = (2)(\pi)(60/s)(2/3 \ H) \approx 251 \ \Omega$.
Then current is: $I_{rms} = V_{rms}/X_L = (120 \ V)/(251 \ \Omega) \approx 0.48$ A.

Transformers—an Application of Inductance

• **Transformers** are used to increase or decrease AC voltage. Transformers are often made up of two coils of wire that are wound onto different segments of the same core. The coils, which are referred to as the *primary input coil* and *secondary output coil*, are electrically insulated from each other. As an AC voltage is applied across the primary coil, a changing magnetic field is generated in the core which induces a voltage in the secondary coil. Transformers are therefore able to increase or decrease AC voltage by causing an induced voltage in the secondary coil using **mutual inductance**. Using *Faraday's Law* the **primary voltage** V_p and the **induced secondary voltage** V_s are:

$$V_p = N_p \Delta\Phi/\Delta t \quad \text{and} \quad V_s = N_s \Delta\Phi/\Delta t$$

where N_p and N_s are the number of loops in the primary and secondary coils. If the coils are the same except the number of loops, these equations combine to what is called the *transformer equation* because it correlates input and output voltage:

$$V_s / V_p = N_s / N_p$$

If the secondary output coil has more loops than the primary input coil, the induced (output) voltage in the secondary coil will be greater than the input voltage of the primary input coil. In this *step-up transformer*, voltage has been *stepped-up*. If the primary input coil has more loops than the secondary output coil, the induced (output) voltage in the secondary coil will be less than the input voltage of the primary input coil. In this *step-down transformer*, voltage has been *stepped-down*. If the primary input voltage and current are $V_p I_p$ and the secondary output voltage and current are $V_s I_s$, we can write:

$$V_p I_p = V_s I_s$$

A step-up transformer increases voltage while decreasing in current, and a step-down transformer reduces voltage while increasing current. This is a useful property for electric power transmission over long distances, since a power plant can immediately *step-up* voltage while lowering current to send electricity which reduces losses due to resistance. Then step-down transformers can be used to lower the voltage and increase current to specified levels near where it is used.

12.4. RCL Circuits

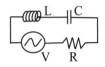
AC circuit with an inductor,
a capacitor, and a resistor.

• An **RCL circuit** is a circuit that includes a resistor, a capacitor, and an inductor. A circuit with only two of the three (a resistor, capacitor, or inductor) is called an RC circuit, an RL circuit, or an LC circuit. When all three elements are in one RCL circuit, the overall resistance to the flow of current is called the **impedance** Z. If a voltage V is applied to an RCL series circuit, then a form of *Ohm's Law* (V = IR) relates V to the current I:

$$V = IZ \quad \text{or} \quad V_{rms} = I_{rms}Z$$

where Z is impedance measured in Ohms. The impedance Z in an RCL circuit consists of R, X_C, X_L, and is found by combining the resistance R, the capacitive reactance X_C, and the inductive reactance X_L. Because of the phase relationships we cannot simply sum the resistances, but in an RCL circuit the resistance and reactances must be added as vectors. Remember, in a circuit with only a resistor, the voltage and current are in phase with each other. In a circuit with only a capacitor, current leads voltage by a 90° phase. In a circuit with only an inductor, voltage leads current by a 90° phase.

To develop the relationship for impedance and find the phase between the voltage and current, we can use an x-y coordinate system and first draw resistance R along the +x-axis. Next, we draw inductive reactance X_L along the +y-axis 90° to R. Finally, we draw capacitive reactance X_C along the −y-axis 90° to R and 180° from the inductive reactance. The vector sum is **impedance** Z.

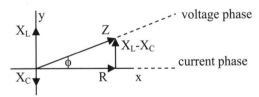

Remember from Section 1.7, the **Pythagorean Theorem** states that the sum of the squares of the lengths of the sides of a right triangle equals the square of the length of its hypotenuse. Using this relation, in an RCL circuit with resistance, capacitance, and inductance in series, **impedance** Z is:

$$Z = [R^2 + (X_L - X_C)^2]^{\frac{1}{2}}$$

From trigonometry (SohCahToa, Section 2.3) we know the angle ϕ is:

$\tan \phi = (X_L - X_C)/R$ or $\cos \phi = R/Z$ or $\phi = \arctan(X_L - X_C)/R = \arccos R/Z$

where the angle ϕ (see above figure) between the impedance Z and the resistance R in the vector diagram represents the phase relationship, or shift, between current and voltage. When phase is depicted as a vector in a plane, zero phase is usually shown as the positive x-axis and associated with the resistor (since the voltage and current of the resistor are in phase). The magnitude is represented as the length of the vector, and its angle ϕ depicts its phase relative to that of current through the resistor. The phase angle ϕ associated with the impedance Z of the circuit is the difference in phase between the voltage and the current. When ϕ is positive, voltage leads current by that angle, and when ϕ is negative, voltage lags current. Similar relations for voltage are:

$$V = [V_R^2 + (V_L - V_C)^2]^{\frac{1}{2}} \quad \text{and} \quad \tan \phi = (V_L - V_C)/V_R$$

• **Example**: What rms current would you expect in an RCL circuit with a rms voltage of 12 V, a 5 Ω resistance, an 8 Ω capacitive reactance, and a 10 Ω inductive reactance?

Begin with: $Z = [R^2 + (X_L - X_C)^2]^{\frac{1}{2}} = [5^2 + (10 - 8)^2]^{\frac{1}{2}} \approx 5.4 \ \Omega.$
Since $V_{rms} = I_{rms}Z$: $I_{rms} = V_{rms}/Z = 12 \ V/5.4 \ \Omega \approx 2.2 \ A.$

• If you follow electrical energy in an AC circuit, power is dissipated in resistors as heat, a capacitor alternately stores charge and then returns it losing no energy, and an inductor alternately stores energy in the magnetic field and returns it to the circuit so none is lost.

12.5. Key Concepts and Practice Problems

• Alternating current (AC) circuits: the voltage V (emf $\mathcal{E}$) oscillates as $\mathcal{E} = V = V_{max} \sin \omega t = V_{max} \sin 2\pi t/\tau = V_{max} \sin 2\pi ft$, where $f = 1/\tau$ is *frequency*, and ω is *angular frequency* $\omega = 2\pi/\tau = 2\pi f$.
• The current through a resistor in an AC circuit is:
  $I = V/R = (V_{max}/R) \sin \omega t = I_{max} \sin \omega t = I_{max} \sin 2\pi t/\tau = I_{max} \sin 2\pi ft$.
• Voltages and currents for AC circuits are measured as rms values:
  root mean square voltage: $V_{rms} = V_{max} / [2]^{\frac{1}{2}} \approx V_{max} (0.707)$;
  root mean square current: $I_{rms} = I_{max} / [2]^{\frac{1}{2}} \approx I_{max} (0.707)$.
• Power in an AC circuit: $P = V_{rms} I_{rms} = I_{rms}^2 R = V_{rms}^2 / R$.

- Current and voltage are in phase in an AC circuit with only resistors.
- Capacitors in AC circuits have a resistance-like effect called capacitive reactance X_C: $X_C = 1/\omega C = 1/(2\pi fC)$, where C is capacitance.
- Voltage across a capacitor: $V_{rms} = I_{rms} X_C$.
- Current through a capacitor: $I_{rms} = V_{rms} 2\pi fC$.
- In a capacitive AC circuit current leads voltage by 90° or $\pi/2$: $V = V_{max} \sin(\omega t)$ and $I = I_{max} \sin(\omega t + \pi/2) = I_{max} \cos(\omega t)$.
- Self-induction of voltage, or current, in a conductor causes opposing current flow and acts to resist sudden changes in the current or flux.
- Mutual inductance occurs when a changing magnetic field from one circuit induces voltage in a nearby circuit.
- In AC circuits with only inductors, current lags voltage by 90° or $\pi/2$: $V = V_{max} \sin(\omega t)$ and $I = I_{max} \sin(\omega t - \pi/2)$ or $I = -I_{max} \cos(\omega t)$.
- Self-inductance $L = \mathcal{E}/(\Delta I/\Delta t)$ or $\mathcal{E} = V = -L\Delta I/\Delta t$ represent induced emf $\mathcal{E}$, or voltage V.
- Inductive reactance X_L is: $X_L = \omega L = 2\pi fL$, where L is inductance.
- Voltage across an inductor in AC circuit: $V_{rms} = I_{rms} X_L$.
- Resistance to current flow in RCL circuits (having resistor R, capacitor C, and inductor L) is impedance $Z = [R^2 + (X_L - X_C)^2]^{1/2}$, and voltage applied to RCL series circuit is $V = IZ$. Phase shift between current and voltage: $\phi = \arctan(X_L - X_C)/R = \arccos R/Z$.

Practice Problems

12.1 (a) Your new electric car is being charged from a 240-V outlet. You measure the current at 32 A. What is the electric power being transferred to the car's batteries? **(b)** What is the instantaneous maximum power?

12.2 (a) Your household electricity is 120 V and 60 Hz. You connect to it a simple circuit having one capacitor and measure a current of 1 A. What is the capacitance of the capacitor? **(b)** If you replace the capacitor with another much smaller one with only one thousandth the capacitance of the first capacitor, what current would you observe?

12.3 (a) Electric motors typically contain coiled wires and create inductance. When you unplug a vacuum cleaner while it is still turned on, you will observe a spark between the plug and socket. What causes the spark? **(b)** AC power with a current of 30 A and 1,200 V enters a transformer having 1,000 loops on the input side and 100 on the output side. What current and voltage would you measure on the output side?

12.4 (a) You have an AC circuit with a resistance of 10 Ω, a capacitive reactance of 15 Ω, and an inductive reactance of 3 Ω. You apply a voltage of 12 V to the circuit. Will voltage lead or lag current and by what phase angle? **(b)** If you double the voltage, what change will you observe in the phase angle?

Answers to Chapter 12 Problems

12.1 (a) $P = V_{rms}I_{rms} = (240\ V)(32\ A) = 7.68\ kW$. **(b)** $V_{max} = V_{rms}[2]^{\frac{1}{2}}$ and $I_{max} = I_{rms}[2]^{\frac{1}{2}}$, so $P_{max} = V_{max}I_{max} = (V_{rms}[2]^{\frac{1}{2}})(I_{rms}[2]^{\frac{1}{2}}) = 2V_{rms}I_{rms} = 15.36\ kW$.

12.2 (a) $I_{rms} = V_{rms}2\pi fC$ so $C = I_{rms}/V_{rms}2\pi f = (1\ A)/(120\ V)(2\pi)(60\ Hz) \approx 2.21 \times 10^{-5}\ F$. **(b)** Since $I_{rms} \propto C$, you would measure 1/1,000 the current, or $10^{-3}A$.

12.3 (a) The motor's induction resists the cessation of current caused by pulling the plug, and the current continues to flow for a short interval with sufficient voltage to jump the gap between the plug and the socket. **(b)** This is a step-down transformer, where $N_s/N_p = V_s/V_p = 1/10 = V_s/1,200V$, so output voltage is 1/10 the input voltage, or $V_s = 120\ V$. Since $V_pI_p = V_sI_s$, output current is 10 times greater, or $I_s = V_pI_p/V_s = (1200)(30)/(120) = 300\ A$.

12.4 (a) $\tan \phi = (X_L - X_C)/R = (3 - 15)/10 = -1.2$, so $\phi \approx -50°$ and, because ϕ is negative, voltage lags current. **(b)** The phase change does not depend on the voltage level.

Chapter 13

WAVES, SOUND, AND LIGHT

"When one door closes another door opens, but we so often look so long and so regretfully upon the closed door, that we do not see the ones which open for us."

"God has strewn our paths with wonders, and we should certainly not go through Life with our eyes shut."

Both attributed to Alexander Graham Bell

13.1. Traveling Waves

• Waves can be divided into two categories: mechanical waves and electromagnetic waves. **Mechanical waves** include waves in water or air, sound waves, and waves on strings. Mechanical waves involve oscillatory motions of matter or particles, and require matter in order to propagate. **Electromagnetic (EM) waves** include visible light, radio waves, microwaves, X-rays, and infrared radiation. Electromagnetic waves involve oscillations of electromagnetic fields and do not require matter to propagate. EM waves can travel through matter or through empty space. Waves not traveling through empty space move through some type of medium. The medium could be air, water, or a string. The speed that a wave travels depends on the characteristics of the medium it is moving through, such as the medium's density or temperature.

A wave is often described as a disturbance traveling from one location to another location through a medium. The **medium** is any substance or material that carries the wave, such as air, water, a string, etc. A medium is usually composed of particles (such as air molecules) which vibrate and can collide or interact with each other, thereby transmitting the disturbance or wave. As a disturbance moves through a medium, energy is transported from one particle to another as it flows from the wave source.

• Elementary wave forms are typically described using sine and cosine functions in the form y = A sin bx or y = A cos bx, which graph in the characteristic crest-and-trough **sinusoidal wave shape**. If the variable A is changed the *amplitude* is affected, and if b is changed the *length* of the repeating pattern is affected.

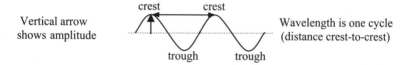

The **energy** in a wave is proportional to the square of its **amplitude**, so a higher amplitude wave has higher energy than an otherwise identical lower amplitude wave: $E \propto A^2$. The **frequency** f refers to the rate at which something occurs or the number of vibrational cycles per time such as cycles per second (Hertz, Hz). So for a wave, frequency is waves per second or vibrations per second. While frequency measures cycles per second, period measures seconds per cycle.

period τ = 1/frequency f   and   frequency f = 1/period τ

The **period** τ of a wave is the time between the passing of wave crests (or troughs) measured at a certain point. During one period the wave crests (or troughs) travel one **wavelength** λ. We can measure wave speed using wavelength and period. The speed of a wave is the distance traveled by a certain point on the wave (such as a crest) per time.

speed = distance/time

For example, if a wave crest travels 10 meters in 5 seconds, the wave speed is 10 m / 5 s = 2 m/s. In terms of wavelength and period, **wave speed** v is also the wavelength/period:

$$v = \lambda/\tau$$

Since frequency f = 1/period τ, **wave speed** v is also wavelength times frequency:

$$v = \lambda f$$

which is referred to as the **wave equation**. Therefore, **wave speed** is:

$$\boxed{v = \lambda/\tau = \lambda f}$$

Wave Pulses—Applying a Single Up-and-Down Motion to a String

Traveling Wave Pulse

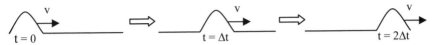

- A helpful model of mechanical wave motion can be demonstrated by applying a single up-and-down motion to a string. If an upward force quickly followed by a downward force is applied to one end of a string, the up-then-down motion will propagate along the length of the string as the up-down force is transmitted to each adjacent section of the string. The speed of the propagating wave pulse along a string can be described by observing the position of the wave pulse at various instants of time (just as we would observe the position of a moving particle in space). For a wave pulse traveling a distance Δx in time interval Δt, the **propagation speed** can be written:

$$v = \Delta x / \Delta t$$

The speed of a propagating wave pulse along a string can be further described by considering which features of the string affect the velocity. If you imagine creating a pulse along a string by applying an up-then-down force, you may expect that the speed of the wave pulse would be affected by the string's density ρ, cross-sectional area A, and tension T with which the string is supported. The speed of a wave pulse along a string can be described using these variables as:

$$v = [T/\rho A]^{1/2}$$

If you substitute the units for these variables in this equation, they reduce to the units of velocity v. Using MKS units we can write these as: $T = N = kg \cdot m/s^2$, $\rho = kg/m^3$, and $A = m^2$, with $v = m/s$. Combining these into the equation gives:

$$v = [T/\rho A]^{1/2} = [(kg \cdot m/s^2)/((kg/m^3)(m^2))]^{1/2}$$

$$= [(m/s^2)/(1/m)]^{1/2} = [m^2/s^2]^{1/2} = m/s \quad \text{where m/s are units of velocity}$$

Traveling Waves—Continuing the Up-Down Motion

- While there are various types of traveling waves, such as ocean waves, let's continue looking at the string model. What happens when we continue to apply an up-then-down motion to the string in a way that one end of the string is subject to a regular oscillating up-down motion? This creates a series of identical pulses along the string which can be described as a sinusoidal traveling wave. If we move the string's end up and down in a simple harmonic oscillating motion, the displacement can be described by a sinusoidal function.

The pulse of each up-down motion travels down the string as a wave. The velocity of each traveling wave is the same as if it was a single pulse and is given by: $v = \Delta x/\Delta t$. The shape of the series of pulses forms a sine wave pattern. The distance between any two successive peaks of the sine wave pattern formed by the string is the **wavelength** λ. Since the motion of the string follows the up-down oscillatory motion driving it, the **period** τ of the wave motion is the same as each oscillatory motion over one complete cycle. This means that during each time interval or period τ the wave moves forward by an amount equal to one wavelength λ. For a time interval $\Delta t = \tau$ and a distance $\Delta x = \lambda$, the wave velocity $v = \Delta x/\Delta t$ is:

$$v = \lambda/\tau = \lambda f \quad \text{where period is } \tau = 1/f \text{ and frequency is } f = 1/\tau$$

Transverse Waves vs. Longitudinal Waves

• The motions of waves can occur as either *transverse waves* or *longitudinal waves*. In the motion of **transverse waves**, particles in the medium (such as in strings or waves on water) oscillate *perpendicular to* the direction of wave propagation. Note that electromagnetic waves are also considered transverse waves, though they can transmit through a vacuum. In **longitudinal waves**, particles move back and forth along, or *parallel to*, the direction of wave propagation. Longitudinal waves can be **compression waves** traveling along a spring or **sound waves** traveling through air. The elastic properties of the medium (such as a string or air) affect how waves propagate since the wave motion depends on the natural restoring forces of the medium. For example, strings provide a natural restoring force for transverse wave propagation, gases and liquids provide a natural restoring force for longitudinal waves, and solids provide a natural restoring force for both transverse and longitudinal waves.

• As one end of a string is oscillated up and down, the string forms transverse waves which propagate, or travel, along its length. Note that while waves travel along the string, the string is displaced up and down, always returning to its original position. If we instead consider a *spring*, waves propagate as *longitudinal waves*, which involve alternately compressing and expanding sections of the spring, thereby creating a variation in the spring's density. This change in the density of coils along the spring's length can be described by a sinusoidal function.

Propagation of *sound waves* also involves alternating *compression* and *rarefactions* as well as a change in density, or pressure, which can be described by sinusoidal functions. It turns out that longitudinal waves vary with distance and time the same way that transverse waves vary, and can therefore be described by the same sinusoidal equations.

13.2. Standing Waves

Reflected Waves and Superposition

• Suppose you have a string connected at one end to a fixed support. When a wave is propagated along the string it will *reflect* back once it reaches the fixed end. This **reflected wave pulse** will have the same size and shape as the initial pulse (except for some loss of altitude due to energy transfer or friction), but the *sign of the displacement* will be reversed from the force the solid support exerts on the wave as it impacts the structure. In other words, after a wave is reflected it will be inverted even though its speed, frequency, and wavelength will be the same. It will also have a slightly lower amplitude due to a loss (transfer) of energy.

• **Reflected waves** that continue to reflect create multiple wave forms propagating along the same string. When, for example, two wave pulses travel along the same string, each propagates independently of the other. As the wave pulses displace the string at a certain location, their combined effect will be the *sum of the two displacements*. The two waves therefore are subject to the principle of **superposition** in which two or more waves crossing the same location superimpose onto one another, adding or subtracting. The waves do not permanently change one another's direction, speed, frequency, or wavelength, though a slight loss in amplitude, or energy, will occur.

When two wave pulses pass one another, the *amplitude displacement is the sum of the two individual displacements*. If these two displacements have *opposite signs*, but the same shape and size, the wave pulses *cancel* at the point or moment they pass (but will continue as they were before passing). If the two pulses have the *same sign*, they will *add* at the moment they pass. If two waves pass each other in opposite directions, then once past they retain their size, shape, and speed (except for frictional losses). Having multiple traveling wave pulses creates **interference**, which is **constructive** when the pulses add and **destructive** when they cancel.

Nodes and Waves on Strings With Both Ends Fixed

• When a wave on a string connected to a support is reflected off that support, the fixed end of the string is motionless. Any point along a string that remains motionless while the rest of the string is moving is called a **node**. A fixed end is always at a node. The wave motion of the string in the figure below has three nodes. **Antinodes** are locations where the wave motion is at a maximum (the figure below shows two antinodes). More generally, in any standing wave pattern, whether on a string or a sound wave in a musical wind instrument, there are points (nodes) along the wave that appear to be standing still and points of maximum displacement (antinodes). The node-antinode pattern alternates.

• What if *both ends of a string are fixed to a support*? Waves can travel in both directions as they reflect off the end points. Imagine a string fixed between two supports and its wavelength λ of wave motion is the length L of the string (solid curve in figure below), so that $\lambda = L$. When the string is in motion, its wave will be reflected until it forms a mirror image (dashed curve in figure). If you could watch the wave leaving the left-hand support at time $t = 0$, it would move along the solid-curve position until it reflected off the right-hand support and formed its mirror image shown as the dashed curve position. As time continues your wave would repeatedly reflect off the two supports and alternate between the solid and dashed curves forming a **standing wave**. If there were no frictional losses, this would continue indefinitely.

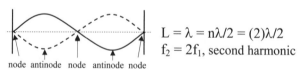

$$L = \lambda = n\lambda/2 = (2)\lambda/2$$
$$f_2 = 2f_1, \text{ second harmonic}$$

node antinode node antinode node

The above figure shows *one complete wavelength* and therefore has $\lambda = L$. If a standing wave displays *one-half of a wavelength* so $L = \lambda/2$, or $\lambda = 2L$, this is the longest wavelength having the lowest frequency that can set up between the two supports. Shown below are additional wavelengths having **nodes** at both ends forming *standing waves*:

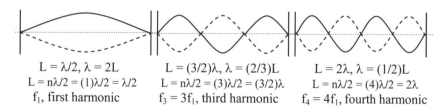

$L = \lambda/2, \lambda = 2L$	$L = (3/2)\lambda, \lambda = (2/3)L$	$L = 2\lambda, \lambda = (1/2)L$
$L = n\lambda/2 = (1)\lambda/2 = \lambda/2$	$L = n\lambda/2 = (3)\lambda/2 = (3/2)\lambda$	$L = n\lambda/2 = (4)\lambda/2 = 2\lambda$
f_1, first harmonic	$f_3 = 3f_1$, third harmonic	$f_4 = 4f_1$, fourth harmonic

The formula relating length L with wavelength λ for a standing wave as shown in the figures is:

$$L = n\lambda/2 \quad n = 1, 2, 3, \ldots$$

• **Standing wave patterns** are produced at certain vibrational frequencies which are characteristic to the medium, such as a string between supports or the air in the cavity of a wind musical instrument. Each characteristic frequency corresponds to a distinct standing wave pattern which is called a **harmonic**. The lowest frequency that a standing wave can exist in is called the **fundamental frequency** f_1, which for a string corresponds to wavelength $\lambda = 2L$, the **first harmonic**. The next lowest frequency for a standing wave has frequency $f_2 = 2f_1$, which corresponds to a wavelength $L = \lambda$ and the **second harmonic**. The standing wave frequencies are $f_1, 2f_1, 3f_1, 4f_1, \ldots$, such that the frequency of a standing wave is an integer multiple of the f_1 fundamental frequency, $f_n = nf_1$. Note that the harmonic number corresponds to the number of antinodes such that the first harmonic has one antinode, the second harmonic has two antinodes, and the nth harmonic has n antinodes. For a standing wave pattern having a node at each end, there are n halves of wavelengths between the nodes, which correspond with the above equation for the nth harmonic: $L = n\lambda/2$.

• In addition to harmonics, sound wave frequencies greater than f_1 are distinguished as corresponding **overtones**. For example, frequency f_2 is the *second harmonic and first overtone*; frequency f_3 is the *third harmonic and second overtone*; frequency f_4 is the *fourth harmonic and third overtone*; and so on. Note that successive overtones count in order even if certain harmonics are absent such that if the second harmonic of the fundamental was missing, the first overtone would match up with the third (or next) harmonic.

• More generally, **standing waves** form when the nodes of a wave motion fall at the two ends of a medium where reflection can occur. A standing wave pattern is an effect of interference that occurs when waves having the same frequency pass through a medium in opposite directions. Standing waves form a vibrational pattern which can be produced when source waves reflect and there is **interference** between the source waves and reflected waves. As mentioned above, standing wave patterns only occur at certain vibrational frequencies which are characteristic to the medium or "container" of the waves (such as wind and stringed musical instruments). These characteristic frequencies are called **harmonic frequencies**, or **harmonics**. For frequencies other than the characteristic or harmonic frequencies, the interference of source and reflected waves create irregular patterns rather than standing wave patterns. Standing waves can be set up for any type of wave motion if proper reflection points or surfaces are present. Waves that can produce standing waves include sound waves, radio waves, and light and other forms of electromagnetic waves.

• **Example**: What is the fundamental frequency of a string stretched 1 m between supports if the wave speed is 2 m/s?

We can use $v = \lambda/\tau = \lambda f$, or $f = v/\lambda$. For the fundamental frequency f_1:

$$f_1 = v/2L = (2 \text{ m/s})/(2)(1 \text{ m}) = 1 \text{ Hz}$$

13.3. Sound Waves

• **Sound waves** in air are longitudinal pressure or compression waves emitted from a vibrating source in which molecules collide with their neighbors as the sound wave travels away from its source. In these longitudinal waves, the medium oscillates in a direction parallel to the sound wave propagation (unlike transverse waves which oscillate perpendicular to the direction of wave propagation). While sound waves can travel through gases, liquids, and solids, they cannot travel through a vacuum. The human ear can detect sound wave frequencies in the range from about 15 to 20 Hz to about 20,000 Hz. Certain species of animals and insects, including bats and dolphins, can detect high frequency sounds up to 100,000 Hz.

The Speed of Sound Waves

• You can actually estimate the speed of a propagating sound wave through air if you stand at a known distance from a large flat wall, then generate a sharp sound just as you begin a stop watch by measuring the time it takes for you to first hear the echo of the sound. If done accurately, you will calculate about 1100 ft/s, or about 330 m/s, for the speed of sound in air on a cold day at a temperature of 0 °C and about 344 m/s at 20 °C. Using $\lambda = v/f$, the **wavelengths at 0 °C** for the lower frequency (long wavelength) and higher frequency (short wavelength) audible sounds in air are about: $\lambda = v/f = (330 \text{ m/s})/(15 \text{ Hz}) = 22 \text{ m}$ and $\lambda = v/f = (330 \text{ m/s})/(20,000 \text{ Hz}) = 0.0165 \text{ m}$ or 1.65 cm.

• **Example**: Your friend wants to know how he could calculate the frequency of a 10-cm sound wave in 0 °C air. What would you say?

You suggest using $f = v/\lambda$ with a more precise value for v of 331.4 m/s.

$$f = v/\lambda = (331.4 \text{ m/s})/(0.10 \text{ m}) = 3,314 \text{ Hz}$$

• The speed that sound travels through a medium depends on the medium's characteristics, particularly its elastic properties. Sound through a **gas or liquid**, where shear stress is absent, has a **wave speed** that depends on the **bulk modulus** B and the **density** ρ and is written: $v^2 = B/\rho$ or $v = [B/\rho]^{1/2}$. In a **gas at constant temperature**, the bulk modulus B and pressure are equal, so wave speed can be written: $v^2 = P/\rho$, or $v = [P/\rho]^{1/2}$. Sound waves traveling through a gas expand and compress adiabatically (constant heat) not isothermally (constant temperature) as heat moves from compressed to rarefied regions. We can model this using the relationship between pressure and volume for an *adiabatic process*: PV^γ = constant, where $\gamma = c_p/c_v = 1.40$, which is the ratio of specific heats in air. In an adiabatic process the bulk modulus is $B = \gamma P$, with P as average pressure in the gas, so the speed of sound in a gas can be written:

$$v^2 = \gamma P/\rho \quad \text{or} \quad v = [\gamma P/\rho]^{1/2}$$

Using the ideal gas law, PV = nRT, the **speed of sound in a gas** is:

$$v = [\gamma P/\rho]^{1/2} = [\gamma RT/M]^{1/2}$$

where M is the molecular mass of the gas in kg/mol.

• **Example**: Now your friend wants to know how to calculate the speed of sound in both 0 °C and 20 °C air. What would you tell him?

You suggest: $v = [\gamma RT/M]^{1/2}$ where R = 8.31 J/mol·K and $\gamma = c_p/c_v = 1.40$, the ratio of specific heats in air. Since air is predominantly a mixture of about 80/20 of N_2/O_2, you say:

$$M = (0.8)(28\text{ amu}) + (0.2)(32\text{ amu}) = 28.8\text{ amu} = 28.8\text{ g/mol} = 0.0288\text{ kg/mol}$$

$$v = [(1.40)(8.31\text{ J/mol·K})(273\text{ K}) / (0.0288\text{ kg/mol})]^{1/2} \approx 332\text{ m/s}$$

$$v = [(1.40)(8.31\text{ J/mol·K})(293\text{ K}) / (0.0288\text{ kg/mol})]^{1/2} \approx 344\text{ m/s}$$

Note that for temperatures other than 0 °C, you can calculate the speed of sound in air by multiplying the speed of sound at 0 °C by $[T/273K]^{1/2}$, or:

$$v = (331.4\text{ m/s}) \times [T \text{ in Kelvin}/273\text{K}]^{1/2}$$

• The speed of sound through a solid material, such as a long bar or rod, can be modeled using Young's modulus Y (discussed in Section 6.2):

$$v = [Y/\rho]^{1/2}$$

• Note that the measure of the speed of sound is often used when reporting the speeds of aircraft and rockets. The *ratio of the speed that an object is traveling to the speed of sound in the same medium* is the **Mach number**, named in honor of **Ernst Mach** (1838–1916).

Seeing Sound Waves

• While some waves are quite visible, other vibrations, such as on a violin string or the sound of our voices in air, cannot be perceived with our eyes. By connecting a microphone to an **oscilloscope**, however, we can translate sound waves into alternating electrical voltage (via the microphone) and display the waves on the oscilloscope screen (as a function of time). When a sound entering the microphone has a pure frequency, the displayed signal is sinusoidal and corresponds to that frequency. When a sound consists of two or more different frequencies, such as a fundamental frequency and its harmonics, the displayed signal shows the superposition of the signals that can be smooth and repeating. When random noise is picked up by the microphone, the display may show jagged lines having no clear repeating patterns.

Beats

• Suppose two pure, steady sound waves having similar frequencies, such as $f_1 = 2,000$ Hz and $f_2 = 2,006$ Hz, are emitted from two sources. While you will hear the average of the two frequencies $(2,000 + 2,006)/2 = 2,003$ Hz, you will also hear a modulation, or **beat**, with the sound intensity at a rate equal to $|f_1 - f_2| = 6$ Hz. In other words the beat you hear is at a frequency equal to the absolute value of the difference in frequency of the two sinusoidal waves:

$$f_{beat} = |f_1 - f_2|$$

• **Example**: Your friend comes to you with two tuning forks—one emitting a pure 500 Hz frequency and one a pure 497 Hz. When sounded simultaneously, how many beats will you hear in a 10 s period?

The beat frequency will be $|f_1 - f_2| = 500 - 497 = 3$ Hz. Since Hertz is per second, you should hear 30 beats in 10 s.

• **Beats** (or *beating*) are caused by the alternating constructive and destructive interference of the waves, which produce alternating soft and loud sound. The following depicts two waves of different frequencies, f_1 and f_2, with time, showing beats. The third and longest curves show the outline of the amplitude variation of the resulting combined waveform of f_1 and f_2 as they add and subtract.

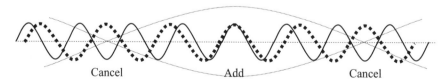

Cancel   Add   Cancel

Music and Sound

• **Musical instruments** are made up of vibrating systems which create sound. Usually an instrument has two or more vibrating systems working together to create its characteristic sound at a high enough volume to be heard by our ears. Sounds made by musical instruments are composed of a fundamental frequency and a mixture of harmonics. The makeup of the harmonics or overtones creates the so-called **timbre** of the sound. The word **pitch** is usually used to describe the sound frequency. In fact, when we hear music we quickly notice its frequency, or *pitch*, and its **loudness**, which is our perception of its volume or intensity.

• The *intensity of a sound wave* is the energy per second (or power) that it transfers to an area of a surface. Sound intensity can be measured in Watts per square meter, $Watt/m^2$. Because the range of sound intensities that we can barely detect to that which we can barely tolerate is so great (about 10^{12} times), a logarithmic scale was developed to measure sound intensity, or loudness. This scale is made up of the unit **Bel** (B) in honor of **Alexander Graham Bell** (1847–1922). This logarithmic scale reflects that for a sound having an intensity level that is 1 B greater than another sound, the ratio of the two intensities is 10. Similarly, for a difference of 2 B, the intensity ratio is 10^2, and a difference of 3 B reflects an intensity ratio of 10^3, and so on. Because a Bel is a large unit, sound intensity is often described in the familiar **decibel** (dB) unit, where 1 B = 10 dB and a difference of 2 B = 20 dB reflects an intensity ratio of 10^2, and 3 B = 30 dB reflects an intensity ratio of 10^3. Decibels are used to give us a relative measure of sound intensity, which can be expressed as the difference in the **sound intensity level** measured in dB for the sounds with intensities I_1 and I_2 as:

The difference in the sound intensity level in dB = $10 \log(I_2/I_1)$

• **Example**: If one sound is twice as intense as the sound before it, what is the difference in the sound intensity level? If a sound is increased by a factor of 10,000, what is the change in decibels?

For sound twice as intense, the difference in sound intensity level is:

$10 \log(I_2/I_1) = 10 \log(2/1) \approx 3$ dB

For sound increased by a factor of 10,000, the difference in sound intensity level is:

$10 \log(I_2/I_1) = 10 \log(10,000/1) = 40$ dB

• Audible sound can be thought of as a pressure wave having ranges in frequency from about 20 Hz to 20,000 Hz and intensity from about 10^{-12} W/m^2 (0 dB) to 10 W/m^2 (130 dB).

Harmonics, Resonance, and Waves in Cylindrical Instruments

• Music is often produced by instruments that employ **resonance** to create characteristic sounds. The *strings of musical instruments* produce standing waves at characteristic frequencies as they vibrate. The longest wave that sets up on a string has two nodes at its attached ends and one antinode at its center having the **fundamental**, or **resonance**. The fundamental wavelength is $\lambda_1 = 2L$, where L is the string length. Additional wavelengths of an **instrument string with a node at each end** are:

$$\lambda_n = 2L/n \quad \text{or} \quad L = n\lambda/2 \quad \text{with } n = 2, 3, \ldots$$

where you can think of n as corresponding to the number of antinodes. The standing waves that can exist are the **harmonic series** for the string.

• Like transverse standing waves on vibrating strings, **longitudinal sound waves in cylinders** produce characteristic frequencies. Cylinders can be open at both ends with an antinode at each end or closed at one end with a node at the closed end and an antinode at the open end. Standing waves in cylinders reflect off the closed end as they do on strings fixed to supports.

• Imagine you have a **cylinder closed at one end only** that is filled with air, and you can control its inside length by sliding a piston at the closed end. A standing sound wave can first occur when the length L is equal to 1/4 of the wavelength of the sound wave, $L = \lambda/4$. As you increase the length inside, **standing sound waves** can occur at lengths, $L = \lambda/4$, $L = 3\lambda/4$, $L = 5\lambda/4$, and so on, as depicted:

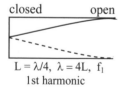
closed   open
$L = \lambda/4, \lambda = 4L, f_1$
1st harmonic

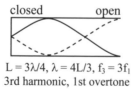
closed   open
$L = 3\lambda/4, \lambda = 4L/3, f_3 = 3f_1$
3rd harmonic, 1st overtone

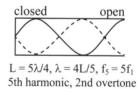
closed   open
$L = 5\lambda/4, \lambda = 4L/5, f_5 = 5f_1$
5th harmonic, 2nd overtone

The intensity of the sound generated in the cylinder is enhanced when the characteristic frequency of the generated sound is the same as the frequency of the wave in the cylinder so a standing wave is produced. This "**resonance**" will occur when $L = \lambda/4$ as well as the additional lengths in the series of one-end-closed cylinders. A **cylinder closed at one end** will **resonate** at the **series of wavelengths** λ:

$$\lambda = 4L/n \quad n = 1, 3, 5, \ldots$$

with corresponding **series of frequencies**:

$$f = v/\lambda = nv/4L \quad n = 1, 3, 5, \ldots$$

so for a fundamental frequency (n = 1) of f_1, the **sequence of standing-wave frequencies** is f_1, $3f_1$, $5f_1$, ... This suggests a cylinder of air with one end closed supports standing waves with *odd* harmonics of the fundamental. The above figure also labels the harmonics and overtones.

• For a **cylinder open at both ends** standing waves will have antinodes at each end.

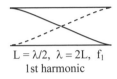

$L = \lambda/2$, $\lambda = 2L$, f_1
1st harmonic

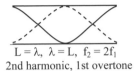

$L = \lambda$, $\lambda = L$, $f_2 = 2f_1$
2nd harmonic, 1st overtone

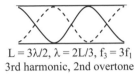

$L = 3\lambda/2$, $\lambda = 2L/3$, $f_3 = 3f_1$
3rd harmonic, 2nd overtone

A **cylinder open at both ends** will **resonate** at the **series of wavelengths**:
$$\lambda = 2L/n \qquad n = 1, 2, 3, \dots$$
with corresponding **series of frequencies**:
$$f = v/\lambda = nv/2L \qquad n = 1, 2, 3, \dots$$
so that for a fundamental frequency (n = 1) of f_1, the **sequence of standing-wave frequencies** is f_1, $2f_1$, $3f_1$, …, suggesting a cylinder of air with both ends open supports standing waves with all integer harmonics of the fundamental.

• **Example**: In an instrument using cylindrical pipes such as organ pipes, the fundamental is the lowest note which corresponds to the smallest part of the wave that fits inside the pipe. As we can see in the drawings above, fitting in more of the wave produces harmonics and additional notes. For a 1.0-m pipe *closed at one end* what is the wavelength of the lowest frequency that can cause resonance? What is the lowest frequency to resonate in the pipe if the air temperature is 20 °C? What is the frequency of the first and second overtones causing sound at 20 °C? What is the lowest frequency to resonate and third overtone if the pipe is *open at both ends*?

The lowest or fundamental frequency for a pipe *closed at one end* requires $L = \lambda/4$, so for a 1-m pipe:
$$L = 1\ m = \lambda/4 \quad or \quad \lambda = 4\ m$$
The lowest frequency to resonate in 20 °C air uses a speed of sound of 344 m/s, so using $v = f\lambda$, the fundamental frequency is:
$$f_1 = v/\lambda = (344\ m/s)/(4\ m) = 86\ Hz$$
The first overtone is the next lowest frequency and third harmonic:
$$f_3 = 3f_1 = 3(86\ Hz) = 258\ Hz$$
The second overtone is the fifth harmonic:
$$f_5 = 5f_1 = 5(86\ Hz) = 430\ Hz$$

If the pipe is *open at both ends* the fundamental frequency requires $L = \lambda/2$, so for a 1-m pipe:

$$L = 1\,m = \lambda/2 \quad \text{or} \quad \lambda = 2\,m$$

Using a speed of sound in 20 °C air of 344 m/s, the fundamental frequency is:

$$f_1 = v/\lambda = (344\ m/s)/(2\ m) = 172\ Hz$$

and the third overtone or fourth harmonic is:

$$f_4 = 4f_1 = 4(172) = 688\ Hz$$

The Doppler Shift

• Waves are often emitted by a source that is moving within and with respect to the medium that is carrying the waves. An example of this is a loud train zooming along its track and emitting noise through the air medium. Imagine standing near a train track and listening to a train approach, pass by, and zoom away. You can hear the frequency of the sound change from higher as it approaches to lower as it passes and moves away. This effect on the frequency of the sound waves is called the **Doppler effect or Doppler shift**, in honor of Austrian physicist, **Christian Johann Doppler** (1803–1853). In essence, a Doppler effect or shift is a change of perceived frequency due to movement of a wave source relative to an observer.

The Doppler shift occurs as the sound waves approaching you with velocity v_S become "bunched up" so that you perceive a higher frequency than you would if the train's engine was running but not moving. You can also imagine "bunching" waves by considering a small object vibrating at a set frequency while floating on a smooth water surface. Then imagine a string is attached to the vibrating object, and you very gently pull it across the smooth surface of the water. The waves caused by the vibration would bunch up in the direction the object is moving and spread out behind it.

• Note that the speed v that waves are traveling depends only on the properties of the medium, such as air or water. Wave speed does not depend on the motion of the source that is emitting the sound, such as the train speed. In other words, sound waves in air will travel at the speed of sound in air at a given temperature (344 m/s at 20 °C) regardless of the speed of a moving object emitting the sound. Nevertheless, the perceived frequency and wavelength of the waves are affected by the motion of the wave-emitting source (such as a train), which produces the Doppler shift. We can see in the figure below that the wave crests are closer together in the direction an object is moving.

• If you, the observer (or listener), are moving toward a stationary object that is emitting sound waves (or other waves), the wave crests will seem to be closer together as you move toward the source. Then, as you pass the source and move away, the wave crests seem further apart (from your perspective) and the sound will shift to a lower frequency to your ears. *A Doppler shift can be perceived when a source of waves moves toward or away from an observer or when the observer moves toward or away from a stationary source.*

The figure shows a stationary train emitting sound and a moving train emitting sound. The sound wave crests in front of the moving train are closer together than those of the stationary train.

• Suppose a train travels from point A to point B while emitting sound waves. From the perspective of the train, the frequency of its sound is f_s. During time period $t = 0$ to $t = t$, the train will emit $f_s t$ number of waves (where $f_s t$ is cycles per second times seconds or equivalently the number of waves). The train's velocity is v_s, and from point A to point B the train travels a distance of $d = v_s t$. The sound waves emitted at $t = 0$ (at point A) will have the most time to travel. These waves move with velocity v and will travel a distance $d = vt$ during the time interval between $t = 0$ and $t = t$ (as the train moves from point A to point B). As the train reaches point B and the final waves have been emitted, the train will have traveled a distance $v_s t$. Therefore the waves emitted from the train occupy the distance $vt - v_s t$. The wavelength λ_L of the waves from your perspective as listener is:

$$\lambda_L = \text{(distance)/(number of waves)} = (vt - v_s t)/(f_s t) = (v - v_s)/f_s$$

The **frequency f_L from your perspective as listener** is:

$$f_L = v/\lambda_L = v/((v - v_s)/f_s) = f_s(v/(v - v_s))$$

or

$$\boxed{f_L = f_s\, v/(v - v_s) \quad \text{as source S moves toward listener L}}$$

This equation highlights the frequency as the train, or wave emitting source S, travels toward the listener L. When the source is moving away from you, $-v_s$ becomes $+v_s$, and the frequency f_L you hear is lowered:

$$\boxed{f_L = f_s v/(v + v_s) \quad \text{as source S moves away from listener L}}$$

• What if *you, the listener, are moving,* and the sound-emitting source (e.g., train) is *stationary?* An equation for frequency f_L from your perspective as listener can also be derived for this scenario. An equation describing the **frequency when you, the listener, are moving** toward (or away from) the sound wave emitting source is:

$$f_L = f_s(v + v_L)/v \quad \text{when L moves toward S, } f_L \text{ higher}$$
or
$$f_L = f_s(v - v_L)/v \quad \text{when L moves away from S, } f_L \text{ lower}$$

• The **Doppler shift** can also occur with waves other than sound, including all mechanical and electromagnetic waves such as water waves and light waves. A familiar use of the Doppler shift is Doppler radar. The Doppler shift of electromagnetic waves is also used to gain information and measure the speed with which different celestial objects in space are moving relative to Earth.

• **Example**: Imagine yourself standing on a street corner when a motorcycle zooms by and the passenger is blowing a 1,000 Hz note on some sort of horn. What frequency do you hear as the motorcycle approaches and after it passes if it is moving at about 45 mph (about 20 m/s) and the speed of sound in air is 344 m/s?

As the source S moves toward you:
$$f_L = f_s v/(v - v_s) = (1{,}000 \text{ Hz})(344 \text{ m/s}) / (344 \text{ m/s} - 20 \text{ m/s}) \approx 1062 \text{ Hz}$$
As the source S moves away from you:
$$f_L = f_s v/(v + v_s) = (1{,}000 \text{ Hz})(344 \text{ m/s}) / (344 \text{ m/s} + 20 \text{ m/s}) \approx 945 \text{ Hz}$$

Shock Waves

• In the previous subsection we discussed a train emitting noise while the train was moving at a speed slower than the speed of its emitted sound waves. What if a train or other source of sound waves is moving at the same speed as the sound waves it emits or at an even greater speed?

When the *source of the waves moves at a speed equal to the speed of emitted waves,* the waves cannot get in front of the moving source, but stack up in a flat planar shape as shown in the figure below. When the *source of the waves moves at a speed greater than the speed of emitted waves,* a **shock wave** forms as the source speeds ahead and "drags" the emitted waves behind it in the shape of a cone as shown in the figure. The edge of the cone forms a large-amplitude supersonic wave front referred to as a **shock wave**. An observer, or listener, hears a **sonic boom** when the shock wave reaches him or her. The speed of a shock wave is usually faster than the speed of sound in that medium and then decreases as its amplitude decreases until its speed is that of the speed of sound.

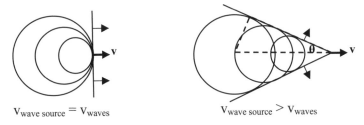

$v_{wave\ source} = v_{waves}$ $v_{wave\ source} > v_{waves}$

If we consider the right-hand figure where the wave source velocity exceeds the velocity of emitted waves, the angle θ is between the conical wave front and the velocity vector of the source. During the time interval when the object emitting waves moves the length of the hypotenuse of the dashed right triangle ($d = v_{object}t$), the initial wave propagates the length of the short leg (opposite to θ) of the triangle ($d = v_{waves}t$). You can determine angle θ from the trigonometric relationship $\sin\theta =$ opposite/hypotenuse:

$$\sin\theta = v_{waves}t / (v_{object}t) = v_{waves} / v_{object}$$

Remember, the ratio of the speed of an object to the speed of sound waves in the same medium is the **Mach number**, $M = v_{object}/v_{waves}$. Therefore, $1/\sin\theta = v_{object}/v_{waves}$ is the Mach number of a moving object. For objects traveling faster than the speed of sound (called supersonic), the air along the wave front is compressed. This high-pressure air forms the **shock wave**. There are actually two shock waves produced by supersonic objects—one at the front of the object and one at the back. First, the air pressure rises along the front shock wave, then drops, and then rises along the back shock wave.

13.4. Light Waves: Interference and Diffraction

Young's Double-Slit Experiment

• Light has characteristics of both a particle and a wave. English physicist **Thomas Young** (1773–1829) demonstrated the wave properties of light in his **double-slit experiment**. This experiment showed that light is subject to constructive and destructive interference, which is a characteristic of waves. In the double-slit experiment, coherent (in phase, single wavelength) light is aimed at a screen which has two parallel narrow slits, A and B. The slits in the screen are separated by distance D. A second screen is placed a distance L beyond the first screen. The light passes through the two slits in the first screen and hits the second screen.

The experiment shows an alternating pattern of bright and dark lines (like stripes) on the second screen, which correspond to constructive and destructive interference such that the light intensity varies from interference. In the actual experimental setup that is depicted below, the distance D between slits must be much less than distance L between screens. Point P corresponds to any point on the second screen where the light from the two slits strikes. The pattern on the right of the drawing portrays what you could see if you were looking at the front of screen 2.

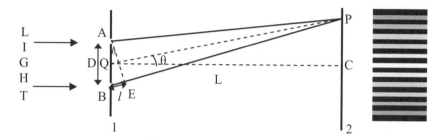

A line drawn from P to Q, which is the midpoint between slits A and B, intersects line QC at an angle θ. Note that when distance L is much greater than D, the angle AEB can be drawn as a right angle forming right triangles AEB and AEP. Also, AE will be approximately perpendicular to PQ, so that angle BAE will be approximately equal to θ. Another way to see this is that since L >> D, the angles that lines AP, BP, and QP make with the horizontal are about equal to θ. Using trigonometry, the light from the lower slit B travels a distance of $l = D \sin\theta$ further than the light from upper slit A in order to reach point P on the second screen. The amount $D \sin\theta$ can be thought of as the amount out of phase the two light beams are when they reach point P. Therefore, the path length difference l is:

$$l \approx D \sin\theta$$

When the **path length difference** l is an integer number of wavelengths (Nλ):

$$l = N\lambda \qquad N = 0, 1, 2, \ldots$$

a bright line due to **constructive interference** will show at point P on the second screen. If we substitute for the difference in path lengths l, a relationship for where the **bright lines in the interference pattern are seen** on the second screen should be:

$$\boxed{N\lambda \approx D \sin\theta \qquad N = 0, 1, 2, \ldots}$$

The bright lines are called the **maxima** and show where the two light waves are *constructively interfering* and are in phase allowing maximum light to strike the second screen. When N = 0, the length of the paths that light travels from each slit to the second screen is the same. In this case, $\theta = 0$ and $l = 0$, and the brightest line, called the *central maximum*, is created in the center of the interference pattern (at C in the figure).

When wavelength λ is equal to an odd number of half wavelengths, **dark lines in the interference pattern are seen**. We can write this as:

$$(N + 1/2)\lambda \approx D \sin\theta \quad N = 0, 1, 2, \ldots$$

Here the two light waves are out of phase by one-half wavelength creating *destructive interference*, causing a dark line on screen 2 at that point P. The dark lines are called the **minima** of the interference pattern.

• When screen spacing distance L is much greater than D, and assuming the length between Q and P in the above drawing is roughly equal to L, we can calculate the spacing between the light interference lines on the second screen. For the spacing between adjacent bright lines on the screen, consider the right triangle PQC. Since sine equals opposite/hypotenuse and $PQ \approx L$:

$$\sin\theta \approx PC/PQ \approx PC/L$$

If we let length PC = x:

$$\sin\theta \approx x/L$$

For bright lines, $N\lambda \approx D \sin\theta$, N = 0, 1, 2, ..., becomes $N\lambda \approx Dx/L$ or:

$$x \approx N\lambda L/D$$

Since there is an integer increase in N between bright lines, we can write the **spacing between adjacent bright lines (or dark lines)** as:

$$x/N = \Delta x \approx \lambda L/D$$

• **Example**: Your friend is studying the interference pattern of the refracted light in the drawing above and wants to create his own double-slit setup. He uses blue light with wavelength 470 nanometer (nm), sets his two screens 1.5 m apart, and carefully creates his parallel slits 0.2 mm apart. What is the spacing between the center bright line and the next maximum line? (Note: one nm is 10^{-9} m.)

"This is easy," you say. "The spacing depends on the light's wavelength λ and the screen separation divided by the slit distance":

$$\Delta x \approx \lambda L/D = (4.7 \times 10^{-7}\,\text{m})(1.5\,\text{m})/(0.2 \times 10^{-3}\,\text{m}) \approx 0.0035\,\text{m} = 3.5\,\text{mm}$$

Therefore, the line spacing is approximately 3.5 mm apart.

Single-Slit and Multi-Slit Diffraction

• What if there is only one slit instead of two slits in the double-slit experiment setup in the diagram above? This would be called **single-slit diffraction**, and we will see a **diffraction pattern** of lines from constructive and destructive interference as we did with double-slit. It will have a stronger, broader central maximum flanked by weaker secondary maxima above and below. Note that **diffraction** bends light around objects. This bending creates interference patterns such as the constructive and destructive interference lines in Young's double-slit experiment or a similar pattern which is created by a single slit.

If you designed a **single-slit experiment**, your equipment would look like the double-slit experiment except the first screen would have only one slit. The distance D between slits in the double-slit is now the size of the one slit and called d. Angle θ would be the angle that a line drawn from the center of the slit to point P would make with the horizontal.

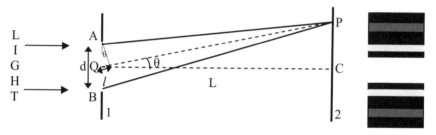

Again, P shows the paths that light through the slit can take to the second screen. Light traveling through the slit will differ slightly in direction, depending on where it passes within the slit. This creates a superposition of the light waves that strike screen 2, causing fuzziness in the maxima flanking the large central maximum. The central maximum at N = 0 is the primary bright prominent maxima. The flanking maxima in each slit of the double-slit experiment, due to diffraction occurring in each slit, were overwhelmed by the maxima and minima of the double-slit interference effects. The light from the ends of the slit will have the largest difference in phase.

• We can develop an equation for the spacing of the interference pattern. Because L >> d in the experimental setup, the light rays that pass through the single slit and hit any point P are virtually parallel. Therefore, line QP is roughly parallel to line AP. If we let angle θ correlate with the first interference minimum above the central maximum on screen 2, then the difference in length of the light path l between light beams QP and AP is one-half the wavelength of the light, or: $l = (1/2)\lambda$. Since $\sin \theta =$ opposite/hypotenuse: $l \approx (1/2)d \sin \theta$. If point P is at the first minimum, the distance between P and the central maximum C (which we will call x) is: $x = L \tan \theta$. Remember $\tan \theta$ = opposite/adjacent, so $\tan \theta = x/L$. Since $L \gg d$, θ is very small, and when θ is small $\sin \theta \approx \tan \theta$. We can then substitute $l = (1/2)\lambda$, $l \approx (1/2)d \sin \theta$, and $\tan \theta \approx x/L$ to obtain an equation for distance between interference minima:

$$\boxed{x = \lambda L/d}$$

The width of the central maximum is roughly 2 times the width of the secondary maxima flanking it.

• What if there are many parallel slits instead of the two slits in the double-slit experiment setup in the diagram above? A **multi-slit system** with a large number of slits spaced distance D apart is called a **diffraction grating**. This system creates an interference pattern which also has maxima at $N\lambda \approx D \sin \theta$ for an integer number of wavelengths $N = 0, 1, 2, \ldots$, where D is the distance between slits called the *grating spacing*. The minima are also at $(N + 1/2)\lambda \approx D \sin \theta$. The pattern has better resolution with less fading than the double slit. Diffraction gratings can be used to examine the spectrum of wavelengths of light composed of more than one wavelength. The positions of interference maxima and minima distinguish the different wavelengths.

• Note that interference and diffraction effects through slits are not restricted to light waves, but occur with other types of waves, such as sound. Interference and diffraction depend on wavelength, and the effects are increased when slit size and spacing are near the wavelength size.

13.5. Electromagnetic Radiation

• **Electromagnetic (EM) radiation** (also referred to as **electromagnetic waves** or **electromagnetic energy**) is produced in various ways including nuclear reactions within the Sun, decay of radioactive substances, changes in the energy levels of electrons, thermal motion of atoms and molecules, and acceleration of electrical charges. We know moving charges (currents) produce magnetic fields and changing magnetic fields produce currents. Just as a traveling wave on a string is produced by applying oscillations to one end, an electromagnetic field can be generated by a changing or oscillating current in a wire antenna. When charges move back and forth, oscillating electric and magnetic fields are produced. For insight into electromagnetic waves, remember that an oscillating electric field **E** produces an oscillating magnetic field **B**, and an oscillating magnetic field produces an oscillating electric field. When EM waves are produced, such as radio waves from an antenna, oscillating electric and magnetic fields perpendicular to one another are produced and travel away from the antenna. The **E** and **B** fields are not only perpendicular to each other, but also to the direction the wave travels (they are **transverse waves**). Similar to the sinusoidal displacement that can be propagated along a string, a sinusoidal variation of the **E** and **B** field vectors can be propagated through a medium or through empty space.

Electromagnetic waves include the visible light we see and the infrared we sense as heat, as well as gamma, X-ray, ultraviolet, infrared, micro-wave, and radio waves. *Electromagnetic waves (EM waves) are transverse traveling waves composed of oscillating electric field E and magnetic field B. They can travel through certain mediums (to the extent they are not absorbed or reflected) or through the vacuum of outer space.*

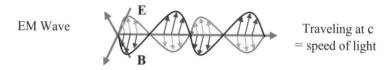

EM Wave   Traveling at c
                                                 = speed of light

We can see that the electric field **E** is in one plane and the magnetic field **B** is in a plane perpendicular to **E**. Electromagnetic waves are **plane waves** with **E** *perpendicular* to **B**, or **E** ⊥ **B**, at every point and for all values as it travels. EM waves are *transverse waves*, such that vectors **E** and **B** are always perpendicular to the direction of propagation of the wave. When EM waves *travel through the vacuum of space* they move at the "speed of light," or $c = 3.00 \times 10^8$ m/s. This is the fastest speed anything has been confirmed to travel. While EM waves carry no mass, as they travel they carry *energy* and *momentum* (which exerts a pressure referred to as the radiation pressure).

• EM waves exist in a broad range, or spectrum, of frequencies referred to as the *electromagnetic spectrum*. The EM spectrum is often identified beginning near 10^3 Hz frequency at about a wavelength of 10^5 m to near 10^{22} Hz frequency at about a wavelength of 10^{-13} m. The **electromagnetic spectrum** can be roughly depicted as follows:

Wavelength (m)

10^3 10^2 10^1 1 10^{-1} 10^{-2} 10^{-3} 10^{-4} 10^{-5} 10^{-6} 10^{-7} 10^{-8} 10^{-9} 10^{-10} 10^{-11} 10^{-12}

radio waves	microwaves	infrared	vis-ible	ultra-violet	X-rays	gamma rays

10^6 10^7 10^8 10^9 10^{10} 10^{11} 10^{12} 10^{13} 10^{14} 10^{15} 10^{16} 10^{17} 10^{18} 10^{19} 10^{20}

Frequency (Hz)

Visible light is between infrared and ultraviolet and ranges from about 400 nm for violet to about 700 nm for red, or 4×10^{-7} m to 7×10^{-7} m. The frequency range for visible light is about 7.5×10^{14} Hz for violet to about 4.3×10^{14} Hz for red. The sequence of colors is *red, orange, yellow, green, blue,* and *violet.* Traditionally this was memorized using the name "ROY G BIV", with the "I" as "indigo". More recently "ROY G BV" is usually used without "indigo" distinguished.

• In EM radiation, the higher the frequency, the shorter the wavelength and the greater the energy. While all EM *waves travel at the same speed in a vacuum*, higher frequency (higher energy) waves oscillate faster. The energy E per photon is E = hf, where h $\approx 6.626 \times 10^{-34}$ J·s $\approx 4.135 \times 10^{-15}$ eV·s is **Planck's constant** and f is frequency. The frequency f is f = v/λ, so when in a vacuum, f = c/λ, where v is speed, c = 3.0×10^{8} m/s is the speed in a vacuum, and λ is the wavelength. Using these equations and the fact that the speed of EM waves is constant in the *vacuum of space*, you can calculate a wave's **frequency** if you know its **wavelength**, and you can calculate its wavelength if you know its frequency. **Wavelength and frequency** distinguish one kind of EM wave from another.

• **Example**: Estimate the frequency of a blue light wave with a wavelength of about 4.7×10^{-7} m. Identify an EM wave with frequency 10^{12} Hz and find its wavelength. Assume the medium is outer space.

$$f = c/\lambda = (3.0 \times 10^{8} \text{ m/s})/(4.7 \times 10^{-7} \text{ m}) \approx 6.4 \times 10^{14} \text{ Hz}$$
$$\lambda = c/f = (3.0 \times 10^{8} \text{ m/s})/(10^{12} \text{ Hz}) = 3.0 \times 10^{-4} \text{ m} \text{ which is infrared}$$

• **Example**: You and your friend are dialing a shortwave radio when he asks, "Which has the shortest wavelength—radio waves, blue light, gamma rays, or X-rays?" He adds, "And what is that frequency?"

"OK," you say, "high-frequency gamma is shortest, with wavelengths near 10^{-12} m. Let's think, the speed of light c is 3×10^{8} m/s and c = fλ, so frequency f = c/λ = $(3 \times 10^{8}$ m/s) / $(10^{-12}$ m) = 3×10^{20} Hz." Impressed, he says, "Wow, you can do that all in your head? Cool."

• Different types of electromagnetic radiation vary according to the amount of **energy** contained in their photons, with radio waves having low energy photons (and long wavelengths), microwaves having slightly more energy, infrared having still more, with increasing energy in visible, ultraviolet, X-rays, and on up to the high energy gamma rays (which have extremely short wavelengths). The EM spectrum is described in terms of energy (in electron volts), wavelength (in meters), and frequency (in cycles/second, or **Hertz**).

While different types of EM waves, energy, or radiation, seem different and they are produced and detected in very different ways, they are all electromagnetic radiation or energy and are fundamentally the same. EM radiation can be described as a stream of photons (massless particles containing energy) or as interfering, diffracting waves (as we saw in our discussion of double, single, and multi-slits). They can all be described by frequency, wavelength, or energy. For EM radiation, wave frequency can be described as the number of wave crests passing a given point per second, where the wavelength is the distance between wave crests. The energy carried is proportional to frequency where $f = c/\lambda$ and energy E per photon is $E = hf = hc/\lambda$ with $h \approx 6.626 \times 10^{-34}$ J·s $\approx 4.135 \times 10^{-15}$ eV·s, which is **Planck's constant**.

13.6. Reflection and Refraction

• **Optics** is in essence a sub-topic of electricity and magnetism since it studies the behavior of light, and light is an electromagnetic wave. Optics studies visible light and how it bounces, bends, and produces visual images. Optical instruments are designed to use the particle and wave natures of light. For example, instruments that diffract light into a spectrum for analysis are observing light's wave properties, and instruments such as detectors in digital cameras use the particle nature of photons.

Reflection

• All objects we can see **reflect** light. Objects generally absorb some of the light that strikes them and reflect certain frequencies, causing them to appear to be a particular color. When light strikes an object with a rough or irregular surface, the reflected light is dispersed in many directions. Conversely, objects with smooth, flat surfaces, such as mirrors, reflect all or most of the incoming, or incident, light. In fact, light has been observed to reflect off a flat, polished surface such that the *angle of incidence* $\theta_{incidence}$ *is equal to the angle of reflection* $\theta_{reflection}$. This is referred to as the **Law of Reflection**:

$$\theta_{incidence} = \theta_{reflection}$$

Note that the angle of incidence $\theta_{incidence}$ and the angle of reflection $\theta_{reflection}$ (as well as angles of refraction) are measured from the perpendicular, or normal, to the reflecting surface. Also, the light ray (or beam) that strikes a reflective surface is the **incident light**, and the light ray (or beam) bouncing off the surface is the **reflected light**.

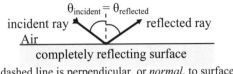

completely reflecting surface
(The dashed line is perpendicular, or *normal*, to surface.)

• Interestingly, an observer perceives that the image of an object reflected in a flat mirror is located behind a mirror and originates from a path that would be the continuation of the line of the reflected ray. Each ray reflected from a point on an object obeys $\theta_{incidence} = \theta_{reflection}$. An observer also perceives that the image of an object is located the same distance behind the mirror's surface as the object is actually located in front of the mirror.

Refraction

• Refraction is observed if you place a straight object, such as a stick or your pencil, into water at a slight angle and see it appear to be *broken* or *bent* at the water's surface. Light that is not reflected from a surface, such as glass or water, may be **refracted** as some of the incident light rays *travel through the surface* into the glass or water. The **refracted rays** are at an angle to the normal that is different than the **reflected rays**. This means as light passes from one transparent medium into another, some rays will be reflected (such that $\theta_{incidence} = \theta_{reflection}$), and some rays will be *refracted* at a different angle to the normal such that $\theta_{incidence}$ is *not* equal to $\theta_{refraction}$.

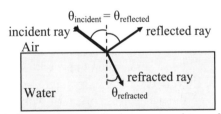

(The dashed line is perpendicular, or *normal*, to surface.)

• Note that the direction of a light ray along a specified path doesn't matter. Whether the light is passing from air into glass or water, or passing from that glass or water into air, the refraction at the interface will be the same. (In the above figure, imagine the arrows for the incident and refracted beams in both directions. The path would not change.) Also note that a ray striking perpendicular to an interface continues perpendicular and does not refract.

• When a beam of light passes from a *less optically dense substance* such as air into a *more optically dense substance* such as water, it will be refracted toward a line drawn perpendicular to the interface. When the ray is traveling from a less dense to a more dense medium, the angle of incidence $\theta_{incidence}$ will be *greater than* the angle of refraction $\theta_{refraction}$. (In the above figure, observe angle $\theta_{incidence}$ is greater than $\theta_{refraction}$.)

• A light beam passing from a *more optically dense substance* to a *less optically dense substance* will be refracted away from a line drawn perpendicular to the interface. When the ray is traveling from a more dense to a less dense medium, the angle of incidence $\theta_{incidence}$ will be *less than* the angle of refraction $\theta_{refraction}$. (In the above figure, imagine the beam traveling in the opposite direction shown by the arrows so that it points from water to air. In this case the depicted $\theta_{incidence}$ angle and the $\theta_{refraction}$ angle are switched. The angle on the less dense air side of the interface is larger.)

• The degree to which a ray is refracted depends on the medium since light travels at different speeds in different media. Light traveling in the vacuum of space travels the fastest (at the "speed of light" c), but when passing through a substance such as air, glass, or water, light travels more slowly. Different substances have an **index of refraction n** given by the following ratio:

n = (speed of light in a vacuum)/(speed of light in the medium) = c/v

Since v is always less than or equal to c, then $n \geq 1$. Examples of n include: $n_{air} \approx 1.0$, $n_{water} \approx 1.3$, and $n_{glass} \approx 1.6$. Index of refraction typically increases with material density. Note that in a medium, but not in a vacuum, the speed of light varies slightly for differing wavelengths, and EM waves with shorter wavelengths move slower than those with longer wavelengths. This also means that EM waves with shorter wavelengths will refract or "bend" more than those with longer wavelengths. For example, blue light refracts at a greater angle than red light. You are probably familiar with the way a glass prism or water droplets disperse sunlight into a rainbow of colors.

• How do we determine the path of a light ray across an interface? The Law of Refraction, or **Snell's Law**, discovered by Dutch scientist **Willebrord Snel van Royen** (or *Willebrord Snellius*) in the early 1600s, relates the *angle of incidence* to the *angle of refraction* for a light ray passing from one optical medium to another. We can easily derive this relationship by considering a beam of light, which is a transverse wave. In the figure below we imagine an expanded version of a ray of light (with width or ray AA') as it traverses into the interface between air and water and bends (as the left side of the wave strikes and begins to slow first).

A light ray leaving air and entering water

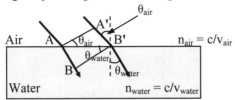

Line AA' and line BB' represent wave fronts as the light is crossing the interface. The light crosses from air with an index of refraction $n_{air} = c/v_{air}$ into water with $n_{water} = c/v_{water}$. The speed in air and water is:

$$v_{air} = c/n_{air} \quad \text{and} \quad v_{water} = c/n_{water}$$

The wave front travels from A to B and from A' to B' during time interval t. Since distance equals rate times time, AB and A'B' are:

$$A'B' = v_{air}t = ct/n_{air} \quad \text{and} \quad AB = v_{water}t = ct/n_{water}$$

Using trigonometry for right triangles:

$$A'B'/AB = \sin\theta_{air} \quad \text{and} \quad \sin\theta_{water} = AB/AB'$$

Rearranging: $AB' = A'B'/\sin\theta_{air}$ and $AB' = AB/\sin\theta_{water}$

Combining $AB' = A'B'/\sin\theta_{air} = AB/\sin\theta_{water}$ with $A'B' = ct/n_{air}$ and $AB = ct/n_{water}$ and substituting for A'B' and AB gives:

$$(ct/n_{air})/(\sin\theta_{air}) = (ct/n_{water})/(\sin\theta_{water})$$

Cancelling ct and rearranging gives: $n_{air}\sin\theta_{air} = n_{water}\sin\theta_{water}$

Or, more generally, for a light ray traveling from medium 1 with index of refraction n_1 across an interface into medium 2 with index of refraction n_2, **Snell's Law** is:

$$n_1 \sin\theta_1 = n_2 \sin\theta_2$$

where θ_1 is the angle of incidence, and θ_2 is the angle of refraction. From Snell's Law and the relationships $n_1 = c/v_1$ and $n_2 = c/v_2$, we can show:

$$(\sin\theta_1)/(\sin\theta_2) = n_2/n_1 = v_1/v_2$$

- **Example**: Your friend sees you standing in waist deep water and begins to laugh, stating how stubby your legs have become. What is he observing? Then he reaches for a water-logged ball that has sunk into the water and his grasp is too high. What happened?

You tell him your "stubby" legs are caused by refraction of light as it passes through the air-water interface. You then laugh at his missed attempt to grasp the ball, explaining that the depth he perceives is less than the actual depth because light rays originating at the ball are refracted away from the perpendicular as they cross out of the water's surface.

• **Example**: You and your friend decide to go out to his swimming pool on a calm dark summer night with two small waterproof laser pointers. If you point one beam into the shallow end of the pool at a 45° angle, what is the refracted angle? Then, you point (with your finger) to the location where your beam enters the water and your friend marks the spot against the pool bottom where the laser light hit. He gently holds his laser pointer on the pool's bottom where your beam hits and aims his laser to where you are pointing (with your finger) at the water's surface. What will the angle of refraction be as your friend's laser beam points up out of the water into the air?

In this section we identified $n_{air} \approx 1.0$ and $n_{water} \approx 1.3$. Now we can find the refracted angle into the water. Using Snell's Law:

$$n_{air} \sin \theta_{air} = n_{water} \sin \theta_{water}$$

$$\sin \theta_{water} = n_{air} \sin \theta_{air} / n_{water}$$

$$\theta_{water} = \sin^{-1}(1.0)(\sin 45°/1.3) \approx 33°$$

Then you find refracted angle up into the air. Using Snell's Law:

$$\sin \theta_{air} = n_{water} \sin \theta_{water} / n_{air}$$

$$\theta_{air} = \sin^{-1}(1.3)(\sin 33°/1.0) \approx 45°$$

Your friend says, "I didn't need to get wet to tell you the angle of the beam *above* the water's surface (45° in this case) is the same whether the light points into or out of the water. Also, the angle below the water's surface is the same (33° in this case) regardless of direction."

• **Example**: Your friend wants to know how to find the angle of refraction if there are three media (air, glass, and water) with two interfaces such as a flat plate of glass floating on a tank of water.

You write Snell's Law for the two refractions and set them equal:

$$n_1 \sin \theta_1 = n_2 \sin \theta_2 \quad \text{and} \quad n_2 \sin \theta_2 = n_3 \sin \theta_3 \quad \text{so} \quad n_1 \sin \theta_1 = n_3 \sin \theta_3$$

"Wow," he says. "Refracted angle θ_3 doesn't care about the intermediate flat plate of glass." "Only if it has parallel surfaces," you reply.

• Note that if media 1 and 3 are the same so that a flat plate with n_2 is immersed in one media ($n_1 = n_3$), a light ray passing through the flat plate will be displaced, but its angle on either side of the plate is the same.

Total Internal Reflection

• When a light ray originating in air strikes a glass surface, part will reflect and part will pass into the glass. When a light ray originating inside the glass strikes the glass/air interface, there may or may not be a ray leaving the glass and passing into the air. If the angle of incidence is greater than a **critical angle** θ_c, **total internal reflection** occurs, and no light ray transmits across the more-dense-to-less-dense (glass-to-air) interface into the air. Light striking the surface of a medium with a lower index of refraction at any angle greater than θ_c is totally reflected. Total internal reflection uses include cutting a diamond to enhance sparkle and sending light waves (information) through fiber optics.

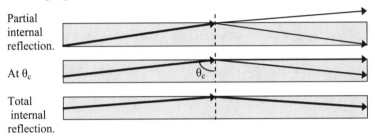

Partial internal reflection.

At θ_c

Total internal reflection.

How do we find this critical angle? If you imagine a simple sine wave plotted vs. time on an x-y coordinate system, you will remember that over time (along the x-axis) the value of $\sin \theta$ moves up and down from -1 to $+1$ along on the y-axis. Because $\sin \theta$ cannot exceed plus or minus 1, if we use *Snell's Law*, there are certain values of n_{glass}, n_{air}, and θ_{glass} in which Snell's Law has no solution for θ_{air} and there is no refracted ray. The **critical angle** θ_c for two media is the smallest angle of incidence for which **total internal reflection** occurs. Using **Snell's Law**:

$$n_{glass} \sin \theta_{glass} = n_{air} \sin \theta_{air} \quad \text{or} \quad \sin \theta_{air} = (n_{glass} \sin \theta_{glass}) / n_{air}$$

and the fact $\sin \theta \le 1$, refraction into a less optically dense medium occurs only if: $(n_{glass} \sin \theta_{glass}) / n_{air} \le 1$. If we set $(n_{glass} \sin \theta_{glass}) / n_{air} = 1$, the critical angle going from more dense glass to less dense air is:

$$\theta_c = \theta_{glass} = \sin^{-1}(n_{air}/n_{glass})$$

Therefore, the **critical angle** θ_c for two media is the smallest angle of incidence for which **total internal reflection** occurs. It is the arcsine of the ratio of the indexes of refraction ($n_{LessDense}/n_{MoreDense}$), or:

$$\boxed{\theta_c = \sin^{-1}(n_{LessDense}/n_{MoreDense})}$$

This equation for **total internal reflection** is supported by observation.

• **Example**: You and your friend, each holding your breath, gently slip into the pool and lie quietly on the bottom pointing your waterproof laser pointers at the surface. What is the maximum angle of incidence that will allow some light from the beam to escape the water, or equivalently, the critical angle for total internal reflection to occur?

Using $n_{air} \approx 1.0$ and $n_{water} \approx 1.3$:

$$\theta_c = \sin^{-1}(n_{LessDense}/n_{MoreDense}) = \sin^{-1}(1.0/1.3) = 50.3°$$

Dispersion

• The index of refraction is somewhat dependent on the wavelength of the incident light. When white light, consisting of a mixture of wavelengths, *refracts*, the different waves vary slightly and **disperse**. This **dispersion** of light creates rainbows as sunlight refracts off water droplets and the longer wavelength light (red) refracts at a smaller angle than the shorter wavelength light (violet), which refracts at a greater angle. Dispersion also occurs when light of many wavelengths passes through a glass prism. This effect occurs since the speed of light in glass depends on the wavelength of light, as it decreases slightly with decreasing wavelength.

Polarization

• Light waves are transverse and therefore oscillate in a plane that is perpendicular to the direction the wave is traveling. Light and other forms of electromagnetic radiation consist of propagating oscillations in an electromagnetic field. When you consider a beam of unpolarized light, its oscillations (which are perpendicular to propagation) can be pointing in all directions within that perpendicular plane. When that beam of light is passed through a *polarizing filter* and becomes **polarized light**, its oscillations will be in one direction within the plane perpendicular to the light's propagation. Light can be polarized by reflection or by passing it through filters or crystals that transmit oscillations in certain directions.

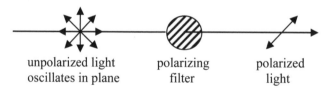

unpolarized light   polarizing   polarized
oscillates in plane   filter   light

13.7. Lenses and Mirrors

• Light reflects off mirrors and refracts through lenses. When light interacts with curved mirrors and lenses, it can be focused to a point (convex lenses and concave mirrors) or diverged from a point (concave lenses and convex mirrors). While focusing light on curved surfaces is more complicated than on flat surfaces, a tiny section of a curved surface is virtually flat so the same principles can apply.

Mirrors and lenses have many characteristics in common with a few differences. Some properties of lenses and mirrors are depicted in the figures of lenses and mirrors below. A perpendicular line drawn through the center, or **vertex** V, of a mirror or lens is called the **principal axis**. Light rays parallel to a principal axis will be reflected, or refracted for lenses, through the **focal point** F. In the opposite direction, a light ray passing through a *focal point* F will be reflected parallel to the principal axis. If you imagine a spherical *mirror* as a slice off a sphere, the center of that slice is the **center of curvature** C, and the radius of the sphere is the **radius of curvature** R. The distance between the vertex and the focal point is the **focal length** f, which is $f = R/2$, or one-half the **radius of curvature** R, for spherical mirrors. Let's explore these principles further, beginning with lenses.

Lenses

• **Lenses** are often circular pieces of glass, having the *center region* either thicker (convex lens) or thinner (concave lens) than the circular edge. *Convex lenses* *cause light rays passing through them to* **converge** *and* *concave lenses* *cause light rays to* **diverge**. A light ray will follow the same path when passing through a lens or system of lenses regardless of which direction the light is traveling. For example, if parallel rays strike a convex lens (on its left side) and converge to the **focal point** F on the right side, then a light source placed at F (on the right) will have its rays radiate out and pass through the lens, exiting as parallel rays along the same path on the left side.

Convex Lenses

• **Convex lenses** have a thicker center and focus light to a *focal point* F. Convex lenses are **converging lenses** with: a **principal axis** which is perpendicular to the center, or **vertex** V; a **focal point** F which lies on the *opposite* side of the lens from an object of interest on the principal axis; and a **focal length** f which is the distance between the vertex V and the focal point F. F' is the focal point opposite to focal point F and also distance f from vertex V. A light ray passing through vertex V of a thin lens (so the ray displacement is negligible) continues without refracting.

• In the figure below of a huge **convex lens**, if your friend stands outside the focal point at a *distance d greater than the focal length* f, or $d > f$, a **real inverted image** is created on the *opposite side* of the lens at distance d' (which is positive). His image will appear upside down with height (negative) h'. The image formed by the converging of the rays is called a *real image* since it can be projected onto a screen or (camera) film. The image can be located by tracing three rays from the top of your friend through the lens. One ray starts parallel to the principal axis, one through the vertex, and one that emerges parallel. They converge to a point determining the location of the top of his image. Note that any two of the rays can determine the position of the image. Note that primes refer to image specifications.

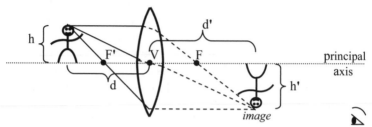

Convex lens with d > f: +d', +f, inverted real image far side.

• Your friend moves closer to the huge convex lens and stands between the lens and its focal point. *Distance d is now less than the focal length* f, or d < f. This time an **upright virtual image** is created behind him on the *same side* of the lens at distance (negative) d'. (Figure below.) The upright virtual image has positive height h'. Tracing the three rays from your friend results in the rays diverging and not passing through the virtual image. Instead, the virtual image is seen by an observer (on the right) as it appears to be behind the lens, on the same side as your friend. The image is also larger than your friend and shows that a convex lens acts as a **magnifying glass** when your friend (or an object) is closer to the lens than the focal point. This virtual image cannot be projected onto a screen since the light rays diverge and do not pass through the image.

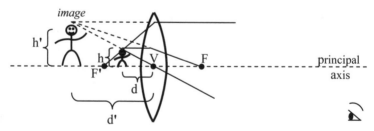

Convex lens with d < f: −d', +f, upright virtual image same side.

Concave Lenses

• **Concave lenses** have a thinner center and divert light away from the *focal point* F. Rays parallel to the axis diverge outward, seeming to originate from the focal point. Concave lenses are **diverging lenses** having a **principal axis** drawn perpendicular through the center, or

vertex V. The **focal point** F lies on the *same* side of the lens as an object of interest and the incident light, making the **focal length** f *negative*. (***Remember, f is distance between the vertex V and the focal point F.***) A light ray passing through V of a thin lens (so the ray displacement is negligible) continues without refracting. When wave fronts in air are perpendicular to the axis of a **concave lens**, the incident rays diverge.

• In the figure below of a large concave lens, if your friend stands a *distance d (outside the focal point) greater than the focal length* f, or d > f, an **upright virtual image** is created on the *same side* of the lens at distance (negative) d'. Also note the figure on the right when he steps inside of F. The upright virtual images in either case have positive height h', and the actual (friend) height h is greater than the image height h'. Rays drawn from the top of your friend diverge.

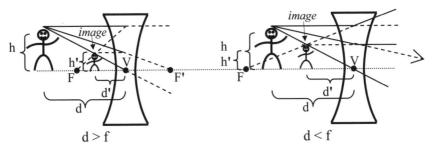

Concave lens: negative d', negative f, upright virtual image on same side.

Applying Equations to Lenses

• The **lenses** described above can be studied using two important equations, which we develop in a few pages after discussing mirrors. These equations provide information about focal length f (which is the distance between the vertex V and the focal point F), distances d and d', heights h and h', magnification M, and whether an image is real, virtual, upright, or inverted. These two **equations for lenses** are:

> **focal length equation**: $1/d + 1/d' = 1/f$
> **magnification equation**: $M = h'/h = -d'/d$

Remember, the primes indicate the image. Lens properties can be summarized as:

Convex lens for d > f: +d', +f, inverted real image far side.
Convex lens for d < f: −d', +f, upright virtual image same side.
Concave lens: −d', −f, upright virtual image same side.

• **Example**: (a) If you place a narrow vase at a distance d of 15 cm and on the left of a convex **lens** having a focal length f of 5 cm, how far is the image from the lens, d'? (b) Now move the vase so that d is 2 cm and determine d'. (c) What is the magnification for parts a and b?

(a) Using $1/d + 1/d' = 1/f$ with d = 15 cm and f = 5 cm, find d':
$$1/d' = 1/f - 1/d = (1/5) - (1/15) = (3/15) - (1/15) = 2/15$$
The image distance d' is 15/2 or 7.5 cm from the lens. Note that f and d' are positive and d > f. (Can you guess the lens image information?)

(b) With d = 2 cm and f = 5 cm, find d':
$$1/d' = 1/f - 1/d = (1/5) - (1/2) = (2/10) - (5/10) = -3/10$$
So the image distance d' is −10/3 or about −3.33 cm from the lens. Note that f is positive, d < f, and d' is negative, so the image is virtual upright and on the same side as the object to a convex lens.

(c) For part (a), we have d = 15 cm and d' is 15/2 cm, so magnification is:
$$M = h'/h = -d'/d = -(15/2)/15 = -1/2$$
The image is half as large as the object.

For part (b), we have d = 2 cm and d' is −10/3 cm, so magnification is:
$$M = h'/h = -d'/d = (10/3)/2 \approx 1.67$$
The image is about 1.7 times larger than the object.

• **Example**: If you have a diverging lens with a negative focal length f of −20 cm and an object distance d of 10 cm, where is the image? What is the magnification?

Since $1/d + 1/d' = 1/f$:
$$1/d' = 1/f - 1/d = -1/20 - 1/10 = -(1/20) - (2/20) = -3/20$$
Therefore, d' = −20/3 ≈ −6.67 cm. This concave lens has a negative sign for d' alerting us that the image is on the same side as the object and is a virtual image. Diverging lenses create virtual images on the same side of the lens as the object. Now, the magnification is:
$$M = h'/h = -d'/d = (20/3)/10 = 2/3$$
So the image size is 2/3 the size of the object.

Mirrors

• Similar to lenses, **mirrors** can be used to *focus an image*. For example, when light rays that are parallel to the principal axis strike a spherical concave mirror, they will reflect to the **focal point** F in front of the mirror. Note that when the surface of the mirror is spherical, focusing of incident parallel rays is most accurate for rays with small angles of incidence. Alternatively, mirrors with *parabola-shaped surfaces* focus all parallel rays accurately.

Concave Mirrors

• Locating an image formed by a mirror is similar to locating an image formed by a lens. A **concave mirror** acts somewhat like the concave side of a shiny soup spoon, which allows you to see an inverted reflection of your face when held at a distance. In fact, concave mirrors form *inverted real images* when an object is located *outside* the focal point F (when d > f), and *upright virtual images* when the object is *inside* the focal point F (when d < f). You can test this by looking into a spherical magnifying mirror and moving it closer and further.

• If your friend, having height h, stands at a distance d *outside of the focal point* F (where d > f) in front of a large *concave mirror*, there will be an *inverted real image* of him with height h' at a distance d' in front of the mirror. (See figure below.) The *real* inverted image is created by light rays and can therefore be projected onto a screen located at d'. Two rays can be traced from the top of his head to locate the image: one ray parallel to the principal axis that reflects off the mirror and crosses through F, and the other ray that passes through F and reflects off the mirror meeting the first ray at the inverted top of his image.

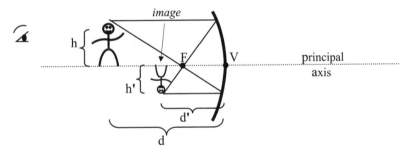

Concave mirror for d > f: +d', +f, inverted real image same side.

• Suppose your friend then steps *inside the focal point* F so that his distance in front of the mirror d is less than the focal length, d < f. (See figure below.) Rays drawn from the top of his head to the mirror will not reflect to a point in front of the mirror (so no real image is formed), but the reflected diverging rays can be extrapolated through the mirror to a point behind the mirror, where they create a large *upright virtual image*. The upright virtual image has height h' and is a distance d' from the vertex V of the mirror. This is similar to what we see when we look into a flat mirror and our image appears behind the mirror, but no light is actually focused on that image. Concave mirrors are often used as magnifying mirrors. Note that the **center** of **curvature C** is equidistant from every point on the **mirror**, and all lines from C strike the mirror perpendicularly.

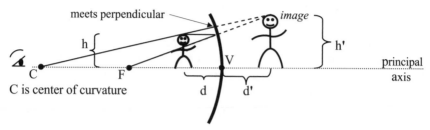

Concave mirror for d < f: −d', +f, upright virtual image far side.

Convex Mirrors

• Now suppose your friend finds a huge convex mirror. He comments that it is like the back of a giant shiny soup spoon. A **convex mirror** creates a smaller *upright virtual image* regardless of how far he (or an object) stands in front of the vertex. Rays drawn from the top of your friend's head reflect off the mirror and diverge (see figure below). Nevertheless, those rays can be extrapolated behind the mirror meeting at the top of the image, with the extrapolated parallel ray passing through focal point F behind the mirror. The upright virtual image is between the back of the mirror and the focal point. Convex mirrors are often used as rear-view mirrors for automobiles because their de-magnification creates a broader view.

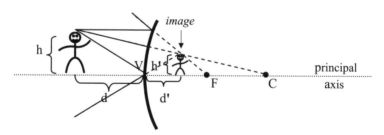

Convex mirror: −d', −f, upright virtual image far side.

Applying Equations to Mirrors

• Mirrors can be described by the same two equations (developed below) that we used for lenses. These equations allow us to gain information about a focal length f (which is the distance between the vertex V and the focal point F), distances d and d', heights h and h', magnification M, and whether an image is real or virtual, and upright or inverted. Again, these two **equations for mirrors** are:

focal length equation: $1/d + 1/d' = 1/f$
magnification equation: $M = h'/h = -d'/d$

Concave mirror for $d > f$: $+d'$, $+f$, inverted real image same side.
Concave mirror for $d < f$: $-d'$, $+f$, upright virtual image far side.
Convex mirror: $-d'$, $-f$, upright virtual image far side.

• **Example**: If your friend's image is inverted and on the opposite side of a lens or a mirror to where he is standing, what type of lens or mirror is it: convex or concave lens or convex or concave mirror?

While both concave mirrors and convex lenses can create inverted images, the images created by concave mirrors are on the same side of the mirror as the person or object. A *convex lens*, however, can create an inverted image on the opposite side.

• **Example**: Your friend stands inside the focal length ($d < f$) of a concave mirror with a focal length f of 4 m, and he is 2 m from the vertex of a mirror. Where is his image and what is its magnification?

First you use $1/d + 1/d' = 1/f$ to calculate d':
$$1/d' = 1/f - 1/d = 1/4 - 1/2 = (1/4) - (2/4) = -1/4$$
Therefore, d' is –4 m, and his image is upright virtual and on the far side. The magnification is found using: $M = h'/h = -d'/d = (4\,\text{m})/(2\,\text{m}) = 2$
Your friend's image is 2 times larger.

Developing Equations for Lenses and Mirrors

• As we saw in the examples for lenses and mirrors, using the equations allows us to quickly calculate focal length f, distances d and d', and heights h and h'. We can use the results to determine whether an image is real or virtual and upright or inverted.

• Let's first look at the equation referred to as the **lens equation, mirror equation**, or **focal length equation**. This equation relates **focal length** f to the **object distance** d and **image distance d'** in **lenses** and **mirrors**:

$$1/d + 1/d' = 1/f$$

Signs of f, d, and d', h, and h' alert us to the properties of lenses and mirrors in that:
Convex lens for $d > f$: $+d'$, $+f$, inverted real image far side.
Convex lens for $d < f$: $-d'$, $+f$, upright virtual image same side.
Concave lens: $-d'$, $-f$, upright virtual image same side.

Concave mirror for d > f: +d', +f, inverted real image same side.
Concave mirror for d < f: −d', +f, upright virtual image far side.
Convex mirror: −d', −f, upright virtual image far side.

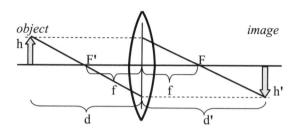

• Using a ray diagram above we can **derive the lens equation**. In the diagram, there are two pairs of **similar triangles**, having the equal corresponding angles in each triangle of a pair. From the triangles, we can write equalities relating ratios of the sides. For the triangles on the object side of the lens: −h'/h = f/(d − f), where h' is negative since h' < 0 because the image is below the lens axis. For the triangles on the image side of the lens: −h'/h = (d' − f)/f. Equating −h'/h gives: f/(d − f) = (d' − f)/f. Cross multiplying and cancelling f^2 gives: dd' − fd' − fd = 0. Dividing each term by dd'f and rearranging gives the **lens or focal length equation**:

$$1/d + 1/d' = 1/f$$

• The equation referred to as the **magnification M equation** describes an image's magnification and applies to **mirrors** and **lenses**:

$$M = h'/h = -d'/d$$

Magnification indicates image size with respect to object size, such that when the absolute value |M| > 1 the image is larger than the object; when |M| < 1 the image is smaller; and when M = 1, which occurs with a flat *mirror*, the image is the same size as the object. Image height h' is positive when the image is upright and h' is negative when the image is inverted. An image will appear larger the closer it is to a *mirror*. An image appears upright with virtual images when M is positive, and the image appears inverted with real images when M is negative.

• We can develop the magnification equation by first considering that **magnification** is *the ratio of the image height (formed by the lens or mirror) to the object height*:

$$M = h'/h$$

The image size depends on where an object is located with respect to the focal point of the lens. Using the triangles in the figure below:

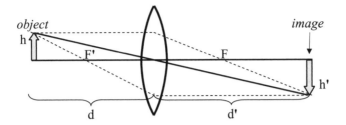

we see that h/d = –h'/d', where h' is negative since the image is below the lens axis. Therefore combining with M = h'/h, the **magnification equation** is:

$$M = h'/h = -d'/d$$

13.8. Key Concepts and Practice Problems

- Mechanical waves involve oscillatory motions of matter.
- Electromagnetic waves involve oscillations of electromagnetic fields and do not require matter.
- A wave is carried by a substance or material called a medium.
- Wave speed is: $v = \lambda/\tau = \lambda f$.
- In transverse waves particles oscillate perpendicular to propagation.
- In longitudinal waves particles oscillate parallel to propagation.
- Reflected waves are inverted with the same v, f, and λ, but less amplitude.
- Superposition: when waves cross and briefly add or subtract amplitude.
- Standing waves form from repeated reflections off endpoint nodes and occur at characteristic frequencies.
- A node point of a wave remains motionless.
- Antinodes are locations of maximum wave (amplitude) motion.
- Length L and wavelength λ for a standing wave: $L = n\lambda/2$, n = 1, 2, ...
- Sound: longitudinal pressure waves oscillating parallel to propagation; emitted from vibrating source; speed depends on medium's properties.
- Beats, or beating, is caused by alternating constructive and destructive interference of waves, producing alternating soft and loud sound.
- Cylinders closed at one end resonate at wavelengths: $\lambda = 4L/n$ for n = 1, 3, 5, ..., with corresponding frequencies: $f = v/\lambda = nv/4L$ for n = 1, 3, 5, ..., and standing-wave frequencies $f_1, 3f_1, 5f_1, ...$
- Cylinders open at both ends resonate at wavelengths: $\lambda = 2L/n$ for n = 1, 2, 3, ..., with corresponding frequencies: $f = v/\lambda = nv/2L$ for n = 1, 2, 3, ..., and standing-wave frequencies $f_1, 2f_1, 3f_1, ...$

- Doppler effect or shift: a change of perceived frequency due to movement of a wave source relative to an observer.
- Doppler frequency f_L of source S from listener's L perspective: $f_L = f_s v/(v - v_s)$ as S moves closer; $f_L = f_s v/(v + v_s)$ as S moves away. $f_L = f_s(v + v_L)/v$ as L moves closer; $f_L = f_s(v - v_L)/v$ as L moves away.
- Shock wave: forms as a wave source speeds ahead of its wave speed and drags the emitted waves behind it in the shape of a cone.
- Mach number is $M = v_{object}/v_{waves}$, or ratio of object speed to speed of sound in the medium, where v_{waves} = speed of sound in medium.
- Light has characteristics of both a particle and a wave.
- Young's double-slit experiment showed light's wave nature through constructive and destructive interference. Line spacing is: $\Delta x \approx \lambda L/D$.
- Electromagnetic radiation or waves are transverse waves of oscillating electric and magnetic fields that move through a medium or a vacuum.
- Law of Reflection: incident angle = reflecting angle, $\theta_{incidence} = \theta_{reflection}$.
- Light incident on a surface can reflect off or refract across the interface.
- Index of refraction n is: $n = c/v$, or
  (speed of light in a vacuum c) / (speed of light in the medium v).
- Snell's Law $n_1 \sin\theta_1 = n_2 \sin\theta_2$ for rays from medium 1 to medium 2 has indexes of refraction n_1 and n_2, incident and refracted angles θ_1 and θ_2.
- Total internal reflection occurs at angle $\theta_c = \sin^{-1}(n_{LessDense}/n_{MoreDense})$.
- Convex lenses cause convergence; concave lenses cause divergence.
- Two equations for lenses and mirrors: focal length equation, $1/d + 1/d'$ $= 1/f$, and magnification equation, $M = h'/h = -d'/d$, where
  Convex lens for $d > f$: $+d'$, $+f$, inverted real image far side.
  Convex lens for $d < f$: $-d'$, $+f$, upright virtual image same side.
  Concave lens: $-d'$, $-f$, upright virtual image same side.
  Concave mirror for $d > f$: $+d'$, $+f$, inverted real image same side.
  Concave mirror for $d < f$: $-d'$, $+f$, upright virtual image far side.
  Convex mirror: $-d'$, $-f$, upright virtual image far side.

Practice Problems

13.1 (a) Two beams of light travel through space. Beam A has twice the frequency of Beam B. Compare their wavelengths. **(b)** Two steel strings vibrate. String A has twice the tension and twice the diameter of String B. Compare the velocities of their wave pulses.

13.2 A guitar string resonates at both 360 Hz and 450 Hz without any intermediate frequencies. Can you compute its fundamental frequency?

13.3 (a) As you drive down a highway at 100 km/hr on a 20 °C day, a train approaches and passes you going in the opposite direction with its horn stuck on. When you first hear the horn, it is a Middle C (261.63 Hz),

and when you last hear it the pitch has dropped to an F (174.61 Hz).
What was the train's speed? **(b)** What is the actual frequency of the horn?

13.4 (a) You see a black-and-white photo of a double-slit experiment
where the distance between the slits is 5 mm, the distance between the
screens is 2 m, and the spacing of the maxima is about 2 mm. What is the
wavelength of the light? **(b)** What color is it?

13.5 (a) A laser emits a beam of light with a wavelength of 550 nm and
an energy intensity of 11 mW. What is each photon's energy? **(b)** How
many photons are emitted each second? (1 mW = 0.001 J/s)

13.6 (a) Diamonds have an index of refraction of 2.42. If light strikes a
diamond at $\theta_I = 45°$, what will be its angle of refraction θ_R? **(b)** What is
the speed of light v_D in the diamond? **(c)** After the light enters the
diamond, what is the critical angle θ_C for it to be internally reflected?

13.7 (a) Sherlock Holmes inspects a hair taken from a crime scene with
his magnifying glass. He holds the lens 4 cm from the hair. It appears 5
times its actual thickness. What is the focal length of his magnifying
glass? **(b)** Your friend is 2 m tall and stands 2.25 m from a concave
mirror with a focal length of 0.75 m. How far from the mirror should you
set a screen to project his focused image? **(c)** How tall will his image be?

Answers to Chapter 13 Problems

13.1 (a) Both beams travel at the speed of light, so $\lambda_A f_A = \lambda_B f_B$. Since f_A
is twice f_B, λ_A must be one half λ_B. **(b)** Since it has twice the diameter,
String A has 4 times the cross sectional area. $v_B = [T_B/\rho A_B]^{1/2}$ and
$v_A = [2T_B/\rho(4A_B)]^{1/2} = [2/4]^{1/2} [T_B/\rho A_B]^{1/2} \approx 0.707 v_B$.

13.2 $f_n = nf_1$ and $f_n = 360$ Hz and $f_{n+1} = 450$ Hz, so $f_{n+1} - f_n = 90$ Hz.
Since 90 Hz $= (n+1)f_1 - nf_1 = f_1$, so the fundamental frequency is 90 Hz.

13.3 (a) The train's speed relative to you is x. For the approaching train
$f_L = 261.63$ Hz $= f_S(v)/(v - x) = 344 f_S/(344 - x)$, so $f_S =$
$(261.63/344)(344 - x)$. For the retreating train $f_L = 174.61 =$
$344 f_S/(344 + x)$, so $f_S = (174.61/344)(344 + x)$. Since $f_S = f_S$ then
$0.76(344 - x) = 0.51(344 + x)$, $1.5(344 - x) = 344 + x$, $x = 68.8$ m/s.
Since you are going 100 km/hr = 27.8 m/s, the train was traveling at
$68.8 - 27.8 = 41$ m/s, or about 148 km/hr. **(b)** Again $f_L = f_S(v)/(v - x)$, so
$261.63 = f_S(344)/(344 - 68.8)$, and $f_S \approx 209.3$ Hz, just above a G Sharp.

13.4 (a) $\Delta x = \lambda L/D$, so $\lambda = \Delta x D/L = (0.002$ m$)(0.005$ m$)/2 = 5 \times 10^{-6}$ m $=$
5,000 nm. **(b)** This wavelength is in the infrared range—not visible to
the human eye (but it can be photographed with the right equipment).

13.5 (a) Each photon's energy $E = hc/\lambda =$
$(6.626 \times 10^{-34} \text{ J·s})(3.0 \times 10^8 \text{ m/s}) / (5.5 \times 10^{-7} \text{ m}) \approx 3.6 \times 10^{-19}$ J.
(b) The number of photons emitted per second $= N$
$= (0.011 \text{ J/s})/(3.6 \times 10^{-19} \text{ J/photon}) \approx 3.1 \times 10^{16}$ photons.

13.6 (a) Using Snell's Law, $n_1 \sin\theta_I = n_2 \sin\theta_R$ or $\theta_R = \sin^{-1}(n_1)(\sin\theta_I)/(n_2)$
$= \sin^{-1}(1)(0.707)/(2.42) \approx 17°$. **(b)** $n_2 = c/v_D$ or $v_D = c/n_2 =$
$(3.0 \times 10^8 \text{ m/s})/2.42 = 1.24 \times 10^8$ m/s. **(c)** $\theta_C = \sin^{-1}(n_{air}/n_{diamond}) =$
$\sin^{-1}(1/2.42) \approx 24°$. The sparkle and depth of a well-cut diamond is
enhanced by all the internal reflection.

13.7 (a) $M = -d'/d$ so $d' = -Md = -(5)(4) = -20$ cm. $1/f = 1/d + 1/d' =$
$(1/4) + (1/-20) = 4/20$, so $f = 5$ cm. **(b)** $1/f = 1/d + 1/d'$, so
$1/d' = 1/f - 1/d = (1/0.75) - (1/2.25) \approx 0.889$, so $d' = 1.125$, so you would
place the screen 1.125 m from the mirror. **(c)** $M = -d'/d = -1.125/0.75 =$
-1.5, so he will appear $(-1.5)(2 \text{ m})$ or -3 m tall (inverted) on the screen.

Chapter 14

INTRODUCTION TO THE "FUN STUFF"

14.1. Special Relativity
14.2. General Relativity
14.3. The Atom and an Introduction to Quantum
14.4. Dark Energy and Dark Matter
14.5. Key Concepts and Practice Problems

"When a blind beetle crawls over the surface of a curved branch,
it doesn't notice that the track it has covered is indeed curved.
I was lucky enough to notice what the beetle didn't notice.
Attributed to Albert Einstein

"Anyone who is not shocked by quantum theory has not understood it."
Attributed to Niels Bohr

• "Modern" physics includes the discoveries of the Twentieth century which went beyond Newtonian explanations of our clearly visible reality into a strange yet very real though unseen reality. Modern physics examines the very small, the very fast, and the very great, using new theories and tools such as relativity and quantum physics.

14.1. Special Relativity

• Published in 1905, Einstein's first theory of Relativity, **Special Relativity**, deviated from the Newtonian construct of space and time as absolute frames of reference. Special Relativity theory asserts that the speed of light is constant regardless of the velocity of an observer, that time slows down for objects traveling near the speed of light, and that objects traveling near the speed of light become shorter and heavier. Special Relativity is restricted to observers in uniform relative motion in the absence of a gravitational field. From the perspective of an *inertial frame* of reference, all *motion* can be described as being relative. For example, if your friend stands next to a train moving 20 mi/hr forward, then relative to you inside the train he is moving in the opposite direction (backward) at 20 mi/hr. Remember, an **inertial reference frame** is any setting in which Newton's First Law—*that every object persist in its state of rest or uniform motion in a straight line unless it is compelled to change*—is valid.

• Let's go back to the 1800s when scientists believed light traveled through an invisible medium in outer space called the "**ether**". This made sense because waves usually travel through some kind of medium, such as air or water. In 1879, Albert **Michelson** and Edward **Morley** measured the speed of the revolving Earth relative to the *ether* of space. *Michelson and Morley* reasoned that if the Earth is moving through some type of ether (similar to a ship through water), one would expect a slight difference in the speed of light depending on whether it is striking the revolving Earth head-on or it is striking the Earth perpendicularly from the side of its orbit. They expected they should easily distinguish differences due to maxima and minima in the interference pattern depending on orientation. Their experiment, however, showed no differences at all, suggesting the speed of light was the same in either (or all) direction.

• In 1905 Einstein proposed the two fundamental postulates of special relativity which enlarged the Newtonian view of reality and shed light on the Michelson-Morley results.

One postulate was that *the laws of physics are the same in all inertial reference frames, so that all observers moving at constant speed should observe the same physical laws.*

Einstein's postulate suggests that the laws of physics are the same in two different inertial reference frames even when the two reference frames are uniformly moving relative to one another at a constant velocity. The laws of physics apply whether you are in a uniformly moving car or train or whether you are standing on the ground. If you are standing inside a train moving at a smooth, constant velocity, from your perspective you are static and your friend standing outside the train appears to be moving backward. From the perspective of your friend standing outside the train, he is static and you (inside the train) are moving forward. Einstein's postulate implies that either perspective is correct and all inertial reference frames are valid.

A second postulate was that *the speed of light in a vacuum is constant at 3.0×10^8 m/s in every reference frame, regardless of the motion of the observer or the source of the light. In other words, the speed of light is the same for all observers, regardless of their motion relative to a source of the light.*

This postulate suggests that whether you are standing or traveling toward a light source, you will measure the light speed to be $c = 3.0 \times 10^8$ m/s relative to yourself. This runs counterintuitive to the behavior of objects moving with respect to one another and how we would combine their velocities using basic vector addition. The consequences of this postulate lead to the unusual outcomes of special relativity, including *time dilation* and *length contraction*.

• These two postulates led to new perspectives including the conversion of mass and energy between one another, the relationship $E = mc^2$, conservation of energy and mass being considered together rather than separately, and motion near the speed of light producing strange time and length effects. For example, imagine if you, while standing on Earth's surface, could view your friend holding a clock on a rocket ship that is traveling near the speed of light. The clock on the ship would appear to run slower than a clock you are holding in your hand. If you and your friend are also each holding yardsticks, his yardstick on the ship would appear shorter to you than your yardstick. From your friend's perspective in the rocket ship, you (standing on Earth's surface) are the one who is moving, and to him your clock is running slower and your yardstick looks shorter. This suggests that space and time are not constant, but depend on a frame of reference. Let's look more closely at *time dilation* and *length contraction*.

Time Dilation

• **Time dilation**, a consequence of Special Relativity, suggests that *time slows at high speeds*. First consider the first postulate of Special Relativity which suggests that absolute speed does not exist—only relative speed. We can see this by considering your perspective from inside a train of your friend who is outside the train. To you, he is moving backward, which is as valid as your friend's perspective that you and the train are moving forward. You in the train and your friend outside the train are each in inertial reference frames, and all the laws of physics apply within your respective frames. Either of you could bounce a ball on the floor, and the fact that your train is moving would not matter—both balls would bounce vertically.

Now let's look at time dilation. The effect of *time dilation* is that when you are inside the train traveling very fast at one-half the speed of light relative to your friend (who is standing outside), it appears to your friend that time is moving slower for you on your train than it is for him. However, from your perspective in your train, it seems that time is moving at its normal speed. From your perspective in the train, you observe your friend moving at one-half the speed of light *relative to you*, so it appears to you that time is moving more slowly for your friend. Time must therefore be relative, and both of you experience time from your respective positions. Scientists have observed time dilation by showing that an atomic clock travelling in a high-speed jet ticks more slowly than its stationary counterpart.

Therefore, according to Special Relativity, the time measured by an outside observer for events occurring on a speeding train or even a rocket occur more slowly than events measured by the person inside the train or rocket. Time is dilating or seeming to expand on the train or rocket from the perspective of the outside observer. A classic depiction of this is a rocket traveling across the sky with an observer on Earth.

Suppose you are in the rocket and your friend is viewing you from the Earth's surface. Now suppose you have a clock in your rocket which measures time by sending a light beam from one side of the rocket to the other, back and forth between two parallel mirrors attached to opposite sides of the rocket interior. Suppose the mirrors are distance D apart. From inside the rocket you measure time intervals by how long it takes the light beam to reflect back and forth across the rocket. From your friend's vantage point on Earth, the reflecting light looks different since he also sees the rocket move across the sky as the light beam reflects. Rather than the beam traveling across the distance D, it also is being moved sideways. The *time you see for a round-trip of the light beam from inside the rocket* is t_0. The time your friend sees from Earth for the same round trip is t as the beam follows the dashed arrows.

To write **time dilation** in equation form, we compare the time measured on the clock in the rocket above from your perspective *on the rocket* t_0 and from your friend's *perspective on Earth* t. Your clock measures time interval t_0, while to him it would take time t.

$$t = t_0 / [1 - (v^2/c^2)]^{1/2}$$

where v is rocket speed and c is speed of light. This shows that the clock appears to tick more slowly to an observer perceiving the clock moving than it does to the observer in the rest frame next to the clock.

- **Example**: Suppose you are in the rocket traveling at one-half the speed of light ($v = 0.5c$) relative to your friend on Earth. If you measure 10^{-8} s on your clock for the light beam to travel from one mirror to the other (distance $D = 3$ m apart) in your rocket ship as shown in the drawing above, what will your friend measure for that same time interval?

$$t = t_0 / [1 - (v^2/c^2)]^{\frac{1}{2}} = 10^{-8}\, s / [1 - ((0.5c)^2/c^2)]^{\frac{1}{2}} \approx 1.15 \times 10^{-8}\, s$$

From your friend's perspective it takes longer for the event to occur, so from his perspective time seems to be running slower for you on the rocket. If instead your friend had the light-beam clock with him on Earth and you were rocketing by, he would measure 10^{-8} s for the light beam to travel between mirrors that were set the same distance D apart, and to you it would seem to have taken 1.15×10^{-8} s for the light beam to travel between the mirrors.

- Note that t and t_0 differ by a factor of $1 / [1 - (v^2/c^2)]^{\frac{1}{2}}$. This factor reduces to the number one for more "normal" travel speeds not near the speed of light c. Therefore, time dilation effects are only noticeable at very high, near-light-speed velocities (which means a normal train would not produce a noticeable time dilation effect). Also note that as v approaches the speed of light c, time appears to slow to a standstill.

Relativistic Length Contraction

- **Length contraction**, another consequence of Special Relativity, suggests that an observer would not only perceive time moving more slowly on a train traveling speed v relative to him at, say, half the speed of light, but he would also perceive the train's length shortening, or contracting. If the train's length at rest is L_0, it will appear to *contract in the direction it is moving* to a length L when it is observed from a reference frame that is moving at speed v relative to the train. The length of the train when it is moving at a speed v relative to the observer is:

$$\boxed{L = L_0[1 - (v^2/c^2)]^{\frac{1}{2}}}$$

This shows that the length of an object appears to be shorter to an observer perceiving the moving object than to a person who is within the rest frame of the object. Note that *this apparent contraction applies to the length parallel to the direction of motion only*.

Train has length L_0 at rest.

Speeding train has observed apparent length L.

Observer on railway platform of length L.

• **Example**: Suppose a train's length at rest L_0 is 50 m. If it could travel 0.4 times the speed of light c, or v = 0.4c, what would be its apparent length to an outside observer?

$$L = L_0[1 - (v^2/c^2)]^{1/2} = (50\,m)[1 - (0.4)^2)]^{1/2} \approx 46\,m$$

Time Dilation and Simultaneous Events

• Special Relativity's *time dilation* can cause two **simultaneous events** to appear not to be simultaneous to different observers. In other words, if time can increase or decrease depending on the reference frame, then two happenings that appear simultaneous to one observer may not appear simultaneous to another observer.

Consider a locomotive that is somewhat longer than a railway platform (as in the above figure). Suppose your friend is standing at the center of the railway platform as you pass by at a high speed riding in the locomotive. Due to the high speed of the locomotive relative to your friend, he observes its length contracted. From your perspective riding in the locomotive, the length of the platform is contracted. Also, suppose a light on the front of your locomotive flashes on as it passes the right side the platform, and a light on the back of your locomotive flashes on as it passes the left side of the platform. Due to apparent length contraction of the locomotive which "shrinks" it to the length of the platform from your friend's perspective, your friend observes the two flashes of light simultaneously from his vantage point at the center of the platform. He then concludes that the two flashes occur at the same time. From your perspective, the right side of the platform passing the front of the locomotive and the left side of the platform passing the back of the loco-motive are two events that do not occur simultaneously because the locomotive is actually longer than the platform. This scenario shows how two events that seem simultaneous for one observer may not be simultaneous for another observer.

Relativistic Velocity Addition

• Simple vector addition of velocity in the easily perceptible Newtonian world shows us that if your friend sees you standing on a train that is moving 16 m/s and you throw a rock forward at 6 m/s in the direction the train is moving, he will observe the rock moving at 22 m/s. Suppose you turn into Superman approaching your friend from your super secret space base riding on your near-light-speed rocket ship. If your ship is moving at *speed* u and you throw a space rock at *speed* v_0 in the direction of motion, he will observe the space rock moving at *speed* v given by:

$$v = (u + v_0) / (1 + uv_0/c^2)$$

This is called **relativistic velocity addition** and shows that speeds of objects traveling near the speed of light do not undergo normal vector addition. Note that this equation can also be applied to velocities in opposite directions by changing signs of velocity values.

• **Example**: If you, Superhero, rocket toward Earth at half the speed of light ($u = 0.5c$) and throw a space rock toward your evil enemy on Earth at 0.6 times the speed of light relative to you ($v_0 = 0.6c$), what is the speed v of the space rock relative to Earth (before impact)?

$$v = (u + v_0) / (1 + uv_0/c^2) = (0.5c + 0.6c) / (1 + (0.5c)(0.6c)/c^2)$$

$$= (1.1c) / (1 + 0.3) = (1.1c) / (1.3) \approx 0.85c$$

Even though the sum of the two velocities ($0.5c + 0.6c = 1.1c$) exceeds c, the actual velocity relative to the Earth is less than c or about 0.85c.

Relativistic Mass Effects

• Another interesting relativistic effect is that as something moves at extremely high relativistic speeds it becomes heavier with increasing speed. This so-called **relativistic mass effect** means that an object traveling at high speed relative to you will have a mass greater than its resting mass. Specifically, for an object with a resting mass of m_0 moving at speed v near the speed of light c, its observed mass when moving at speed v relative to you is m:

$$m = m_0 / [1 - (v^2/c^2)]^{1/2}$$

This equation suggests that as the speed of an object approaches the speed of light $c = 3.0 \times 10^8$ m/s, its mass approaches infinity. It would therefore require infinite energy to accelerate a massive object to light speed. This, in practice, prevents a massive object from being able to travel at the speed of light relative to us. Obtaining high speeds for objects with non-zero resting masses requires large amounts of energy. Relativistic mass effects have been interpreted to suggest nothing can travel faster than the speed of light. However, these effects do not necessarily preclude objects that *always* move faster than c. Such theoretical particles are referred to as **tachyons**. Note, for light itself, the equation's denominator is zero, so by convention, the resting mass of light is zero.

Relativistic Energy Effects

• Very high speeds also affect energy. Approaching the speed of light, the mass of an object increases, so its kinetic energy also increases. The kinetic energy at near light speeds has **relativistic energy effects** which are expressed:

$$KE = mc^2((1 / [1-(v^2/c^2)]^{\frac{1}{2}}) - 1)$$

This equation suggests that *as speed* v *approaches the speed of light* c, *kinetic energy approaches infinity*, and it would therefore require infinite energy to accelerate large objects to light speed. For this reason traveling faster than the speed of light seems impractical. Note that for slower, more "normal" v values, determining KE using the above equation converges with values obtained using the more familiar $(1/2)mv^2$.

• If you graph kinetic energy vs. velocity at normal speeds, the curve begins at the origin and rises as a parabola. Alternatively, if you graph kinetic energy vs. velocity at relativistic speeds, the curve begins at the origin and starts to rise as a parabola. Then as the curve approaches the *speed of light* c it rises more steeply approaching *but never reaching* c, with increasing KE occurring due to greater mass more than from speed.

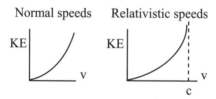

Mass-Energy Equivalence

• Einstein's discovery of the relativity of space and time revealed a **mass-energy equivalence**, that *matter and energy are not just inter-related but have an equivalence*. Using principles of relativity, Einstein therefore developed the equation showing that **mass and energy can be converted from one to another**. Mass can be converted to energy E as:

$$E = m_0c^2$$

where m_0 is the resting mass and c is the *speed of light*.

Relativistic Momentum

• At very high speeds the equation for **momentum** p, where p = mv is:

$$p = mv / [1 - (v^2/c^2)]^{\frac{1}{2}}$$

The effect occurs near the speed of light c when the denominator becomes very small. At lower speeds the denominator is essentially equal to one and we have the Newtonian relationship with mass m and velocity v:

$$p = mv$$

The Twin Paradox

• Suppose with clock in hand you rocket off Earth, traveling very fast to your space base and then return to Earth. Suppose also that your twin remains on Earth. According to relativity, your clock runs slow compared to your twin's clock, so when you return to Earth you will be younger than your twin. This is because not only does your physical clock run more slowly, but your internal (aging) clock also presumably runs more slowly. Your experience in your rocket ship seems normal to you, and it does not seem to you that time runs more slowly. In fact, from your vantage point, except for the effects of your rocket accelerating and decelerating, you are stationary and it seems that your twin moves relative to you at a high speed. From your perspective your twin's clock runs slower, and therefore he should be younger when you rejoin each other on Earth.

How can you and your twin both wind up younger? This is referred to as the **twin paradox**. It can be resolved by the fact that as you travel to your space base and back, you experience high g-forces during your launch, turn around at your base, and the deceleration and landing back on Earth. You do not remain in one inertial reference frame during your trip. It is therefore not valid to analyze the twin paradox from your reference frame.

While the Earth is in a circular orbit around the Sun and rotating on its axis, these accelerations are small enough that Earth may be treated as an inertial reference frame for your twin. This means your twin remains in one inertial reference frame. Because you do not remain in one inertial reference frame during your trip, there is not a symmetrical situation between you and your twin. In order to create the twin paradox, we must assume that you have been in a single inertial frame throughout your trip out to the base and back home. Since this assumption is not true, there is no real paradox. Therefore, your clock is the one that runs more slowly and you age less than your twin.

14.2. General Relativity

General Relativity Principles

• While Special Relativity applies to situations in which objects are moving at uniform velocities, **General Relativity** expands to situations involving accelerated motion and gravity. During the years leading up to 1916 when his landmark paper on General Relativity (GR) was published,

Einstein published several related papers and fine-tuned and developed his theory. General Relativity is an expansion of Special Relativity that provides a new theory of **gravity** and a description of **spacetime** which treats its geometry as a 4-dimensional continuum. We usually think of time and space being separate, but in GR we can treat time as a 4th dimension, giving us *spacetime*. Einstein theorized that *gravity is not a force but rather a manifestation of curved space and time*, or spacetime.

Space and time had been assumed to be absolutes before the 20th century, but in GR they are viewed as dynamic. In GR theory, matter causes spacetime to curve, and that curving affects the behavior of matter. The laws of Euclidean geometry are not valid in GR. In Newtonian physics, when a particle has no forces acting on it, it will continue in a straight line, but in curved spacetime it will move along curved paths.

The Newtonian theory of gravity describes gravity as a force such that two bodies are believed to exert a pulling force on each other. Newton's perspective on gravity describes not only the fall of apples to Earth but also the orbital motions of planets and moons. GR describes such effects differently. For example, a sun causes the spacetime surrounding it to curve, which in turn affects the motion of nearby planets, creating their orbits. In GR, gravity is described in terms of the dynamic characteristics of space and time such that it is curved by the effects of matter which in turn can affect the behavior of that matter.

• A classic depiction of space curved by mass uses a 2-dimensional (2D) slice of space and shows a bowling ball resting on a stretched rubber sheet.

This drawing is referred to as an **embedding diagram** and shows how a 2D slice through 3D space can be curved. The idea of embedding is that it describes a region of curved space by modeling it as a curved surface. We can see that the bowling ball deforms the surface of the sheet. If you add a golf ball to the sheet, it will roll toward the bowling ball. Einstein viewed the "attraction" of a small object to a larger one as the smaller object moving through space that has been warped by the larger object rather than the objects being drawn together by an attractive *force*.

• In curved space strange things happen. For example, suppose you and your friend begin walking north from two different points on the equator, and your paths begin parallel to each other. As you continue toward the North Pole, with each of you walking straight, you will at some point no longer be walking parallel to each other. Eventually you will meet each other at the North Pole. The "geodesics" you were walking along did not remain parallel to each other. (A **geodesic** is the shortest path between two points. On a sphere it is a path along a great circle.) On a sphere parallel lines curve inward toward each other, having *positive* curvature. Alternatively, on a saddle-shaped object parallel lines curve outward, having *negative* curvature. In curved space paths that remain parallel are not the shortest-distance paths between two points. Note that in a tiny region of curved space we can often use assumptions that apply to flat space. On a microscale, parallel lines remain parallel.

• While in Newtonian physics space and time are viewed as separate, in GR, time is considered a fourth dimension. Since it is difficult to visualize a 4-dimensional Universe with 3 space dimensions plus a time dimension, we can at least imagine 2 space dimensions as a surface plus one time dimension. For example, consider an x-y-z coordinate system where x and y are 2-dimensional space drawn horizontally and z is time drawn vertically. We can depict a single particle at rest as a vertical line, and a particle traveling at a constant velocity is a diagonal line.

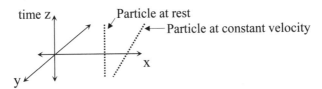

• In the Newtonian world mass appears as either **inertial mass** in F = ma when we consider motion or **gravitational mass** in F = mg. In his **Principle of Equivalence**, Einstein hypothesized that these two kinds of mass are equivalent. That means in Newton's equation ma = mg, the first m is inertial mass and the second m is gravitational mass, and they are equivalent. When a given mass is examined in two different states—one where the mass is acted on by gravity or another where the mass is in a state of inertia such that it resists forces and accelerations—Einstein's Principle of Equivalence says the mass is the same. An example of this is a spinning top which has the same mass whether it is falling off a table and acted on by gravity or whether it is spinning on the table in a state of inertia.

Because of the Principle of Equivalence, an experiment conducted in a uniformly accelerating reference frame with acceleration "a" would play out in an indistinguishable manner from the same experiment conducted in a non-accelerating reference frame within a gravitational field where the acceleration of gravity g = −a.

What we see is that a uniform gravitational field, such as near the Earth, is equivalent to a uniform acceleration. This suggests that if you were standing on Earth's surface feeling the effects of gravity or standing in an elevator in outer space that is smoothly accelerating upward at the rate of gravity (9.81 m/s²), you would feel the same downward pull of "gravity".

• Imagine you are in an elevator that is at rest relative to the Earth's gravitational field. The gravitational force on your body (your weight) pushes you down onto the floor of the elevator as the elevator floor pushes up on you with the same force. You feel this reaction force as your weight. Uh oh, if the elevator's cables break, you will become weightless. Since inertial and gravitational mass are equivalent, all objects fall freely with the same acceleration. The elevator and its contents are accelerating toward Earth's center at the same rate. This means that the elevator floor is no longer supplying a force on you, and you are at rest relative to the elevator. If you let go of the coffee cup you are holding, it will appear to float since it is also in free fall and is accelerating toward Earth's center at the same rate as you and the elevator.

This scenario suggests that within a **frame of reference** that is in free fall, gravity has no effect. Apparently, Einstein reasoned, if gravity can be made irrelevant, then it may not be an actual force but rather related to motion in spacetime. Furthermore, he thought any experiment performed within the gravity-less elevator should have the same results as experiments done in any region void of gravity's effects.

Field Equations

• GR theory is expressed in a series of **Field Equations**, which are complex and suggest strange phenomena such as black holes and gravitational waves. *Einstein's Field Equations* describe the gravitational field around a mass and, as such, how a mass curves spacetime and how curved spacetime affects movement, stretching, or shrinking of a mass. The equations describing GR required many years to understand and to begin to solve.

At first the equations were only applied to simple systems. **Karl Schwarzschild** applied Einstein's GR theory to a symmetrically spherical star and the curving of spacetime. He soon thereafter described the spacetime curvature inside a star. Schwarzschild's work led to *Schwarzschild geometry* and a description of a **singularity** with infinite spacetime curvature inside—what was later described as a black hole. (See discussion of black holes later in this section.)

General Relativity Predictions

• Newtonian physics and General Relativity yield similar predictions when you are modeling situations with speeds much slower than the speed of light and when gravitational fields are weak. When speeds are very high or gravitational fields are great, the predictions diverge and GR is more accurate. The following sections highlight some of the predictions of General Relativity.

Gravitational Time Dilation and Length Change

• We learned that a prediction of Special Relativity is that time is variable and traveling clocks appear to run slow relative to a stationary observer. This effect is most dramatic at speeds approaching the speed of light. General Relativity predicts that gravity also appears to slow or *dilate* the flow of time. While **gravitational time dilation** is generally negligible since gravity is usually weak, when spacetime is curved by a massive object, time dilation becomes significant. For example, an observer far away from a black hole would perceive time passing extremely slowly for an object being drawn into the hole.

• Similar to *length contraction* of objects at high speeds in Special Relativity, GR also predicts a **gravitational length change**. In General Relativity, gravity (which is the curvature of spacetime by matter) appears to *stretch or shrink distances* or lengths of an object depending on its orientation in the gravitational field.

General Relativity Theory Survives 1919 Eclipse

• GR's prediction of the bending of light near large masses was observed as starlight deflected around the Sun during a 1919 solar eclipse. The measurements made supported Einstein's GR predictions. During the eclipse, the Sun was silhouetted against the Hyades star cluster which had known star positions. With Sir Arthur Eddington on an island off the west coast of Africa and a group of British scientists in Brazil, they measured a number of stars in the Hyades cluster, proving that the star's light bent precisely according to Einstein's predictions as it grazed the Sun. The starlight's position seemed to shift. *This apparent displacement of light was due to the light following the warped or curved space near a massive object (the Sun).* The light did not deviate from its path, but followed the curvature of spacetime. The eclipse results were the first big test of Einstein's theory, and brought him fame and his theory respect.

This effect of light curving around large masses is called **gravitational lensing**. Stars or objects in a region of space behind a massive object have an apparent shift in their locations. When there is a cluster it can appear magnified. Lensing by the Sun and other similar masses is referred to as *weak gravitational lensing*, as they cause a shifting or a magnification of objects behind the lensing mass. If the lensing object is massive enough, it can create multiple images of objects behind it. This creation of multiple images is called *strong gravitational lensing*.

Mercury's Orbit

• A second test of Einstein's theory was an explanation of the slight alteration in **Mercury's orbit around the Sun**. Astronomers had been aware of a defect in Mercury's orbit. While Newton's Laws generally describe the orbits of the planets, a small discrepancy existed between theory and observation in the orbit of Mercury. Mercury is the planet closest to the Sun and experiences the strongest gravitational field from the Sun. It is therefore in a region where spacetime is warped the most by the Sun's mass. Mercury precesses such that the long axis of its ellipse revolves around the Sun slightly faster than predicted by Newton's theory. In other words, Mercury's elliptical path shifts slightly with each orbit so that its closest point to the Sun, or perihelion, shifts forward with each lap. While Newton's theory predicted an advance half as large as what was observed, Einstein's theory predicted what was observed.

Gravitational Redshift

• A **Gravitational Redshift** occurs when the frequency of a photon appears to shift to lower energy as it climbs out of a gravitational field, or "well". This is explained as occurring as a photon expends energy to rise out of a gravitational well. Since the speed of light is maintained during the climb, energy is lost through a frequency change rather than velocity. As a photon's energy decreases, its frequency decreases. This corresponds to an observed increase in, or lengthening of, the photon's wavelength or a shift toward the red end of the electromagnetic spectrum (remember $\lambda = c/f$). This means the wavelength of light or any other form of electromagnetic radiation that is observed to pass against a gravitational field will be shifted toward redder regions of the spectrum. Also, note that when a photon is observed to "fall" into a gravitational field or well, it will gain energy, and its observed wavelength will be shortened or shifted toward the blue end of the spectrum, called Gravitational Blueshift.

The Gravitational Redshift of photons is also described in terms of the gravitational slowing of time. Since gravity changes the flow of time, as predicted by GR, a Gravitational Redshift occurs as the oscillations of light waves slow or become redder when pulled by gravity.

In 1960 a Gravitational Redshift was observed when a beam of high energy gamma rays was very slightly redshifted as it climbed out of Earth's gravity up an elevator shaft at Harvard University. The observed redshift was very close to the value calculated by GR theory. In the 1960s scientists at Princeton University measured a redshift of sunlight which also matched GR theory. Since then numerous Gravitational Redshifts have been measured for light from high-mass heavenly bodies and compared with GR theory's predicted values.

Black Holes

• Extremely massive or dense objects generate strong gravity. GR predicts unimaginably compact objects with gravity so strong that nothing, including light, can escape. These objects are called **black holes**. Einstein's GR predicted both *singularities* and *black holes*.

Let's see how a black hole could occur. First, if we consider the "gravitational field" around a mass such as the Sun, by GR we are considering the curvature of spacetime caused in its vicinity. In the vicinity of a massive object such as the Sun, light (and anything else) is warped or bent toward it. If we next consider something smaller but more dense and massive, such as a neutron star, gravity is very strong in its vicinity, and the curvature of space and time is greater. For an object even smaller and denser, spacetime is warped even more, and light shined outward from such an object cannot escape its gravity. Such an object is appropriately called a **black hole**. There is a spherical region around a black hole that demarcates where light can no longer escape called an **event horizon**.

When any object gets squeezed into a sphere that is equal to or smaller than a certain radius called the **Schwarzschild radius**, it will become a black hole. The characteristic Schwarzschild radius for a given object depends on the object's mass m and can be expressed as: $r_{sch} = 2Gm/c^2$, where G is Newton's gravitational constant. The *event horizon* of a given black hole is a sphere the size of the *Schwarzschild radius*. Once an object becomes dense enough to form a black hole, it continues collapsing and becoming denser until it eventually forms *an infinite point of density and curvature* called **singularity**. If you could observe a black hole forming, you would see the original object, such as dying star,

Master Math: Essential Physics

collapse to the size of its Schwarzschild radius and disappear. The object would continue collapsing, but no light would escape, and you could not see it anymore.

• **Example**: Your friend asks you to what size our sun would need to collapse in order to form a black hole. (Assume $G = 6.67 \times 10^{-11} \, m^3/kg\cdot s^2$, $m_{sun} = 1.99 \times 10^{30} \, kg$, and $c = 3.00 \times 10^8 \, m/s$.)

You say the Sun would need to be squeezed into a sphere the size of the Schwarzschild radius or smaller:

$r_{sch} = 2Gm/c^2 = 2(6.67 \times 10^{-11} \, m^3/kg\cdot s^2)(1.99 \times 10^{30} \, kg) / (3.00 \times 10^8 \, m/s)^2$

$\approx 2{,}949.6 \, m$ or about 3 kilometers

Gravitational Waves

• As predicted by General Relativity theory, **gravitational waves** are disturbances or ripples in the fabric of spacetime created by the motion of matter. Vigorous movements of large masses create a spacetime disturbance, or ripple, which will propagate outward as gravitational waves. Gravitational waves pass through space or matter and weaken with distance traveled from a source. As gravitational waves travel through something, it minutely stretches and shrinks.

Gravitational waves require not only a massive object but also some sort of vigorous movement. Candidates for production of gravitational waves include events such as the supernova collapse of stellar cores into neutron stars or black holes, the collisions and combining of neutron stars or black holes, the close unsteady orbiting of two neutron stars or black holes around each other, and the remnants of gravitational radiation created during the birth of the Universe.

An Expanding Universe

• Finally, on the largest scale, the General Relativity equations predict a Universe that is not static and therefore either expanding or contracting. This prediction, in essence, results from the effect of gravity. Einstein was apparently troubled by the thought of a non-static Universe and ineffectively attempted to make modifications to his theory to allow static solutions. In 1929 Edwin Hubble observed that all distant galaxies appeared to be moving away, which was in accordance with Einstein's General Relativity prediction of an expanding Universe.

• Once again, Relativity theory was proven correct, which has happened consistently since Einstein developed it. Relativity has changed our view of the Universe including: gravity being defined as a manifestation of curved space and time, time and space no longer considered fixed, the bending of light and slowing of time near large masses, the accurate calculation of Mercury's precession, and the prediction of black holes.

14.3. The Atom and an Introduction to Quantum

• An **atom** is the fundamental building block of elements. It consists of a central nucleus having neutrons and protons surrounded by one or more negatively charged electrons. Scientists throughout history have been preoccupied with the search for the building blocks of matter and the physical world around them. Many believed there was some sort of inseparable fundamental entity of matter. The root of the ancient Greek word which translates "atom" means "not divisible."

Plum Pudding Atom

• In 1897 *J. J. Thomson* discovered **electrons** and the fact that they are a fundamental particle of matter, have a negative electric charge, and possess a very small mass. It became clear that **atoms** were not the smallest existing particles. Also, for matter to be electrically neutral, both negative and positive charges had to be present. In 1904 Thomson proposed the **plum pudding model of the atom**, which used the concept of plums scattered in a plum pudding to illustrate how small negatively charged electrons are scattered about a positively charged *medium*. The negatively charged electrons balanced what he imagined to be a positively charged *medium*, giving an atom a neutral charge. He also suggested that the geometry of the atom was spherical.

Rutherford Model

• As a result of experiments conducted from 1909 to 1911, **Ernest Rutherford** showed that atoms have *defined nuclei* rather than a "pudding" distribution. Rutherford showed that an atom's positive charge was concentrated in a tiny spot smaller than the atom itself and that most of the atom's mass was also concentrated in its positively charged center called a **nucleus**. He found that most of the volume of an atom was empty space.

Rutherford made these discoveries in his **gold foil experiment** in which charged **alpha particles** were directed at a thin sheet of gold foil. (*Alpha* *particles* are large positively charged particles consisting of two protons and two neutrons.) It turned out a majority of the alpha particles passed through the foil without hitting anything, suggesting that most of an atom's volume was empty. Some alpha particles were deflected, however, revealing the localized, small, positively charged (nuclear) masses within the atoms. In keeping with the idea that atoms have a stable structure, and since the positive charge appeared to be localized, it was theorized that the electrons must be symmetrically distributed to create counterbalanced forces with each other.

Prior to the experiment, it had been expected that the smaller, negatively charged electrons would cause a slight deflection in the path of the larger positively charged alpha particles so that the distribution of electrons in the atoms could be analyzed. Rutherford had assumed at the time of his experiments that atomic structure fit the plum pudding model. To his surprise, a small number of alpha particles were bounced back toward their source, while most alpha particles were not significantly deflected, revealing the empty space. *Rutherford's model* therefore showed that an atom's mass is primarily concentrated in a nucleus (composed of protons and neutrons), with the remaining mass composed of electrons which orbit the nucleus and take up most of the volume. The electrons are held in orbit around the nucleus by the electromagnetic force.

This new model began to be compared to planets orbiting the Sun. The problem was that according to classical electromagnetic theory, a charge moving or being accelerated in a circular path would lose or release energy as electromagnetic radiation, so in Rutherford's model there was nothing to prevent the Coulomb attraction from causing the electrons to lose energy and spiral into the nucleus. Also without explanation in the Rutherford model was the observation that light seemed to travel in certain discrete frequencies.

Bohr Model

• In the late 19th century it was known that the spectrum of each pure element was unique, and the values of the line spectra for many elements were identified. To develop his theory **Niels Bohr** (1885–1962) brought together Rutherford's work on the discovery of the atomic nucleus as well as the available information on the existence of line spectra from chemical elements, particularly work by *Johann Jacob Balmer* for the hydrogen spectrum. Bohr also had Planck's and Einstein's work on the early development of *Quantum Theory*, including the idea of energy traveling in distinct quanta.

In 1913 Bohr devised a workable model of the atom, which is still a convenient model of the hydrogen atom. He addressed the question of why electrons do not spiral into their nuclei using a model having specific orbits in which the electrons do not lose energy or fall into the nucleus. The **Bohr model** showed electrons circling the nucleus at specific levels or **orbitals**. Electrons exist at definite energy states or levels and can move from one energy state to another. *Energy is released or absorbed as electrons change states in the form of electromagnetic radiation.*

Bohr's model was *based on the hydrogen atom* with one proton and one electron and was somewhat consistent with a planetary model. Bohr's model required that the electron only orbit at certain "allowed" radii. It also stipulated that when an electron is in an allowed orbit it is not radiating (and losing) energy. The allowed orbits are given by:

$$mvr = nh/2\pi \quad \text{where the integer n can be 1, 2, 3, ...}$$

and where mvr = *angular momentum* L, m = electron mass, v = electron velocity (speed), r = radius of its orbit around the nucleus, and $h \approx 6.626 \times 10^{-34}$ J·s $\approx 4.135 \times 10^{-15}$ eV·s is **Planck's constant**.

Bohr proposed that when an electron absorbs energy from incident electromagnetic radiation, it makes a "quantum jump" up into a higher energy allowed state, which is an allowed orbital radius with a higher n value. An electron in a higher energy state can also spontaneously quantum jump down (or fall) into a lower energy state with a smaller n value, while emitting its lost energy as a single photon of electro-magnetic energy. The lowest orbital radius is called the electron's **ground state**. A state with energy greater than that of the ground state is referred to as an **excited state**.

Since the electron jumps down to *specific orbits*, it can only emit *specific frequencies* consistent with that orbital change. This means each kind of atom can only emit photons of *certain frequencies*. For example, if the Sun is largely composed of hydrogen, then most of the light we measure from the Sun is at the allowed frequencies for energy jumps in hydrogen atoms. When an electron jumps down an orbit level, a photon is emitted with an energy equal to the energy lost by the atom due to the transition. The energy of the emitted photon is related to its frequency f by: E = hf. The energy change for the transition is:

$$\Delta E = hf = E_i - E_f$$

where E_i is the energy of the atom with the electron in its initial higher orbit, E_f is the energy of the atom with the electron in its final lower orbit, and $h \approx 6.626 \times 10^{-34}$ J·s $\approx 4.135 \times 10^{-15}$ eV·s is *Planck's constant.*

Similar to a planetary orbit, the **centripetal force** on an electron is directly proportional to its mass times its velocity-squared and inversely proportional to the radius of its orbit: $F_c = mv^2/r$. The centripetal force is also related to the *electrostatic force,* $F_E = Kq_1q_2/r^2 = K(Ze)(-e)/r^2$, where Z describes the number of protons in the nucleus, Ze is the electric charge of the nucleus, and $-e$ is the electric charge of the electron. The **centripetal force for an electron** in a hydrogen atom is being supplied by the electrostatic force between it and the proton and can be written:

$$F_c = F_E = m_e v^2/r = K(Ze)(-e)/r^2$$

where K is **Coulomb's Constant** $\approx 9.0 \times 10^9$ N·m^2/C^2, v is the electron velocity, and r is the electron radius. In the force equation above, $m_e v^2/r = K(Ze)(-e)/r^2$, we can cancel the r's and write: $m_e v^2 = -KZe^2/r$.

The **total energy** of this system is the sum of kinetic and potential, so we can write:

$$E = KE + PE = (1/2)m_e v^2 + (-KZe^2/r)$$

Substituting $m_e v^2 = -KZe^2/r$ gives the **total energy of the electron**:

$$E = (1/2)KZe^2/r - KZe^2/r$$

$$E = -K(Ze^2)/2r$$

which relates energy to the radius r of the orbit. Note, the negative sign reflects the electron is trapped in an energy well about the nucleus and would require energy to be freed from orbit.

To determine r, Bohr found that an electron is allowed only to be in a discrete state or orbit, so the **angular momentum** L = mvr of the electron is an integer n multiple of $h/2\pi$. Now we can write: $L = m_e v_n r_n = n(h/2\pi)$, for n = 1, 2, 3, . . . where h is *Planck's constant.* The subscripts on v_n and r_n show that they correspond to a certain value of n where n is referred to as the **principal quantum number**. The angular momentum is restricted, or **quantized**, to certain values so that the energy is restricted to certain values and radii r_n. If we substitute the *quantization of angular momentum,* $m_e v_n r_n = n(h/2\pi)$ solved for $v_n = nh/2\pi m_e r_n$, into the equation for centripetal force, $m_e v^2 = KZe^2/r_n$ (leaving off the negative sign since we are describing the attractive electrostatic force), we can solve that combined equation for the **electron radius** r_n as:

$$r_n = n^2 h^2 / 4\pi^2 m_e KZe^2$$

Therefore, r_n exists at certain (integer) values of n, and r is proportional to n^2, with each higher radius value of the electron further from the nucleus. For a *hydrogen atom in its ground state* with n = 1 and Z = 1: $r_n = 0.53 \times 10^{-10}$ m, or 0.53 Angstroms.

When we know the radius of an electron we can find its **energy**. Using the above energy equation, $E = -K(Ze^2)/2r$, and substituting for r:

$$E = -2\pi^2 K^2 Z^2 e^4 m_e / h^2 n^2$$

The **energy for a hydrogen atom in its ground state** with n = 1 and Z = 1, works out to:

$$\boxed{E_n = E_1/n^2 = -13.6 \text{ eV}}$$

This corresponds to the energy to free the electron from orbit. Since an electron jumps to a higher energy level by multiples of n, for a **hydrogen atom with any n value, its energy** becomes:

$$\boxed{E_n = -(1/n^2)13.6 \text{ eV}}$$

The energy in the n = 2 state is: $E_2 = -(1/2^2)13.6$ eV = -3.40 eV

When an electron jumps to a lower energy state, a hydrogen atom emits a photon. That means an electron at the n = 2 state that jumps to the ground state at n = 1 will emit a photon with energy:

$$E = E_2 - E_1 = (-3.40 \text{ eV}) - (-13.6 \text{ eV}) = 10.2 \text{ eV}$$

Since energy also equals frequency times Planck's constant, or E = hf, the *frequency f of the emitted photon* from n = 2 to n = 1 is:

$$f = E/h = 10.2 \text{ eV} / (4.135 \times 10^{-15} \text{ eV·s}) \approx 2.47 \times 10^{15} \text{ Hz}$$

Since frequency f equals the speed of light c divided by wavelength λ, or $f = c/\lambda$, the wavelength λ of that emitted photon from n = 2 to n = 1 is:

$$\lambda = c/f = (3 \times 10^8 \text{ m/s}) / (2.47 \times 10^{15} \text{ Hz}) \approx 1.21 \times 10^{-7} \text{ m}$$

Therefore, a photon with an energy of 10.2 eV has a wavelength of about 1.21×10^{-7} m, which is in the ultraviolet part of the spectrum. This means that for an electron to jump up from n = 1 to n = 2 in the Hydrogen atom, it must absorb an ultraviolet photon having a wavelength of 1.21×10^{-7} m. Also, when an electron jumps down from n = 2 to n = 1, it emits a 1.21×10^{-7} m photon of ultraviolet light.

Note that the absorption of a photon is the reverse of emission. A photon having energy equal to the difference in energy between two states of an atom can be absorbed by that atom, which will cause the atom to be in a higher energy state. A photon having energy that is not equal to the difference in the energy between two states of an atom will not be absorbed. This selective absorption of light energy is reflected in spectroscopy data.

• In summary, the **Bohr atom** is described as having a central nucleus made up of neutrons and protons surrounded by "orbiting" electrons. The protons carry a positive charge and have a mass of about 1 atomic mass unit or about 1.67×10^{-27} kg. Neutrons are electrically neutral and about the same mass. Electrons carry a negative charge and have mass of only 0.00055 amu or about 9.1×10^{-31} kg. The number of protons (designated Z and called **atomic number**) in an atomic nucleus determines the element of the atom. The number of neutrons is sometimes designated N and called the *neutron number*. The total number of neutrons N and protons Z in the atom is often referred to as the **mass number** A, where Z + N = A. **Elements** are atoms delineated by their number of protons. Elements are arranged in the *Periodic Table of Elements* according to increasing Z. A given element usually has a particular number of neutrons, but it can vary. An element with a non-standard number of neutrons is called an **isotope** of the element. There may be one or more isotopes of a given element.

Quantum Mechanics

• The Bohr model was only able to describe very simple atoms, particularly hydrogen. A new model was needed. Work of physicists such as Louis de Broglie, Max Planck, Albert Einstein, Erwin Schrödinger, and Werner Heisenberg contributed to what became Quantum Mechanics or Quantum Physics. **Quantum Mechanics** provided a new description of the structure and behavior of matter and electromagnetic radiation (including light) and a new mathematical model of the atom and subatomic particles. Quantum Mechanics revealed that energy is emitted (or absorbed) from matter in packets called quanta and theorized that light and other electromagnetic radiation is composed of photons, or quanta. The energy of a photon is also discrete and indivisible. It was found that photons have properties of both particles and waves. In fact, electromagnetic radiation as well as matter both have properties of particles and waves.

Going back to the 18th century, light was viewed as consisting of particles. Beginning about 1800 evidence for a wave nature of light began to accumulate. A wave theory was supported by Young's double-slit experiment and also single-slit diffraction experiments (discussed in the previous chapter). By the end of the 19th century, the wave theory of light dominated. Some inconsistencies were evident with a pure wave nature of light, however, since all objects radiate electromagnetic energy as heat. In fact, an object or body emits radiation at all wavelengths. Scientists were unable to accurately calculate the energy distribution or spectrum for the radiation from a radiating object, referred to as a *blackbody*, using wave theory. A **blackbody** is defined as an ideal body or surface that absorbs and re-emits all incident light or electromagnetic radiation. **Blackbody radiation** refers to the spectrum of light emitted by any heated object, such as a toaster element or the filament of an incandescent light bulb.

In 1900 **Max Planck** calculated a blackbody spectrum that matched experimental results. He assumed that electromagnetic radiation is emitted and absorbed in *discrete packets called quanta* and that *the energy E of a packet is directly proportional to the frequency f* of the radiation, $E = hf$. Planck determined the constant of proportionality $h \approx 6.626 \times 10^{-34}$ J·s $\approx 4.135 \times 10^{-15}$ eV·s (now called **Planck's Constant**) by comparing his theoretical results with the experimental data.

• In 1905 **Einstein** further challenged the concept that light always behaves as a continuous wave and expanded on Planck's work to explain the *photoelectric effect*. The **photoelectric effect** occurs when an electron in an atom absorbs sufficient energy in the form of electro-magnetic radiation (light, X-rays, UV, etc.) to be ejected from its orbit. This can occur, for example, when light strikes a sheet of metal and the surface atoms absorb the energy from the electromagnetic radiation in sufficient amounts to eject electrons from their orbits. These electrons are called **photoelectrons**. (The photoelectric effect was observed as early as 1887 by Heinrich Hertz, but Einstein's later work was the basis for his 1921 Nobel Prize.)

Einstein described the inconsistencies of light behavior by suggesting that in certain situations light behaves as particles, or light quanta (photons), and each photon carries a discrete amount of energy. A beam of light is therefore made up of many photons which together are observed as a continuous wave. The energy in a beam of light is the sum of the individual energies of the photons.

In the photoelectric effect the kinetic energy of the emitted electrons depends on the frequency f of the incident electromagnetic radiation, but not its intensity (as a wave-theory of light would predict). There is also a **threshold frequency** f_0 for a given metal or material below which no electrons are emitted. This means low frequency light may be insufficient to free an electron from a given material. This would not be predicted from a pure wave nature for light. Emission of an electron is also observed almost instantly with the presence of incident light since a sufficient amount of light (to eject the electron) arrives all at once as a photon. Einstein suggested these observations revealed that light is composed of particles or photons which have energy defined by **Planck's relationship** $E = hf$, where Planck's formula determines the amount of energy in a given quantum. Einstein further suggested that an atom in a metal must absorb either a discrete quanta, meaning a whole photon, or else absorb nothing.

When a photon is absorbed by an atom on a metallic surface, some of the energy of the absorbed photon frees an atomic electron, and the remaining energy is converted into the kinetic energy of the emitted electron. The absorbed energy required to free an electron from its atomic orbit is called the *work function* ϕ of the metal or other material. Different materials have a **work function** given by:

$$\phi = hf_0$$

where h is Planck's constant and f_0 is the threshold frequency. The remaining energy that is converted into the kinetic energy of the emitted electron is: $KE = (1/2)m_e v^2$, where m_e is the electron's mass and v is its velocity. The **energy** of this process can be written:

$$E = hf = \phi + KE = \phi + (1/2)m_e v^2$$

When the frequency f of incident electromagnetic radiation is less than the threshold frequency f_0, where $hf_0 = \phi$, no electrons are emitted. This means the energy of the incident light must have a higher energy than the work function of the material for the light to free electrons. When an electron is freed its kinetic energy can be expressed as: $KE = hf - \phi$.

• As we saw in the photoelectric effect, the particle nature of electromagnetic waves was revealed since light was absorbed as discrete photons. In 1923 **Louis de Broglie** brought together the evidence that electromagnetic radiation has characteristics of both a particle and a wave by stating that **matter also has both particle and wave properties**. In other words, he suggested that matter *particles can have wave properties*. This is referred to as **wave-particle duality**. Analogous to photons, de Broglie proposed that matter particles such as electrons can behave as waves, and their wavelength λ is correlated with the linear momentum p of the particle by:

$$\lambda = h/p = h/mv$$

This is referred to as the **de Broglie wavelength**. Because Planck's constant, $h \approx 6.626 \times 10^{-34}$ J·s $\approx 4.135 \times 10^{-15}$ eV·s, is very small, a de Broglie wavelength is relevant or noticeable for small masses of atomic or subatomic sizes.

For example, remember *angular momentum* L *of the electron* from the Bohr model above is L = mvr = nh/2π, where the integer n can be 1, 2, 3, …, m = electron mass, v = electron velocity (speed), and r = radius of its orbit around the nucleus. Rearranging mvr = nh/2π, gives mv = nh/2πr. From the de Broglie's formula above, λ = h/mv, so mv = h/λ. Combining mv = nh/2πr with mv = h/λ, gives: h/λ = nh/2πr, or:

nλ = 2πr   where λ is the de Broglie wavelength of an electron

This says that the de Broglie wavelength of an electron is an integer multiple of 2πr, which is the *length of one orbit by the electron*. This means that an electron must orbit the nucleus of the atom at the radius that will allow the completion of integer n wavelengths. The de Broglie wavelength, therefore, accounts for the observation that electrons orbit a nucleus at specific radii. We can think of a planetary model of the atom, but with electrons as waves. What *orbital radii* are allowed? Remember, for waves in a defined space, standing waves can set up to "fit" that space. These standing waves are the allowed waves. In a circular space or orbit, the standing waves that can exist could look somewhat like the following (shown as dotted):

The circumference is positive integer n times the wavelength λ. The values of n shown are 1, 2, 3, and 4, respectively, and represent the electron in its n energy state completing n cycles in its orbit around the nucleus. Interestingly, these standing wave states correspond to the "allowed" electron orbits in the Bohr model. The integer n is called the *principal quantum number*. It labels the overall **orbital** of the atom describing the size of the orbital, or state, of the electron for the hydrogen atom and indirectly describes the energy of an orbital.

• In 1926–1927 a new mathematical framework for describing the atomic world that incorporated de Broglie's work was developed by physicists including German physicist Werner Heisenberg and the Austrian physicist Erwin Schrödinger.

Werner Heisenberg is credited with determining that the *position* and *momentum* of a particle cannot be precisely measured at the same time. This is called the Heisenberg **Uncertainty Principle** and is stated: *The* *more precisely the position is determined, the less precisely the momentum is known in this instant, and vice versa.* While simultaneously measuring both position and momentum of large objects does not generally require micrometer accuracy, measuring the position and momentum of an electron-size object does require accuracy on a tiny scale. For example, if you measure an electron's position, the energy from your measuring tool (electromagnetic radiation) will affect the tiny electron's movement, and you cannot accurately detect its motion. If you know the precise position of a particle, you cannot know its speed, and if you know its speed, you cannot know its exact position. The more precisely you measure position or momentum, the less precisely the other is known. **Heisenberg's Uncertainty Principle** can be written:

$$\Delta x \Delta p \geq h/4\pi$$

where the particle's *position* uncertainty is Δx, its *momentum* uncertainty is Δp, (remember p = mv), and h is Planck's constant. The Uncertainty Principle is also written:

$$\Delta x \Delta p \geq \hbar/2 \quad \text{where h-bar is } \hbar = h/2\pi$$

The Uncertainty Principle affects our understanding of our ability to accurately observe the very small, since observation influences the observed. This makes obtaining experimental data a challenge.

• In 1926 **Erwin Schrödinger** published papers providing the foundations of *Quantum Wave Mechanics*. His work involved a partial differential equation—a foundational equation of Quantum Mechanics—called the **Schrödinger equation**. Integrating Louis de Broglie's ideas that particles of matter have a dual nature and can behave as waves, Schrödinger described such a system in his *wave equation*. The solutions to Schrödinger's wave equation are wave functions that provide the *probability* of an event.

The Schrödinger equation has two general forms, one time dependent and one time independent. The **time-dependent Schrödinger equation** describes how the wave function of a particle evolves with time and the dynamic behavior of the particle. The three-dimensional **time-dependent** Schrödinger equation can be written:

$$-(\hbar^2/2m)\nabla^2\Psi(x,y,z,t,) + U(x,y,z)\Psi(x,y,z,t,) = i\hbar(\partial/\partial t)\Psi(x,y,z,t,)$$

where $\Psi(x,y,z,t,)$ is the time-dependent wavefunction, $\hbar = h/2\pi$ is called h-bar and is often used in applications involving angular momentum (h is Planck's constant), Δ is the del operator of partial derivatives, $U(x,y,z)$ is the potential energy function for the potential energy at position x, y, z, and i represents the imaginary part. Note that $i\hbar(\partial/\partial t) = E$ is called the *energy operator*, so the term $i\hbar(\partial\Psi/\partial t)$ is sometimes written $E\Psi$.

The **time-independent Schrödinger equation** can usually be derived from the time-dependent equation. Information it provides includes the allowed energies of a particle. The **time-independent Schrödinger equation** for three dimensions can be written:

$$-(\hbar^2/2m)\nabla^2\Psi(x,y,z) + U(x,y,z)\Psi(x,y,z) = E\Psi$$

where $\Psi(x,y,z)$ is the time-independent wavefunction.

Note that the square of the wave function, Ψ^2, was initially interpreted by Schrödinger as indicating that the electron was spread out in space with its density at point x, y, z given by the value of Ψ^2 at that point. Physicist **Max Born** instead proposed that Ψ^2 gives the probability of finding the electron at x, y, z, which is held as the better interpretation. The distinction is that for a small Ψ^2 value at a particular position, the first interpretation suggests a small fraction of an electron will always be detected there. Born's interpretation suggests instead that *either the whole electron will be detected at a location or nothing will be detected.*

Setting up, solving, and analyzing the solutions to the Schrödinger equation is a fundamental part of Quantum Mechanics referred to as *Wave Mechanics*. We have come from the idea of an electron as a point particle moving along a set orbit around the nucleus to wave mechanics. **Wave Mechanics** describes clouds, or orbitals, having probable locations of electrons in different states, with the **wave function** providing a way to calculate the probability of finding an electron or particle at a point in space.

• Schrödinger's wave equation was not, however, consistent with Special Relativity, as it has a nonrelativistic expression for the kinetic energy. In 1928 physicist **Paul Dirac** combined Quantum Mechanics and Relativity in a wave equation for the electron, which included an additional quantum number with values +1/2 and −1/2 corresponding to an additional form of angular momentum. Over the years Quantum Mechanics progressed. In the 1940s **Quantum Electrodynamics** (QED) brought more elucidation with the work of **Richard Feynman** and others. QED provided a new framework for processes involving the transformations of matter into photons and photons into matter. QED studies the interaction of light with matter. We have just scratched the surface of Quantum Theory. The nature of light, matter, and the subatomic world remain mysterious.

14.4. Dark Energy and Dark Matter

Dark Energy

• The cosmos is composed of about 4% **ordinary matter** which is the detectable matter of atoms such as in stars, planets, and living creatures; about 22% **dark matter** which is believed to have a *gravity* that holds together galaxies and clusters of galaxies; and about 74% of the strangely repellant **dark energy** which pushes everything out causing the acceleration in the expansion of the Universe. **Dark energy** is believed to be a form of energy that produces a repulsive, opposite-to-gravity type of force, causing the Universe's expansion rate to accelerate.

Many physicists theorize the Universe began instantaneously as a "Big Bang," emerging extremely hot and dense, and then for a tiny fraction of a second expanding at faster than the speed of light, which is called "**inflation**". During the inflation expansion, density fluctuations occurred. The gravity of highly dense regions drew in *dark matter* and *ordinary matter* in a mixture of subatomic particles. Within the regions of dark matter, galaxies formed.

Dark energy opposes gravity. In the early Universe, gravity dominated dark energy. Then, long after the Big Bang when space had expanded and matter became diluted, the gravitational attractions weakened and dark energy began to dominate gravity. Some physicists imagine a distant future with dark energy's continued increasing dominance over gravity resulting in galaxies being spread farther and farther apart.

• Dark energy was experimentally revealed in 1998 using distant exploding stars referred to as type Ia **supernovae**. All type Ia supernovae explode with about the same energy and therefore brightness, emitting constant, intrinsic, measurable light called "**standard candles**." Their reliable brightness allows their distance from Earth to be calculated. Astronomers can also examine how long ago supernovae exploded by measuring how much their light has been shifted to longer, redder wavelengths by the expansion of space. Such observations showed that the rate of expansion of the Universe is increasing and that **dark energy** is pushing objects out at accelerating speeds.

Our understanding of the cosmos has also been enhanced through measurements of the so-called afterglow of the big bang radiation referred to as the **cosmic microwave background** (CMB). Studies of the detectable radiation from the Big Bang reveal temperature variations of about one part in 100,000. These temperature variations from one point to another correlate hotter spots with denser regions of the early Universe. CMB data provide information about the amounts of ordinary and dark matter in the early Universe and the geometry of space, including that the Universe is "flat". Measurements from the CMB suggest that about 70% of what makes up the Universe is dark energy.

Recent results from NASA's space-based Galaxy Evolution Explorer and the Anglo-Australian Telescope on Siding Spring Mountain in Australia confirmed that dark energy is a smooth, uniform force that now dominates the effects of gravity and drives the accelerated expansion of the Universe. The new data support the idea that dark energy fits the **cosmological constant** in Einstein's equations and theory. These new observations are based on detailed measurements of the distances between pairs of galaxies. Another "standard" measurement was also employed. *Sound waves* from the nascent Universe made imprints in the patterns of galaxies, which caused pairs of galaxies to be separated by about 500 million light-years. This acts as a "**standard ruler**" which was used to determine the distance from the galaxy pairs to Earth. Galaxy pairs closer to us appear farther from each other. Similar to the supernovae "standard candles," the newer data combining distances and speeds with which pairs of galaxies are moving away from us also confirmed the acceleration in the rate of expansion of the Universe.

• We know dark energy exists by its effects on the expansion of the Universe. What dark energy actually is remains unclear. Different theories have been propounded, such as the idea that Einstein's description of gravity is wrong, and gravity, rather than dark energy, becomes repulsive at long distances, pushing everything apart. Data has not supported this idea. Another theory is that *dark energy is an intrinsic property of space.* **Einstein** suggested that empty space may not be void of everything and it may be possible for more space to come into existence. Einstein's theory of gravity contains a **cosmological constant** and predicts that "empty space" can possess its own energy. This energy would be an inherent property of space and therefore would not be diluted as space expands. More energy would appear as more space comes into existence. This would allow the Universe to increase its expansion rate. The *cosmological constant* is not well understood. Even Einstein struggled with it, as he initially intended it to explain why the fabric of what he believed was a static Universe did not collapse under gravity. At that time **Edwin Hubble** had not yet discovered that the Universe is expanding. Once that was discovered Einstein rejected his constant.

Today physicists are trying to gain insight into dark energy by examining how its density changes as space expands. Its density should not become diluted if dark energy is an inherent property of space. Conversely, its density should become more dilute if dark energy is something that exists within space. Observational data are being used to examine these questions. Whether dark energy is mathematically equivalent to Einstein's *cosmological constant* or some as yet undiscovered dynamic field, dark energy is a very real part of our mysterious world.

Dark Matter

• Combining theory and observation, physicists estimate the Universe is made up of about 74% repellant **dark energy** (discussed above), about 4% **ordinary matter** which includes detectable mass made from atoms, and about 22% **dark matter**. Dark matter accounts for nearly 90% of all matter yet it does not emit, absorb, or reflect light or other electro-magnetic radiation, making detection very difficult. Dark matter makes itself known, however, by its gravitational effects.

The idea of **dark matter** was proposed decades ago when it became obvious that the ordinary mass contained in stars, cosmic dust, and all other forms of detectable matter could not possess enough gravity to hold together a spinning galaxy. The amount of detectable ordinary matter was so far from adequate it became obvious that there must be some type of invisible matter responsible for most of the gravitational effects in the Universe.

Dark matter is believed to have contributed to the birth of massive galaxies in the early Universe. In fact, the current view of **galaxy formation** is that they form as dark matter clusters together under its own gravity. The cluster then draws gases, dust, stars, and smaller galaxies into a flat wide spiral form, which we call a **spiral galaxy**. The dark matter cluster remains in a spherical **halo** shape, surrounding the formed galaxy. Even after a spiral galaxy forms, new stars continue to develop throughout the flat disk as neutral hydrogen gas showers onto the spiral arms from the dark matter halo. **Elliptical galaxies** can form when two galaxies of about the same mass collide.

• We know dark matter exists due to its gravitational effects, but what is it and what do we know about it? We know it makes up the *halos* that surround galaxies and clusters of galaxies. It played a role in the formation of the structure of the Universe. Dark matter is called **"cold"** since its particles are relatively slow-moving (non-relativistic).

Dark matter is not simply dark clouds of ordinary matter or it would be detectable by the absorption of radiation passing through it. It is believed not to be antimatter since it does not create the characteristic gamma rays observed from matter-antimatter annihilations. Since dark matter interacts with ordinary matter gravitationally, it is believed it must possess enough mass to cause the observed gravitational effects on galaxies and clusters of galaxies.

It has been speculated that dark matter could be composed of difficult-to-detect ordinary matter in the form of such objects as compact brown dwarfs or in small, dense clumps of heavy elements. These ordinary matter candidates are called **MACHOs**, or **MAssive Compact Halo Objects**, and can include black holes, neutron stars, and brown dwarfs. While neutron stars and black holes can be dark (without enough detectable light), since they are a product of some type of supernovae, they are not believed to be common enough to provide the halo of dark matter around galaxies. The population of brown dwarfs also may not be large enough to explain all of the dark matter. Astronomers have been detecting MACHOs using *gravitational lensing* of light from distant objects. (Note that **brown dwarfs** are very small "failed" star-like objects that do not have sufficient mass to sustain fusion of hydrogen, as does a star, and are not luminous enough to be detectable by telescopes.)

In the early 1980s, scientists thought dark matter might consist of nearly massless particles called **neutrinos**. Then cosmologists showed that neutrinos were not a likely candidate since neutrinos whiz along at near-light-speed and are therefore referred to as "hot". Scientists believed dark matter needed to consist of some type of as-yet unobserved particle that is larger, slow-moving and therefore "cold". Scientists believe the evolution of structure in the Universe indicates that dark matter cannot be fast moving, since fast-moving particles would prevent the clumping of matter observed in the Universe. A group of researchers used large-scale simulations to show that only "cold" dark matter particles could create the distribution of galaxies that are observed.

Currently, a leading candidate for dark matter is a particle that is not ordinary matter, but rather an exotic particle. This popular candidate particle is believed to be composed of some type of elementary particle referred to as **WIMP**, or **Weakly Interacting Massive Particle**. While WIMPs are not ordinary matter, they are considered subatomic particles. They are described as "Weakly Interacting" since they are able to pass through ordinary matter without having effects. They are called "Massive" because they must have some type of mass in order to affect gravity. Cosmologists imagine that dark matter particles could have been created just after the Big Bang.

Candidates for WIMPs have included neutrinos, axions, and neutralinos. **Neutrinos** not only move fast, as mentioned above, but also are believed to have insufficient mass to make them a good dark matter candidate. **Axions** are particles speculated to be responsible for why neutrons do not possess an electrical dipole moment. They have very little mass, but should have been plentifully produced in the Big Bang. **Neutralinos** are electrically neutral massive (having mass) particles which are part of a group of particles proposed under the theory of supersymmetry. Supersymmetry endeavors to unify all the known forces in physics. Observations have so far not detected axions or neutralinos.

• Astronomers and particle physicists are actively working on detecting and identifying dark matter through astrophysical observations in space as well as in underground particle detectors. The search is complicated because these still somewhat theoretical WIMPs can zoom through ordinary matter without leaving a trace.

How have physicists endeavored to detect dark matter? In their search for dark matter particles, physicists have taken several approaches. One approach is to attempt to directly detect WIMPs. Since our galaxy lies within a vast dark matter *halo*, physicists can potentially directly detect particles using highly sensitive techniques. Theoreticians had predicted we should see a seasonal modulation in the presence of measurable dark matter particles (providing we could detect them) because of the relative motion of the Earth and Sun as we move with respect to the plane of our galaxy. As the Sun orbits the Milky Way's center on the outskirts of one of its spiral arms, the solar system drives through our galaxy's halo of dark matter particles. At the same time, the Earth is also orbiting the Sun. During the winter, the Earth moves mostly opposite to the Sun's motion through the galaxy. However, during the summer, the Earth moves in roughly the same direction as our Sun's orbit around the Milky Way's center. When the Earth is moving with the Sun, their alignment increases the Earth's net velocity through our galactic halo of dark matter particles. This means that during the summer, there should be a peak in dark matter particles available for detection as we crash through them at a greater speed. Conversely, during the winter, there should be a dip in detectable particles as the Earth travels opposite to the Sun's orbital motion.

Because of this seasonal modulation, scientists are attempting to directly measure dark matter particles and see if their data fits with the annual modulation. The main problem is that dark matter particles should only barely interact with ordinary matter, so sensing their presence requires extremely sensitive detectors. Scientists have set up detectors deep underground where there are very low levels of cosmic rays and ordinary radiation, but where dark matter can penetrate. Scientists have designed detectors made up of certain atoms, so that when a dark matter particle strikes an atomic nucleus, the recoiling nucleus can produce a tiny pulse of electricity, light, and heat. This signal reveals that a dark matter particle has struck. Some researchers have, in fact, detected the expected seasonal fluctuation in the scintillation signals, suggesting they have detected the type of dark matter particle they were looking for (WIMPs). Other researchers dispute the results. There is controversy at this time regarding the validity of data by different research groups.

Another method of measuring dark matter is through indirect detection of gamma rays or other familiar particles which presumably result when two dark matter particles collide. Researchers are searching by peering into the skies toward such places as the center of our galaxy for excess gamma rays, signaling dark matter particle collisions. Instruments used for these endeavors are both orbiting and ground based. An additional method of measuring dark matter is to attempt to blast dark matter particles into existence using an accelerator. Eventually it will become evident through a preponderance of evidence what data is valid and what is the nature of dark matter. This is an extremely exciting, unfolding story!

14.5. Key Concepts and Practice Problems

- Special Relativity applies to observers in uniform relative motion in the absence of a gravitational field. According to Special Relativity: the laws of physics are the same in all inertial reference frames; and the speed of light is constant regardless of the observer's velocity.
- At near-light speeds time dilates (slows), length contracts in the direction of motion, events can appear not simultaneous, objects get heavier, speed vectors don't add normally, and KE approaches infinity.
- Mass and energy can be converted from one to another: $E = mc^2$.
- General Relativity includes accelerated motion and gravity. It provides a description of spacetime with its geometry as a 4-dimensional continuum spacetime and a new theory of gravity as a manifestation of curved space and time (spacetime) rather than a force. Matter causes spacetime to curve, and that curving affects the behavior of matter.
- General Relativity theory is expressed in a series of Field Equations and predicts: the shifting of Mercury's elliptical path; black holes;

singularities; the expanding Universe; and gravitational time dilation, length changes, lensing, redshifts, and waves.
- An atom, the fundamental building block of elements, consists of a nucleus of neutrons and protons surrounded by one or more electrons.
- Rutherford's gold foil experiment showed atomic mass concentrated in the nuclei of protons and neutrons with electrons orbiting, filling the volume.
- The Bohr model based on the hydrogen atom: electrons circling the nucleus at specific allowed orbitals. Electromagnetic energy is released or absorbed as electrons change states: $\Delta E = hf = E_i - E_f$.
- Energy for ground state hydrogen atom: $E_n = E_1/n^2 = -13.6$ eV.
- Quantum Mechanics: energy is emitted (or absorbed) from matter in packets called quanta, and light and other EM radiation are composed of photons, or quanta, which have properties of particles and waves.
- Photoelectric effect: electron absorbs proper EM energy to be ejected.
- Louis de Broglie stated matter has both particle and wave properties.
- Heisenberg Uncertainty Principle: can't know both position and speed.
- Schrödinger wave equation describes wave function of particles.
- The cosmos: 4% detectable ordinary matter of atoms, stars, planets, creatures; 22% dark matter which holds galaxies together and makes up halos around galaxies; and 74% repellant dark energy which pushes everything out causing acceleration of Universe's expansion.

Practice Problems

14.1 (a) The nearest star, Proxima Centauri, is about 4 light years away, so traveling there and back at 0.8 the speed of light (2.4×10^8 m/s) would take about 10 years. If you volunteer for such a journey, how much "younger" would you and your clock be upon your return? Assume your acceleration and deceleration times are negligible compared to 10 years. **(b)** If your mass is 75 kg before the trip, what would your mass become during your trip and what would it be after your return home?

14.2 (a) How small a ball would the incredible shrinking man have to shrink into before collapsing and forming a black hole, assuming his mass remains 80.0 kg throughout? **(b)** Our author speculates that time is accelerating along with the increase in the rate of expansion of the universe (not just that it *seems* faster as we age due to our perspective). What arguments can you make for and against this theory? How might you measure time acceleration?

14.3 (a) How can an astronomer determine the elements present in a star by evaluating the spectrum of the light it emits? **(b)** Does the Heisenberg Uncertainty Principle imply that a small moving particle does not in fact at an instant in time possess a specific location and velocity? **(c)** If Mr. Bolt has a mass of 85 kg and is sprinting at 11 m/s, what is his wavelength?

14.4 Do the observed effects of the accelerating expansion of the Universe and the holding together of spiral galaxies mean that dark energy and dark matter exist, or could there be other explanations?

Answers to Chapter 14 Problems

14.1 (a) Each tick of your clock t compared to an at rest clock's tick t_0 would take $1/[1 - (v^2/c^2)]^{1/2} = 1/[1 - (0.8/1)^2]^{1/2} \approx 1.67$. Therefore time would pass at a rate of $1/1.67 \approx 0.6$ the rate it would if you had not taken the trip, and you would return 4 years younger. **(b)** During the trip your mass $m = m_0/[1 - (v^2/c^2)]^{1/2} = 75/[1 - (0.8/1)^2]^{1/2} = 125$ kg. After decelerating to a stop on your return, your mass returns to 75 kg.

14.2 (a) $r_{sch} = 2Gm/c^2 = 2(6.67 \times 10^{-11} \text{ m}^3/\text{kg·s}^2)(80)/(3 \times 10^8 \text{ m/s})^2 \approx 1.186 \times 10^{-25}$ m. So he would have to form a ball of diameter 2.37×10^{-25} m. By comparison, a hydrogen atom has a diameter 10^{-10} m and a proton about 10^{-15} m. **(b)** Suppose time is created by the steady expansion of the Universe. Since the rate of expansion is accelerating, and space and time are linked as spacetime, then shouldn't time also be speeding up? Can the speed of light remain constant if the expansion of space is accelerating but time is remaining constant? Might the interval of a "second" become shorter as spacetime expands? What do you think? How else could time be created?

14.3 (a) The amount of energy emitted as electrons change between different energy levels is unique for each element. Therefore photon emissions occur at specific identifiable frequencies, giving each element an identifiable electromagnetic signature. **(b)** The fact there are limits on the accuracy of simultaneously measuring both location and velocity does not necessarily imply that a particle does not possess a specific location and velocity, only that we cannot measure it with precision. Of course, attempting to ascribe a particular location in either space or time to a particle that in some ways behaves like a wave may be futile. **(c)** $\lambda = h/mv = (6.626 \times 10^{-34} \text{ J·s}) / (85 \text{ kg})(11 \text{ m/s}) \approx 7.1 \times 10^{-37}$ m, a vanishingly short distance. (Note Joule = kg·m^2/s^2)

14.4 Within a cause-and-effect Universe, the effects imply natural causes, but unless and until the existence of the hypothesized dark energy and dark matter are observed and measured, the field remains open for new or modified theories and explanations. Many times so-called "settled science" has later been discovered to be incorrect in whole or in part.

INDEX

*"The first gulp from the glass of natural sciences will turn you into an atheist,
but at the bottom of the glass God is waiting for you."*
Attributed to Werner Heisenberg

Other Books by These Authors ~ See GlacierDog.com

MASTER MATH Books
BASIC MATH & PRE-ALGEBRA, ALGEBRA, TRIGONOMETRY, GEOMETRY, PRE-CALCULUS, and CALCULUS

These **best-selling** books explain principles, definitions, and operations of each subject, provide step-by-step solutions, and present examples and applications. These comprehensive books teach in a way that is easy for students of all ages to understand!

ARROWS THROUGH TIME
A Time Travel Tale of Adventure, Courage, and Faith

"A must read for anyone fascinated by time travel and its possibilities. An intriguing, insightful, and inspiring tale of faith, fate, and heroism with compelling characters and surprising twists. This fun, fast-moving adventure will take you on an unforgettable journey."

The 3:00 PM SECRET 10-DAY DREAM DIET

The 10-Day Dream Diet shows you how to achieve your dream body step-by-step. It provides a simple, straight-forward weight-loss formula. It has daily menus, simple exercises you can do in minutes, nutrition tips, and revealing Dream questions to guide you to your dream life.

The 3:00 PM SECRET
Live Slim and Strong Live Your Dreams

Living The 3:00 PM SECRET is so easy and such a positive experience, it seems like magic! The 3:00 PM SECRET is unique because of its motivation and simple, innovative approach to eating. This book develops the philosophy and practical steps that will make you slim, strong, and healthy.